D0205870

AN INTRODUCTION TO
ASPECTS OF

THERMODYNAMICS AND KINETICS

RELEVANT TO
MATERIALS SCIENCE

E.S. MACHLIN

Howe Professor of Metallurgy

Columbia University, New York, N.Y.

Figures 2.10, 3.8 and the Fe-Ni phase diagram in Figure 3.9 are reproduced with the permission of McGraw-Hill, Inc. and are taken from M. Hansen,Constitution of Binary Alloys, McGraw-Hill, 1958.

Published by:
GIRO PRESS
P.O. Box 203
Croton-on-Hudson, N.Y. 10520

ISBN 1-878857-02-9

PRINTED IN THE UNITED STATES OF AMERICA
95 94 93 92 91   5 4 3 2 1

CONTENTS

XI. Growth of Phases (Diffusion or Reaction Control.) 269

XII Morphological Instability and Growth via Cellular Segregation 309

Index 337

PREFACE

This book is based on a set of notes developed over a period of many years for an introductory course that the author taught to seniors and entering graduate students in materials science. Lately, this course has also been taken by students at IBM Fishkill Facility, who have a varied background, in a program leading to a master's degree in materials science.Our pedagogical philosophy in the teaching of materials science courses has been to take a materials generic approach to each of the core courses in this program. Thus, this book is based on such an approach, which is incidentally consistent with a recommendation in the recent report entitled "Materials Science and Engineering for the 1990's:Maintaining Competitiveness in the Age of Materials" of the Committee on Materials Science and Engineering of the National Research Council. Reference to specific materials, however, will be found in both the text and the problems, inasmuch as each material is unique in its properties.

Because this book is on the introductory level, it has not been possible to provide a detailed discussion of each topic covered in depth. Indeed, it is possible to write a book on the subject of each chapter. Thus, in many places detailed derivations of equations used in the text are absent or provided in appendices. Emphasis has been placed upon providing the student with a physical understanding of the phenomenon under discussion, with the mathematics presented as a guide to such an understanding. Thus, the researcher seeking comprehensive knowledge of a field will need to search further. It is hoped that the references and bibliography listed at the end of each chapter will aid in this search.

The problems have been used not only to provide practise in quantitative application of principles, but also to give examples of applications of the general subject matter to problems having current interest and to emphasize the important physical concepts.

Except for the last chapter, part of the section on DIGM and section A1.2 in the third chapter, the material in this book consists of material already discussed in the literature. The treatment in the last chapter of cellular segregation problems came about because of my discontent with the current status of theory in this area and because I thought that there must be a common element to all the cellular segregation processes, which I believe I have found in Langer's suggestion of thermally induced drift in the interlamellar spacing.

I-THERMODYNAMICS OF PHASES HAVING CONSTANT COMPOSITION

INTRODUCTION

Applications of thermodynamics and kinetics to materials is the theme of this book. Materials consist of phases or mixtures of phases. By 'phase' is meant the homogeneous configuration of atoms or molecules corresponding to a liquid, or to some crystalline solid, such as a solid having the body-centered-cubic (bcc) structure, or to an amorphous solid, such as glass, or to a vapor. A phase has a set of properties including thermodynamic properties, such as entropy, enthalpy and Gibbs free energy. The phase may or may not be in an equilibrium state. In the equilibrium state the thermodynamic properties are not functions of its past history, but are unique. A state of equilibrium is defined by the existence of a stationary state in one of the thermodynamic potentials. Stationary states defining stable equilibrium are described in Table 1.

Table 1.   Stationary States Corresponding to Stable Equilibrium

Thermodynamic Potential	State[+]	Constrained Variables
S,Entropy	max.	none**
G,Gibbs free energy	min.	P,T
F,Helmholz free energy	min.	V,T
H,Enthalpy	min.	P,S
E,Internal energy	min.	V,S

[+]   This condition refers to the stationary state of the thermodynamic potential (minimum or maximum.)

**   Closed system.

The set of constraints that apply to the Gibbs free energy are those applicable to most of the phenomena we shall study. Consequently, this thermodynamic potential will occupy an important role in this book.

1. THERMODYNAMIC POTENTIALS.

It is useful to obtain some physical insight into the materials aspect of the thermodynamic potentials. Since the Gibbs free energy of a phase is given by

$$G = E+PV-TS = H-TS \qquad (1.1)$$

then a study of the material origins of the energy and the entropy will be helpful in solving problems involving the equilibrium between phases at constant temperature and pressure.

Materials may exist in either the solid, liquid or vapor states. When in the vapor state, and approximating the ideal gas, a monatomic gas consists of atoms that do not interact and that move freely throughout the container that defines the volume of the gas. In the solid state the atoms are confined to potential wells of known coordinates and oscillate in these wells. The liquid state is more akin to the solid state in that the atoms oscillate in potential wells, but there is a free volume, much smaller than in a gas, in which these atoms can accomplish a gaslike motion, as well. These aspects of the states of materials affect their thermodynamic properties in that the configurational and thermal contributions to the latter differ in type in the different states.

1.1. Energy

Energy determines the stable crystal structures of solids at $0°K$. Although first-principle calculations of the energy are possible, it is unusual for such calculations to be sufficiently accurate to distinguish between the stability of competing polymorphs. Thus, recourse to empirical methods is necessary in attempts to predict the stable structures. One successful method is based on the concept of structure maps. This method deduces from theory the variables likely to affect the relative stability of different crystal structures and then uses these variables as coordinates. It is found that the stable crystal structures of compounds having a given stoichiometry each occupy a space in this coordinate system that is unique to that crystal structure. The appropriate variables differ dependent upon the type of binding.[1,2] Also, much use has been made of interatomic potential models to describe the configuration dependent energy of various solids (see Bibliography for references to interatomic potentials.)

Prediction of phase diagrams requires quantitative values for the energies of phases. The major source for such values is experimentally derived data (see Bibliography for list of data banks.)

1.1.1. Lattice stability energy

The difference in cohesive energy between two competing polymorphs, at a given temperature and composition, is called the lattice stability energy. This energy enters into calculations having as their objective the prediction of phase diagrams, a subject that will be considered in detail later in this book. Appendix 1 to this chapter lists values of lattice stability and cohesive energies for elements and crystalline compounds obtained by several different methods. A generalization applicable to these data is that the lattice stability energy is about one order of magnitude smaller than the respective cohesive energies. Another generalization is that the closest packed phases are often the most stable polymorphs at 0 °K, although there are exceptions to this rule due to the effect of unusual factors, such as the effect of spin order in ferromagnetism and the like.

1.2. Entropy

We show in Appendix 2 that entropy has a statistical significance and can be given by the following equation:

$$S = k \ln W \qquad (1.2)$$

where k is Boltzmann's constant and W is the number of different ways that the system having entropy S can be arranged at constant energy. This is the fundamental relation that we shall use to describe the entropy associated with various systems. It is worthwhile to the uninitiated reader to become familiar with this relation and its derivation. In quantum statistical mechanics the corresponding relation is

$$S = k \ln N \qquad (1.2b)$$

where N is the number of quantum states accessible to the system.

Another way of defining the entropy is also described in Appendix 2 based upon a "canonical" ensenble of systems. The result is

$$S = -k\Sigma_i [p_i \ln p_i] \qquad (1.2c)$$

where k is Boltzmann's constant and p_i is the fraction of the total very large number of specimens of the same system that are in the ith quantum state, distinguishable from the jth quantum state. If there are g distinguishable quantum states, all of which are equally probable, in the ensemble of specimens, then $p_i = 1/g$, which, when substituted in equation 1.2c, yields the following statistical definition of entropy

$$S = k \ln g \qquad (1.2d)$$

We shall make use of equations 1.2 and 1.2d in most of our evaluations of the entropy of a system.

Conceptually, we may compartmentalize the various contributions to the entropy of a substance as follows:

1) Configurational entropy. This contribution refers to the distinguishable ways the atoms can be arranged on the lattice sites for the case of a monatomic crystalline solid, or for a molecular crystal, the distinguishable ways of arranging the orientations of the molecules. Thus, for a monatomic crystalline solid, the only entropy of configurational origin that may exist is contributed by point defects, such as vacant sites or atoms occupying interstitial sites. The configurational contribution to the entropy in a multicomponent phase is considered later in this chapter. Since it is apparent that there are many distinguishable ways of arranging the different atoms on lattice sites in a solid solution or on the instantaneously frozen set of sites of atom centers in a liquid solution it is understandable that the configurational entropy of such solutions will be positive definite. Polymers have configurational entropy even for the case of a polymer corresponding to a linear chain of a single type mer. This configurational entropy depends upon the stationary states of the polymer, which in reality may be complex. At constant energy, it is apparent that there are many distinguish-

able configurations of a linear chain polymer of a single type mer. A derivation, following Kittel, of the configurational entropy of an unreal model polymer in which the mers are confined to a plane is given in Appendix 3.

2) Thermal entropy. At any finite temperature, the atoms of a solid are in continuous oscillating motion about their equilibrium positions. Thus, at any time there exists uncertainty concerning the exact positions of the atoms. From the theory of such thermal motion-the theory of phonons-certain conclusions can be drawn concerning the influence of material properties, such as cohesive energy, on the entropy of thermal origin. For Einsteinian monatomic solids (i.e. consisting of $3N$ independent harmonic oscillators), the thermal entropy is given by[3]

$$S_T = 3Nk(\theta_E/T)[(1 - e^{-\theta_E/T})^{-1} + \ln(1 - e^{-\theta_E/T})] \qquad (1.3a)$$

where the Einstein characteristic temperature, $\theta_E = h\omega/2\pi k$. Also, ω is the frequency of the harmonic oscillator, h is Planck's constant, k is Boltzmann's constant, and T is the absolute temperature.

For $T \gg \theta_E$, this expression simplifies to

$$S_T = 3Nk[1 + (\theta_E/T)\ln(\theta_E/T)] \qquad (1.3b)$$

Thus, the thermal entropy of a monatomic solid should be a monotonic increasing function of T. Further, by dividing the temperature by a number characteristic to each solid it should be possible to make the thermal entropy-temperature functions of different monatomic solids superpose. This characteristic number is, of course, the Einstein characteristic temperature, θ_E, roughly equal to the Debye temperature.(See Table 1.1.)

Some insight, from the viewpoint of a materials scientist, can be achieved by noting that the frequency of a harmonic oscillator is given by $(K/M)^{1/2}$, where K is the force constant and M is the mass of the oscillator. Dimensional analysis suggests that the force constant of the "spring" should be proportional to the quotient of the cohesive energy divided by the square of a characteristic distance, such as the interatomic distance, d, i.e. $K \gg E/d^2$ or $K \gg E/\varpi^{2/3}$, where ϖ is the atomic volume. A similar relation can be derived on the assumption that the energy of the solid is given by a sum over interatomic potentials.

The Einstein model is a poor approximation of the vibrational modes in a solid. A more realistic, but still inadequate, model was suggested by Debye. In a Debye solid the density of normal modes depends upon the square of the frequency up to some maximum frequency determined by the constraint that the total number of normal modes equals 3N. Although Debye's model was an improvement on Einstein's it is still not realistic in that the density of normal modes departs drastically from the frequency squared dependence in real crystals. The atomic nature of each crystal exerts a significant effect on the normal mode-frequency spectrum, which is investigated in books discussing phonons (see bibliography.) Although we shall require little use of the results of this discipline in the remainder of this book, a very brief discussion of Einstein and Debye theories is given in Appendix 4.

The main results of the theories of lattice vibrations applicable to the present subject are as follows. The thermal entropy is a continuous monotonic function of the temperature, increasing with the temperature. It is also a function of the atomic mass and the cohesive energy, increasing as the former increases and decreasing as the latter increases for a given type of bonding and crystal structure. Variation of the bonding exerts a strong effect as will be discussed later. Further, temperature dependent soft phonon modes may influence the onset of displacive phase transitions.

The entropy of thermal origin plays a controlling role in many polymorphic transitions. As we will note later, with increasing temperature, a first-order polymorphic transition is between a phase that has a lower entropy to one that has a higher entropy. In view of the relation between thermal entropy and cohesive energy discussed above then for transitions controlled by thermal entropy the transition is between a phase of higher cohesive energy and one of lower cohesive energy.

3) Electronic entropy. Because electrons at the Fermi level in metals have unoccupied states into which they can be excited by thermal energy and because of the corresponding uncertainty in the distribution of such electron states there is a contribution to the total entropy of electronic origin in metals, which is absent in insulators. This contribution is generally small and usually can be neglected in problems that will be encountered in this text. Van Vechten[4] has suggested another contribution to the electronic entropy of metals absent in insulators, to be discussed later.

4) Magnetic entropy. The atoms of many elements have net magnetic moments. These moments may have their vectors oriented randomly, as in a paramagnetic substance, or may have them aligned along a specific direction, as in a ferromagnetic substance. In the latter case the

Table 1.1. Approximate Debye θ Values for the Elements

Ar	85	Cd	175	Ge	360	α Mn	~380	Pd	290	(white) Sn	170	W	310
Ag	220	Co	119	Gd	170	β Mn	~380	Pr	141	Sr	140	Y	214
Al	395	h.c.p. Co	390	Hf	200	f.c.c. Mn	~355	Pt	233	Ta	230	Zn	235
As	285	f.c.c. Co	385	Hg	100	b.c.c. Mn	~370	Rb	61	Te	180	h.c.p. Zr	260
Au	177	Cr	490	Ho	162	Mo	380	Re	275	Tb	170	b.c.c. Zr	212
B	1250	Cs	45	In	129	Na	150	Rh	370	Th	140		
Ba	110	Cu	315	Ir	285	Nb	250	Ru	400	Tm	167		
Be	920	Dy	157	K	100	Nd	145	Sb	210	h.c.p. Ti	365		
Bi	120	b.c.c. Fe	432	Li	360	Ne	63	Sc	~400	b.c.c. Ti	300		
C (diamond)	1860	f.c.c. Fe	420	La	132	Ni	390	Si	625	Tl	94		
C (graphite)	~400	Er	167	Lu	166	Os	250	Sm	150	U	160		
Ca	230	Ga	240	Mg	318	Pb	88	(grey) Sn	260	Va	335		

From: R.J. Weiss, Solid State Physics for Metallurgists, Pergamon Press, 1963.

magnetic entropy is zero because there is no uncertainty governing the orientation of the magnetic moment vectors. In the former case, it is

$$S_{mag} = Nkln(2J+1) \tag{1.4}$$

where J is the quantum number of net spin on an atom.[5]

The problem of describing the magnetic entropy for the case of partial ordering of the spins is known as the Ising problem. An analogous problem exists in the determination of site occupation probabilities by different components of alloys. We shall postpone an analysis of the Ising problem to a later section.

Magnetic entropy exerts a controlling effect in the low temperature polymorphic transition found in iron and in many alloys, and intermetallic compounds.

The free energies, enthalpies and entropies of many elements and stoichiometric solids have been measured and are collected in various data banks. The bibliography at the end of this chapter provides a list of these sources.

There is an interesting experiment that the reader can perform to be convinced of the reality of separation of the entropy into configurational and thermal contributions. The human lip is a sensitive thermometer. We suggest that you use it to determine changes in the temperature of a wide rubber band as it is stretched and unstretched rapidly. The rapid stretching or unstretching of a rubber band approaches an isentropic process in which the total entropy is conserved because there is no transfer of heat to or from the rubber band in the period during which the rapid stretching or unstretching is carried out. From Appendix 3 we can determine that the stretching of an elastomer, such as a rubber band, should decrease the configurational component of the entropy. Since the stretching is isentropic, then the thermal entropy must increase to compensate for the decrease in the configurational entropy. But, the thermal entropy is a single valued function of the temperature, increasing with an increase in the temperature. Thus, the temperature of a stretched rubber band just after stretching should be higher than just before the stretching process. Similarly, the temperature of the rubber band just after rapid unstretching should be lower than it was just before the unstretching process. Carry out the experiment and convince yourself that the above analysis is valid.

2. POLYMORPHISM.

2.1. Phase transitions on varying the temperature

Let us now consider several aspects of the concepts discussed above. From the values listed in Appendix 1 for non-transition elements it is possible to deduce that usually the energy of the fcc structure (the ideally close-packed structure) is more stable than that of the bcc structure (a less close-packed structure). (In this context, the term "packing" is meant to correspond to nearest-neighbor coordination number rather than atomic volume. The lower this coordination number the more open is the "packing" in the present usage.) First-principle calculations of cohesive energy are not yet sufficiently accurate to distinguish between the energies of the fcc and bcc crystal structures, at equilibrium, and consequently it is not possible to provide a physical basis for the empirical rule that the energy at $0°K$ of the fcc structure is more stable than that of the bcc structure for most of the elemental solids. We shall assume the validity of this rule. Then, equation 1.3 yields that the entropy of thermal origin is higher for the bcc crystal structure relative to that for the fcc structure. This result is illustrated in Figure 1.1a. As shown there, the energy and entropy of each phase has been assumed to be independent of temperature (i.e. the entropy equals $- [\frac{\partial G}{\partial T}]_P$ and hence a constant entropy corresponds to a constant slope of the line representing the dependence of the free energy G on temperature T.)

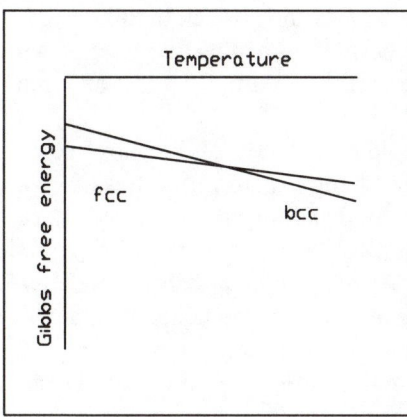 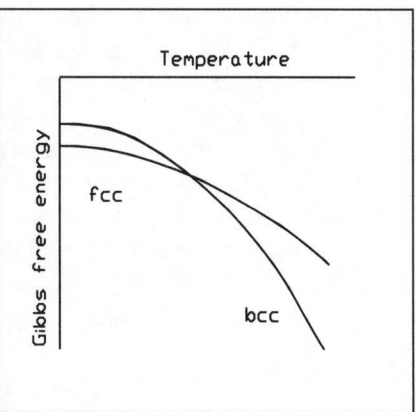

Figure 1.1.   (a)   (b)

In general, this assumption is not representative of the real behavior. The entropy of a phase increases with increasing temperature and the dependence of free energy on temperature corresponds more to that shown in Figure 1.1b. The specific heat is proportional to the curvature of the $G(T)$ line, i.e. $C_p = -T[\frac{\partial^2 G}{\partial T^2}]_P$. Since, the specific heat of materials has positive definite values, then the curvature of the $G(T)$ line must be negative.

We may deduce, from the knowledge that the free energy is a minimum for stable equilibrium, that the fcc structure is the relatively stable one at temperatures below T_E and that the bcc structure is stable relative to the fcc structure at temperatures exceeding T_E. At the latter temperature, the bcc and fcc structures are in equilibrium with each other. This conclusion can be generalized to imply that the close packed structure is stable at low temperature, whereas the less close-packed structures become stable at higher temperatures, for the case of metals that are not covalently bonded. This "rule" is also applicable to ionically bonded solids[2], where it is stated in terms of coordination number. The latter increases with increasing close-packing. Again, as suggested above, covalently bonded solids do not obey this "rule".

Because $S_{fcc} \neq S_{bcc}$ at T_E and $S = -[\frac{\partial G}{\partial T}]_{P...}$, the transformation that occurs at T_E in Figure 1.1 is a first-order transformation, e.g. the discontinuity occurs in the first derivative of the free energy. Higher order transformations occur in Nature. The ferromagnetic-paramagnetic transformation in bcc iron is a second order transformation, for example, as is the order-disorder transformation in beta brass and NH_4Cl. In this case, the discontinuity at T_E occurs in the 2nd derivative of the free energy with respect to an intensive parameter, such as temperature, i.e. C_p. (See Figure A.4.1 in Appendix 4.)

It is possible to supercool a liquid below the equilibrium freezing point temperature with no discontinuous change in its properties, as well as to superheat a solid above its melting point again with no discontinuous change in its properties. Thus, at the equilibrium melting (=freezing) point, the temperature at which the free energies of the solid and liquid are equal, there must be a discontinuous change in the entropy and hence the transformation solid to liquid (and vice versa) is of first order.

Liquids have a higher entropy than the corresponding solid states, in part because there is an additional contribution to the thermal entropy in liquids not present in solids. This added factor corresponds to the uncertainty in specification of the positions of the potential wells about which the atoms oscillate, due to rapid diffusion of these wells. For a monatomic material, this difference in entropy, the "entropy of fusion", can be derived and the derivation due to Fowler and Guggenheim is given in Appendix 5. The entropy of fusion of metallic elements, and compounds that upon melting retain their molecules in undissociated form, is about 2 cal/mol/deg C (Richard's rule), which is just the value derived in Appendix 5. A consequence of this result is that the vibrational frequency spectrum of these atoms is not too different in the liquid and solid states.

Covalently bonded elements (C,Si,Ge) and compounds which dissociate into atoms or ions rather than molecules upon melting exhibit higher values than the Richard's rule value (i.e.6-6.8 cal/mol/°C) for the entropy of fusion.

The increment in the entropy of fusion for covalently bonded elements, such as C, Ge and Si, can be explained to a first approximation using the concept relating the thermal entropy to the force constant of the "bond". Liquid Ge and Si are known to be metallically bonded. Hence, it is not the entropies of the liquids that are abnormal in the entropy of fusion of the covalently bonded elements. Rather, the thermal entropy of the covalently bonded solids must be abnormal relative to metals. Indeed, for the elements Ni and Si, which have about the same melting point, the entropies of the liquids are about equal at the melting point, whereas those for the solid state at this temperature are about 20.6 and 14.8 cal/mol/°C, respectively.

Van Vechten[4] has provided an explanation for the major contribution to the entropy of fusion of covalently bonded elements in terms of the significance of covalent bonds. As he noted, in order to form the bonding states of a covalently bonded solid the correlated phases of orbitals on neighboring atoms must add coherently. However, in the liquid phase, the phases of orbitals are randomized so that both bonding and anti-bonding states are equally likely to be occupied. There are 4N sp^3 type hybridized orbitals centered on N atoms. There is only one way these orbitals can be arranged to form the covalently bonded solid. However, in the liquid there are 2^{4N} choices of bonding or anti-bonding state allowed. This leads to an increase in the entropy upon melting of covalently bonded elements equal to k4N ln2 = 5.55 cal/mol/°C. The entropy of fusion of such covalently bonded elements includes an additional term corresponding to that for metals, equal to kN lne or 2 cal/mol/°C. Also, the existence of unoccupied states into which electrons can be easily promoted by thermal excitation in the metallic liquid, and the improbability of this excitation in the solid contributes a small positive term to the entropy of fusion.

Amorphous solids or glasses are in a "frozen-in", non-equilibrium, state rather than an equilibrium state. Nevertheless, this class of polymorphs have an energy and entropy that characterize them thermodynamically. Such solids usually have a larger specific volume, a more positive energy and a larger entropy than their crystalline counterparts. However, there may be exceptions to the latter statement in certain cases, which we will consider in a later chapter. The entropy of the amorphous solid is not as high as that for the liquid because the diffusional contribution to the thermal entropy, that contribution due to diffusion of the potential wells, is larger in the liquid (it is absent in glasses below the "glass" temperature). The latter effect is often neglected.

2.2 . Phase transitions on varying the pressure

It is also possible to vary the pressure P. Hence, by $[\frac{\partial G}{\partial P}]_T = V$, and the fact that the molar volumes of competing phases are not equal, in general, we obtain the possibility of having polymorphic transitions with increasing pressure. This concept is described graphically in Figure 1.2. As shown, with increasing pressure the phase having the smaller molar volume will be stabilized. This transformation is first-order because the first derivative of the free energy with respect to an intensive variable, in this case pressure, is discontinuous, i.e. V is discontinuous at the transformation pressure. Below the transition pressure, the phase with the larger molar volume is stable (it has the more negative values of Gibbs free energy),

while above the transition pressure, the phase with the smaller molar volume is the stable phase. These curves also have a curvature to them given by
$$[\tfrac{\partial^2 G}{\partial P^2}]_T = [\tfrac{\partial}{\partial P}][\tfrac{\partial G}{\partial P}] = [\tfrac{\partial V}{\partial P}] = -\beta V,$$
where ß is the compressibility. Since ß and V are positive quantities, the curvature must be negative in sign.

2.3. P-T diagram

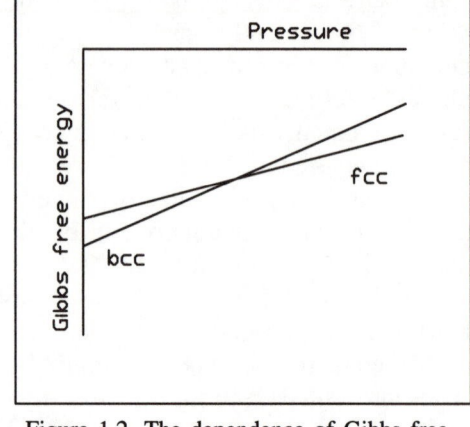

Figure 1.2. The dependence of Gibbs free energy, G, on pressure, P.

Thus, the equilibrium situation of a single component system is described in a P-T diagram, such as that shown in Figure 1.3. The existence of a number of stable solid phases in different regions of the P-T diagram is termed polymorphism. Polymorphic transitions may occur on changing the temperature or the pressure. It is often desirable, in the search to synthesize some particular phase, to be able to predict the corresponding P-T diagram. With the current rate of progress in the development of first-principle methods of calculating energies of phases and possible entropies of phases it is likely that this objective will be achieved before the end of this century.

There is a relationship called the Clapeyron equation that relates the slope of a phase boundary in a P-T diagram to the changes in molar enthalpy and volume and the transformation temperature. It is derived as follows. At equilibrium between two phases ' and " the molar free energies of the two phases must be the same,

$$G' = G'' \tag{1.5}$$

Substituting from equation 1 we are then able to write

$$H' - T_E S' = H'' - T_E S''$$

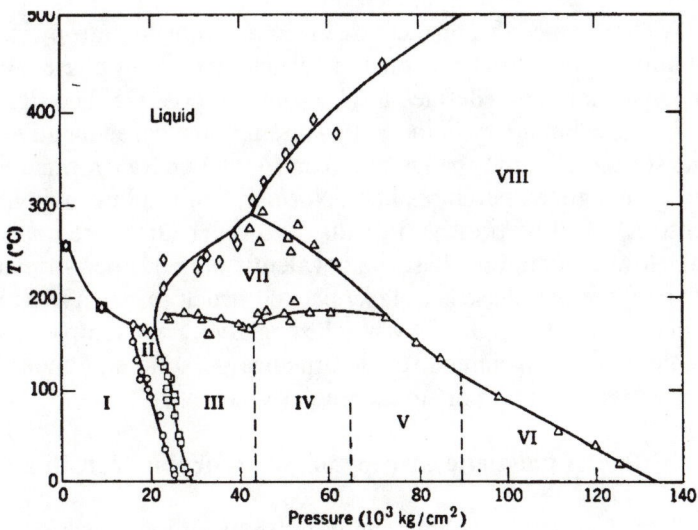

Figure 1.3. Phase diagram of bismuth. From H. M. Strong, *Am. Scientist*, **48**, 58 (1960).

or

$$\Delta H_t = T_E\, \Delta S_t \qquad (1.6)$$

Along the phase boundary both T and P must be varied in such a way that equation 1.5 is satisfied, or that dG'=dG". But for the single component system

$$dG = -SdT+VdP.$$

Thus, we may write

$$-S'dT+V'dP = -S"dT+V"dP$$

or

$$(\partial P/\partial T)_E = (S'-S")/(V'-V") = \Delta S_t/\Delta V_t$$

We can substitute for ΔS_t from equation 1.6 to obtain

$$(\partial P/\partial T)_E = \Delta H_t/(T_E\Delta V_t) \qquad (1.7)$$

Equation 1.7 is the Clapeyron equation.

We have noted that on increase in temperature the entropy increases while within a given phase or upon transformation to a new phase. Thus, the heat of transformation as defined in (1.7) must be positive. The sign of the slope of a phase boundary in the P-T diagram, must consequently depend upon the sign of ΔV, the change in molar volume on transformation from the low to the high temperature phase. Normally, this volume will increase and hence ΔV will be positive in value. However, there are many cases where the low temperature phase is a covalently bonded open structure and the high temperature phase is a closer packed structure, such as a metallic phase. In this case, the sign of ΔV will be negative. For example, covalent Si is stable in the open diamond cubic structure as a solid and upon melting becomes a metallic liquid of smaller molar volume.

2.4. Effect of magnetic entropy on polymorphic transitions

In the foregoing, we have suggested only one reason why we expect the energies(entropies) of different polymorphs of a substance to differ. Namely, we have suggested that the cohesive energy (entropy) is a function of the close-packing or coordination number. Actually, this is only one of the factors that affect the cohesive energy and entropy. For example, the state of electron spins leading to atomic magnetic moments can affect these quantities sufficiently so as to bring about polymorphic transitions in iron. Indeed, the stable low temperature polymorph of iron is bcc. Above 910 °C and below 1390 °C the fcc structure is the stable structure for iron, and above 1390 °C the bcc structure again becomes stable up to the melting point of iron.

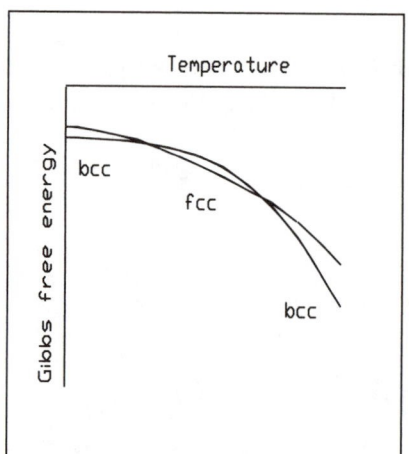

Figure 1.4. Schematic dependence of Gibbs free energy versus temperature for iron.

The clue to understanding this behavior was first provided by Zener[6], who noted that the curvature in the plot of $(G_{bcc} - G_{fcc})$ versus temperature exhibited a maximum in the vicinity of the Curie temperature for bcc iron. Thus, one may interpret the low temperature transition as follows. At low temperature bcc iron is ferromagnetic and hence the atomic magnetic moments are fully ordered in this structure. This contribution to the stability makes the bcc structure stable relative to the more close packed fcc structure at T=0°K. However, in the fcc structure, this ordered state of spins does not exist and hence there is a magnetic contribution to the entropy of the fcc structure. Thus, the magnetic entropy in the bcc structure and its total entropy is less than that for the fcc structure. Consequently the free energy versus temperature curves start out as shown in Figure 1.4. The higher entropy of the fcc structure brings about the first transition at 910 °C. The complete loss of short-range magnetic order of electron spins in the bcc structure does not occur until the temperature exceeds 910 °C. Because in the absence of any magnetic spin order, such as is the case at very high temperature above the Curie temperature, the bcc structure has a higher entropy than the fcc structure, as already noted, it will eventually become more stable than the fcc structure and this fact is responsible for the transition that occurs at 1390 °C. The latter transition obeys the general rule, which has its origin in the dependence of entropy on coordination number, while the low temperature transition has its origin in the different magnetic properties of bcc and fcc iron.

The rapid change in entropy that occurs below the first transition temperature (910 °C) is indicated by the rapid change in the specific heat of iron below this temperature that is shown in Figure A.4.1 in Appendix 4.

2.5. Other contributions to polymorphic transitions

Zener[7] has also suggested that some crystal structures have low resistance to certain crystal shear displacements. For example, the bcc structure will have negligible resistance to a $(110)[1\bar{1}0]$ shear if many-body terms are absent in the bonding. (Nearest-neighbor interatomic distances are not distorted to a first approximation by such a shear displacement.) This factor may contribute significantly to the greater thermal entropy of the bcc structure relative to the fcc one. Similar 'soft mode' phenomena may be operating in other crystal structures tending to stabilize one structure at the expense of another with increasing temperature. Such 'soft mode' induced transitions have been found in some systems that also exhibit superconducting transitions.

3.THERMODYNAMIC STABILITY AND METASTABILITY.

We have used the term 'equilibrium' without defining it. We are, of course, discussing thermodynamic equilibrium. Hence, thermodynamically, equilibrium is defined in terms of the appropriate thermodynamic potential. For a closed system, equilibrium corresponds to the situation where the entropy of the system is at a maximum, i.e.$dS=0$ and $d^2S<0$. For the case of a system at constant temperature and pressure, the system is at equilibrium when $dG=0$ and $d^2G>0$. But these conditions can be satisfied at any minimum in the free energy for the latter case or maximum in the entropy for the former case. That there may be different minima in the free energy, for example, is illustrated in Figure 1.5. As shown, at point 'a' there is a relative minimum while at point 'b' there is an absolute minimum in the dependence of Gibbs free energy on the coordinate describing the state of the system. The former represents a condition of metastable equilibrium while the latter that of stable equilibrium. The two are differentiated by the fact that a large perturbation in the reaction coordinate describing the system will lead to a spontaneous decrease in the free energy, whereas for the case of stable equilibrium, no matter how large this perturbation may be, the free energy of the system will always increase.

Is a "frozen-in" amorphous solid or glass in a metastable state? In the absence of thermal energy to overcome activation energy barriers to diffusion, the "frozen-in" solid may be considered to be in a metastable state. However, in the absence of thermal energy to allow for the interchange of atoms between phases it is not possible for the "frozen-in" solid to be in metastable equilibrium with another phase. A partial metastable equilibrium may be possible, which does not involve the exchange of atoms, such as mechanical equilibrium, but this is not usually an interesting case.

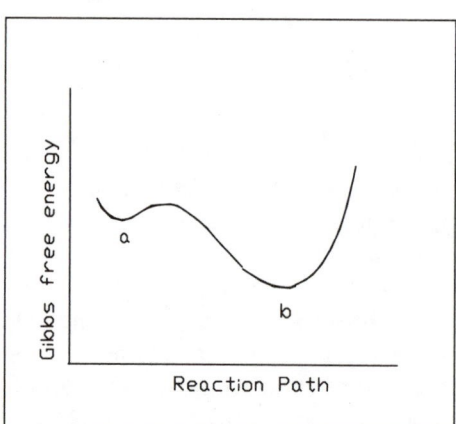

Figure 1.5. Illustrating metastable and stable equilibrium.

4. SOME THERMOPHYSICAL PARAMETERS.

At equilibrium, there are a variety of derivative properties of the thermodynamic potentials. For example, the bulk modulus is given by $-V[\frac{\partial^2 G}{\partial V^2}]_T$. A listing of the relations between derivatives of thermodynamic parameters and physical properties is given in Table 1.2.

Table 1.2.   Summary of Thermodynamic Relations[a]

X	Y	Z	$\left(\dfrac{\partial Y}{\partial X}\right)_Z$	X	Y	Z	$\left(\dfrac{\partial Y}{\partial X}\right)_Z$
T	V	p	αV	T	p	V	α/β
T	S	p	C_p/T	T	S	V	$C_p/T - \alpha^2 V/\beta$
T	V	p	$C_p - \alpha p V$	T	U	V	$C_p - \alpha^2 VT/\beta$
T	H	p	C_p	T	H	V	$C_p - \alpha^2 VT/\beta + \alpha V/\beta$
T	F	p	$-\alpha pV - S$	T	F	V	$-S$
T	G	p	$-S$	T	G	V	$\alpha V/\beta - S$
p	V	T	$-\beta V$	T	p	S	$C_p/\alpha VT$
p	S	T	$-\alpha V$	T	V	S	$-\beta C_p/\alpha T + \alpha V$
p	U	T	$\beta pV - \alpha VT$	T	U	S	$\beta p C_p/\alpha T - \alpha pV$
p	H	T	$V - \alpha VT$	T	H	S	$C_p/\alpha T$
p	F	T	βpV	T	F	S	$\beta p C_p/\alpha T - \alpha pV - S$
p	G	T	V	T	G	S	$C_p/\alpha T - S$

[a] From J. Lumsden, *Thermodynamics of Alloys*, Institute of Metals, London, 1952,

* U in the table corresponds to our internal energy E.

5. ISING PROBLEM, LONG-RANGE ORDER (SUPERLATTICES) OF SPINS OR OF ATOMS ON LATTICE SITES.

Consider the atom positions in the common crystal structures, such as the fcc, the bcc and the hcp structures, of binary solid solutions or the equivalent of a distribution of up and down spins on these lattice sites. At stoichiometric compositions, the unlike atoms (spins) can be rearranged on these atom positions to form what are called superlattices. Figure 1.6 illustrates some of the possible superlattices in the fcc and bcc structures for various stoichiometric compositions.

If the unit cell of a superlattice is translated along each of the unit cell

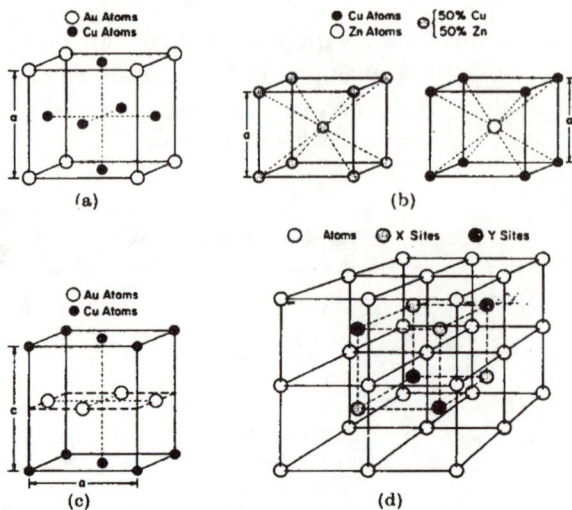

Figure 1.6.   Various types of ordered superlattices: (a) Cu$_3$Au, (b) disordered and ordered structure of b brass, (c) CuAu, (d) the structure of Fe$_3$Al and FeAl.

dimensions to build up a macroscopic crystal, the latter will be said to be a long-range ordered structure. Such long-range ordered structures can transform to the disordered parent solid solution on increasing temperature. This transformation can be of nth order, as measured by a discontinuity in the nth derivative of the free energy with respect to temperature or to any other independent variable.

Let us first consider the significance of long-range order.   Long-range order may be differentiated from short-range order as follows. Consider the unit cell of the long-range ordered superlattice. Translation of this unit cell along each of the axes of the unit cell by the corresponding lattice parameter results in the construction of the three dimensional ordered crystal. The lattice sites corresponding to a particular atom of the unit cell make up a lattice of sites that are occupied by only that type atom in the long-range ordered crystal. The degree of long-range order may thus be related to the fractional occupancy of a lattice of sites by the type of atom that occupies this type site in the long-range ordered unit cell. A typical definition of the long range order parameter is then

$$\eta = (\rho*-c)/(1-f*)$$

where $\rho*$ is the fraction of the total number of * lattice type sites occupied by the atom that occupies this type site of the superlattice unit cell, c is the concentration of this type atom in the alloy and f* is the fraction of the total number of atoms per superlattice unit cell corresponding to the site in question. In the event that there are more than two types of atoms (i.e. a ternary or higher system) then more than one long-range order parameter is required to describe the long-range order in the long-range ordered phase.

Short-range order corresponds to the number of unlike atoms that are nearest-neighbors to a given atom relative to this number in the completely disordered alloy. Thus, the short-range order parameter can be described by

$$\sigma = (N_{AB}(sro)-N_{AB}(random))/(N_{AB}(lro)-N_{AB}(random))$$

where N_{AB} refers to the number of unlike atom bonds in the short-range ordered condition (sro), the number in the perfectly long-range ordered condition (lro) or the number in the random disordered condition (random). Another definition of short-range order parameter involves the correlation parameters for different coordination spheres about an atom. For example,

$$^{LL*}\varepsilon_{AB}(\rho_l) = {}^{LL*}p_{AB}(\rho_l)-{}^{L}p_A{}^{L*}p_B$$

defines the correlation parameter $^{LL*}\varepsilon_{AB}(\rho_l)$ in terms of the probability, $^{LL*}p_{AB}(\rho_l)$, that sites L are occupied by the atom A and that sites L* located a distance ρ_l from atom A (in the lth coordination sphere) are occupied by atoms B, while $^{L}p_A$ and $^{L*}p_B$ are the probabilities for the occupation of L and L* sites, as defined for the case of the long range order parameter, by atoms A and B, respectively.

If the probabilities of occupation of sites of a given type were independent of the arrangement of atoms on the other sites, then the correlation factor would equal zero because the first term on the right hand side of the above equation would equal the second term. When the correlation factor is not equal to zero then there can be both short and long range order in a long-range ordered phase. For disordered alloys in which the long-range order parameter is equal to zero the existence of short-

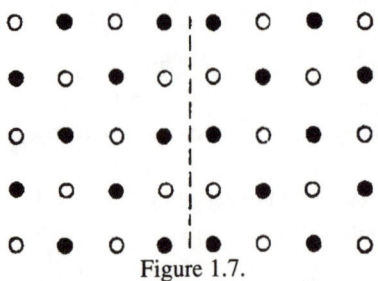

Figure 1.7.

range order is related to a non-zero value of the correlation factor. The short-range order parameter in disordered alloys can be related to the correlation factor, as follows.

$$\alpha(1) = 1 - p_{AB}(\rho_l)/(C_A C_B)$$
$$= -\varepsilon_{AB}(\rho_l)/(C_A C_B)$$

There are two manifestations of partial long-range order. Both correspond to a value of η less than unity but greater than zero. However, one achieves this result in the absence of anti-phase boundaries and the other brings this about by the presence of anti-phase boundaries between perfectly short-range ordered domains. Figure 1.7 illustrates a case where there is complete short-range order in each domain, but where the long-range order parameter equals zero. It is believed that most ordering systems exhibit both types of partial long-range order at equilibrium.

Superlattices will generally transform from the long-range ordered state to a disordered state as the temperature is increased. This transformation can be either of first-order or of higher order in the discontinuity in the derivative of the free energy with respect to temperature. In the copper-gold system, the order-disorder transformations that occur are all first-order ones. However, that in the copper-zinc system, at the equiatomic composition, is a second-order transformation. In the former, the long-range order parameter changes discontinuously at the transformation temperature, while in the latter there is a continuous decrease in the long range order parameter as the temperature is increased. The reason for these differing behaviors is related to the crystal structure of these superlattices. This subject has been considered in detail by Landau and Lifshitz. Let us now examine examples of the various types of superlattices.

It may not be apparent but for the unit cells shown in Figure 1.6 the arrangement of atoms on the basis lattice is such as to maximize the number of bonds between unlike atoms. However, although for many cases the driving force for long range order may be a tendency to maximize the number of bonds between unlike atoms, other factors contribute to the formation of a superlattice arrangement of the atoms. For example, consider the Cu-Pt superlattice which consists of (111) planes containing only Cu atoms alternating with planes containing only Pt atoms. In this case, the number of bonds between unlike atoms is the same in the superlattice arrangement as in a random arrangment of the atoms on the fcc basis lattice. Thus, models of the thermodynamics of the long-range order phenomenon which limit their description of the energy to a sum over bond energies are likely to be insufficient descriptions in some cases. Modern models of the thermodynamics of partially disordered superlattices thus allow for the possibility of including many-body terms in the energy.

We have given no indication how to express the energy or entropy of a partially long-range-ordered solid solution on the long-range order parameter. This is not a simple matter and has not been rigorously solved for three dimensions. However, good approximations have been developed. They are of two kinds. One makes use of computer simulation to achieve a non-analytic description of the energy and entropy, and the other of analytic methods of approximating the energy and entropy. Reviews of the latter may be found in the article by De Fontaine .

In the simplest treatment of the order-disorder transition, the Bragg-Williams model, it is assumed that the distribution of atoms on one sub-lattice is independent of the distribution of atoms on the other sub-lattices. Thus, short-range order is neglected in the Bragg-Williams model. The result is a poor quantitative description of the order parameters. However, a review of the procedure used to evaluate the free energy of the partial long range ordered system is instructive and is given below.

Consider, the beta-brass system. In this case, there are two simple cubic sub-lattices, each one occupied by one of the two components, Cu or Zn, in the completely long range ordered superlattice. The nearest-neighbor sites of one lattice belong to the other lattice. An assumption in this model is that the energy is described as a sum over nearest-neighbor bond energies. There are four classes of such bonds: A/I-A/II, A/I-B/II, B/I-A/II, B/I-B/II, where A,B denote the two types of atoms and I,II denote the corresponding two sub-lattices occupied by these atoms, respectively in the superlattice. Denoting the corresponding number of such bonds by $N(1)$, $N(2)$, $N(3)$ and $N(4)$, respectively, then

$$N(1) = Nz\rho(1-\rho)$$

$$N(2) = Nz\rho^2$$

$$N(3) = Nz(1-\rho)^2$$

$$N(4) = Nz\rho(1-\rho)$$

where, N equals half the total number of atoms in this equiatomic system, z is the coordination number (number of nearest-neighbors, equal to 8 in this case), and ρ is the fraction of A(B) atoms on the I(II) sub-lattice. Thus, the energy of the partially long-range ordered system is

$$E = Nz\rho(1-\rho)(E_{AA} + E_{BB}) + Nz(\rho^2 + (1-\rho)^2)E_{AB}$$

The entropy of configuration corresponding to the Bragg-Williams approximation is based upon the assumption that the atoms on a given type of lattice sites are distributed randomly and is then

$$S = k\ln W = k\ln\{N!/[(N\rho)!(N(1-\rho))!]\}^2$$

Using Stirling's approximation for the factorial of large numbers, then

$$S = -2Nk(\rho\ln\rho + (1-\rho)\ln(1-\rho))$$

Since the energy of the superlattice is given by NzE_{AB} and the entropy of configuration of the superlattice is zero then the disordering energy and entropy are

$$\Delta E = -2Nz\rho(1-\rho)[E_{AB} - (E_{AA} + E_{BB})/2]$$

$$\Delta S = -2Nk(\rho\ln\rho + (1-\rho)\ln(1-\rho))$$

Minimizing the free energy $\Delta F = \Delta E - T\Delta S$ yields the equilibrium value of the occupation probability, ρ, and by its relation to the long-range order parameter, η, the equilibrium value of the latter. Long-range order disappears above a critical temperature, equivalent to the Curie temperature in ferromagnetism, given by

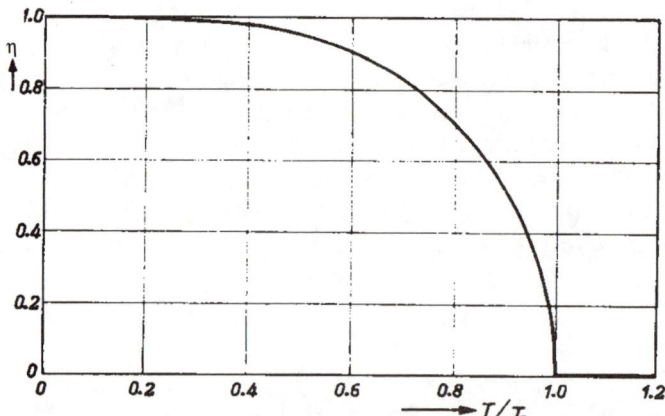

Figure 1.8. Long-range order parameter as a function of reduced temperature in the BraggWilliams model.FromJ.D. Fast, Entropy, Philips Technical Library,1962.

$$kT_C = -4[E_{AB} - (E_{AA} + E_{BB})/2]$$

Figure 1.8 shows the equilibrium value of the long-range order parameter as a function of the reduced temperature in the Bragg-Williams model. The transition at T_C is second-order. That this model is grossly inadequate needs to be restated, because it appears to describe the partial long-range order in beta brass, at least qualitatively, fairly well. Much better quantitative models and computer models have been developed, but their description is outside the scope of the present volume.

6. SUMMARY

Summarizing, in this chapter we have introduced the concept that the most stable phase, at constant temperature and pressure, has the most negative value of the Gibbs free energy. Also, we have recalled that at equilibrium any system, at constant temperature and pressure, has its lowest possible value for the Gibbs free energy. We have found it possible to partition the various thermodynamic potentials into various categories according to their origin. For example, the entropy can be partitioned into thermal and configurational contributions. Also, at equilibrium, a system constrained to a constant composition proceeds, from low to high temperature (from low to high pressure), through a sequence of phases in such a way that the entropy (volume) continually increases (decreases). If the increase in entropy (decrease in volume) is discontinuous at distinct temperatures (pressures) corresponding to transitions from the regime of stable equilibrium of one phase to that of another, the transitions are of first-order. For first-order transitions, it is possible to represent the regimes of phase stability in a P-T phase diagram. The Clapeyron equation yields the dependence of the transition temperature on pressure for polymorphic transitions. We have also defined the state of thermodynamic equilibrium and have been introduced to the concepts of long- and short-range order, the equilibrium values of which are determined by minimizing the free energy with respect to the order parameter.

Appendix 1.

The table below lists the energies of elements at 0 °C in cal/gatom.

AG	-68400.	K	-21500.	SC	-88000.
AL	-77500.	LA	-102000.	SN	-72000.
AM	-66000.	LI	-38400.	SR	-39100.
AU	-87300.	MG	-35600.	TA	-186800.
BA	-42500.	MN	-66700.	TC	-125000.
BE	-77900.	MO	-157500.	TH	-136600.
BI	-47210.	NA	-25900.	TI	-112700.
CA	-42200.	NB	-173000.	TL	-43000.
CD	-26750.	NI	-102800.	U	-125000.
CE	-97000.	NP	-105000.	V	-123000.
CO	-101600.	OS	-187000.	W	-200000.
CR	-95000.	PB	-46800.	Y	-98000.
CS	-18700.	PD	-91000.	YB	-40000.
CU	-81100.	PR	-80000.	ZN	-31200.
EU	-42000.	PT	-135200.	ZR	-146000.
FE	-99500.	PU	-92000.	IN	-59100.
GD	-84000.	RB	-19500.	SI	-108000.
HF	-160000.	RE	-187000.	GE	-90000.
HG	-15320.	RH	-133000.	GA	-69000.
IR	-159000.	RU	-153000.	SB	-62000.
				AS	-69000.
				P	-79800.

Lattice Stability Parameters (1)
(cal/g-atom)

ELEMENT	$F_{bcc}-F_{fcc}$	$F_{fcc}-F_{hcp}$	$F_{bcc}-F_{hcp}$
Ag	900−0.45T		
Al	2410−1.15T		
Au			
Ba			−1040−0.8T
Be	635−1.15T	+465+0.43T	1100−0.72T
Ca	58−0.08T		
Cd			−700+0.6T
Co	+1390+1.05T	110−0.15T	1500+0.9T
Cr	−2500−0.15T	500+0.15T	−2000
Cs			
Cu	1500−0.8T		
	850−0.2T	−150−0.3T	700−0.5T
Fe	−1303(0°K)		
Hf	1030−0.9T	800	1830−0.9T
Hg			
Ir	1650+1.05T	−150−0.15T	1500+0.9T
K			
Li			
Mg	635−1.15T	465+0.43T	1100−0.72T
Mn	420−0.3T		
Mo	−2500−0.15T	500+0.15T	−2600
Na			
Nb	−2150−0.85T	650+0.05T	−1500−0.8T
Ni	1350+0.8T	−250−0.3T	1100+0.5T
Os	1000	120+0.8T	1120+0.8T
Pb		−600+0.1T	
Pd	1350+0.8T	−250−0.3T	1100+0.5T
Pt	1350+0.8T	−250−0.3T	1100+0.5T
Rb			
Re	150+0.1T	250+0.3T	400+0.4T
Rh	1650+1.05T	−150−0.15T	1500+0.9T
Ru	1000	120+0.8T	1120+0.8T
Sn			
Sr	200−0.23T		
Ta	−2150−0.85T	650+0.05T	−1500−0.8T
Tc	150+0.1T	250+0.3T	400
Ti	240−0.9T	800	1040−0.9T
Tl	14−0.22T	76+.04T	90−0.18T
V	−2150−0.85T	650+0.05T	−1500−0.8T
W	−2500−0.15T	500+0.15T	−2000
Zn	250−0.2T	440−0.4T	690−0.6T
Zr	230−0.9T	800	1030−0.9T

APPENDIX 1

Successive tables in this appendix list values of lattice stability energies for compounds.

Thermodynamic Data for Transition from Rock Salt to Cesium Chloride Structure

Compound	P (kbar)	T (K)	$-\Delta V^\circ$ (cm^3 mol^{-1})	ΔG (cal mol^{-1})	$\Delta H^\cdot$ (cal mol^{-1})	ΔS° (cal mol^{-1} K^{-1})
NaCl	290 ± 25[a]	298	0.864	6067[f]	6514	+1.50[g]
KF	17.1[a]	298	3.747	1551	1274	−0.924[h]
KCl	19.3[a]	298	4.315	2015	2022	+0.025[e]
KBr	17.4[a]	298	4.517	1902	1884	−0.060[g]
KI	17.8[a]	298	4.489	1934	1994	+0.200[g]
RbCl	5.2[a]	298	6.045	761	892	+0.440[g]
RbBr	4.5[a]	298	6.029	656	555	−0.340
Rb5	3.4[a]	298	7.478	2260	2156	−0.350[g]
CsCl	0.001[a]	745	8.115	0	−760	−1.060[h]
CsBr	0.001	(1153)[r]	(9.41)[r]	0	−1175	−1.019[h]
AgF	26[a]	373	1.647	1033	1168	+0.361[h]
BaS	65 ± 5[b]	298	4.826	7590	7352	−0.225[h]
BaO	145 ± 35[c]	298	3.351	12390	12311	−0.263[h]
CaO	650 ± 50[d]	298	1.844	29006	29012	+0.021

Thermodynamics of Phase Transitions in Pyroxenes and Pyroxenoids

Transition	Change in tetrahedral repeat	ΔH^c (cal mol^{-1})	ΔS° (cal K^{-1} mol^{-1})	ΔG° (cal mol^{-1})	ΔV° (cc mol^{-1})
MgSiO$_3$(ortho → clino)[a]	∞ → ∞			<300	+0.03
FeSiO$_3$(ortho → clino)[b]	∞ → ∞	−40	−0.097	+85	−0.06
CoSiO$_3$(ortho → clino)[b]	∞ → ∞	0	−0.25	+318	−0.23
MnSiO$_3$(rhodonite → pyroxmangite)[c,d]	5 → 7	60	−0.246	+373	−0.391
MnSiO$_3$(pyroxmangite → clino)[c,d]	7 → ∞	−211	−0.635	+846	
Fe$_{0.5}$Ca$_{0.5}$SiO$_3$(hedenbergite → bustamite)[e]	∞ → 3	400			−2.35
CaSiO$_3$(wollastonite → II)[f]	3 → peculiar chain silicate	+1580	+0.30	+1198	−0.77
CaSiO$_3$(wollastonite → pseudowollastonite)[f]	3 → 3	+1370	+0.99	+110	+1.00
Fe, Mg disorder in orthopyroxene[g]	∞ → ∞			+3600	
Ca, Mg disorder in diopside[h]	∞ → ∞	1500			

Thermodynamics of Transition from Rock Salt to Nickel Arsenide Structures

Compound	ΔV (cm^3/mol)	ΔV (%)	ΔG (cal)	P (kbar)
MnTe	−0.77	−2.233	0[a]	0.001[a]
CaTe	−1.089	[b]	6447[c]	248[d]
MnSe	−0.826	−3.378	860[e]	90[e]
FeSe			−1940[c]	Negative
CoSe			−784[c]	Negative
NiSe			−1415[c]	Negative
MgS	−0.594	[b]	9813[c]	691[d]
MnS	−0.603	[b]	1972[c]	137[d]
FeS			−248[c]	Negative
MgO	−0.313	[b]	24888[f]	3327[d]
MnO	−0.371	[b]	8041[f]	907[d]
FeO	−0.339	[b]	2319[f]	286[d]
CoO	−0.326	[b]	1443[f]	185[d]
NiO	−0.308	[b]	1497[f]	203[d]
ZnO	−0.326	[b]	6014[f]	772[d]
CaO	−0.467	[b]	28163[f]	2523[d]

Thermochemical Data for Transitions among Olivine (α), Spinel (γ), and Modified Spinel (β) Phases[a]

$\alpha \rightarrow \gamma$	ΔV°_{298} (cm^3)	ΔH°_{986} (cal)	ΔS^c (cal/deg)
Mg$_2$SiO$_4$	−4.133	5133 ± 1000	−5.593 ± 1.0
Fe$_2$SiO$_4$	−4.353	703 ± 198	−4.774 ± 0.2
Co$_2$SiO$_4$	−3.916	2690 ± 300	−3.142 ± 0.3
Ni$_2$SiO$_4$	−3.420	1430 ± 700	−1.389 ± 0.7
Mg$_2$GeO$_4$	−3.520	−3030 ± 530	−2.790 ± 0.5
$\alpha \rightarrow \beta$			
Mg$_2$SiO$_4$	−3.244	4011 ± 1000	−4.149 ± 1.0
Co$_2$SiO$_4$	−2.901	2150 ± 390	−2.164 ± 0.3
$\beta \rightarrow \gamma$			
Mg$_2$SiO$_4$	−0.889	1122 ± 1000	−1.444 ± 0.8
Co$_2$SiO$_4$	−1.015	540 ± 430	−0.978 ± 0.3

APPENDIX 2.

Entropy of gases.

To relate entropy to statistically based quantities we follow the lead of Fast (ENTROPY, Philips Technical Library, 1962) who noted that from thermodynamics the change in entropy upon increase in volume of a gas of n molecules from V_1 to V_2 is

$$\Delta S = nkln(V_2 / V_1) \tag{A2.1}$$

and upon mixing of n_1 molecules of gas 1 of volume V_1 and n_2 molecules of gas 2 of volume V_2 by removing a partition separating the two volumes is

$$\Delta S = k\{ n_1ln[(V_1+V_2)/V_1]+n_2ln[(V_1+V_2)/V_2]\} \tag{A2.2}$$

Now consider that the volume, V_1, containing gas is partitioned to produce many equal volumes, the number of which is much larger than the number of molecules in the volume, and the size of which is such that no cell contains more than one molecule. As a result of the constant thermal motion of the molecules, different cells become occupied with time. Suppose there are N_1 cells and n molecules. Thus, n cells are occupied and N_1-n are unoccupied. The total number of different configurations is then

$$W_1 = N_1!/[n!(N_1-n)!]$$

because the first molecule can fill N_1 boxes, the second N_1-1, and so on, but not all these configurations are different because there are n indistinguishable molecules and N_1-n indistinguishable cells.

Using Stirling's approximation for large numbers, $lnN! = NlnN-N$ then

$$lnW_1 = -nln(n/N_1) - (N_1-n)ln[(N_1-n)/N_1]$$

Suppose the volume containing the n molecules were increased to V_2. This is equivalent to increasing the number of cells to N_2, each cell maintaining the same cell volume, (i.e. $V_2 /V_1 =N_2 /N_1$). Then, obviously

$$lnW_2 = - nln(n/N_2) - (N_2-n)ln[(N_2-n)/N_2]$$

and carrying out the algebra we find that

$$ln(W_2/W_1) = nln(V_2/ V_1) \tag{A2.3}$$

Now let us consider n_1 type 1 molecules in volume V_1 and n_2 type 2 molecules in volume V_2, which are allowed to mix by removing a partition between the two volumes.

Before the mixing we had

$$\ln W_1 = - n_1 \ln(n_1 / N_1) - (N_1 - n_1)\ln[(N_1 - n_1)/ N_1]$$

$$\ln W_2 = - n_2 \ln(n_2 / N_2) - (N_2 - n_2)\ln[(N_2 - n_2)/ N_2]$$

Thus, before the mixing the total number of different configurations is the product of W_1 and W_2. After the partition is removed we have $n_1 + n_2$ molecules distributed over $N_1 + N_2$ cells and the number of different configurations is given by

$$W' = (N_1 + N_2)!/\{ n_1! \, n_2! \, [N_1 + N_2 - (n_1 + n_2)]! \}$$

We are interested in the ratio $W'/(W_1 W_2)$ which can be shown to be given by

$$\ln[W'/(W_1 W_2)] = n_1 \ln[(V_1 + V_2)/V_1] + n_2 \ln[(V_1 + V_2)/V_2] \qquad (A2.4)$$

for the same approximation $n_1, n_2 \ll N_1, N_2$ and $V_2/V_1 = N_2/N_1$. Comparing (A2.4) with (A2.2) and (A2.3) with (A2.1) allows us to write

$$\ln[W_{final}/W_{initial}] = \Delta S/k$$

If the initial state is taken as the reference state at 0 K for equilibrium at complete order the third law of thermodynamics yields $S_o = 0$. Hence,

$$S = k \ln W$$

Although this relation was developed on the basis of change in the entropy due to change of volume of a gas or to mixing of two gases, it is more general in application. In the next section we derive the relation between entropy and statistical parameters for a canonical ensemble of systems, which need not be limited to gases.

Canonical ensemble and entropy.

Let us consider an ensemble of N total number of systems immersed in a constant temperature bath with which each system can exchange only energy subject to the constraints that the total energy of the ensemble is constant and N is constant. This type of ensemble is called "canonical". Let n_i be the number of systems in the ensemble in the ith quantum state. A complexion is a way of arranging the members of the ensemble of systems, each member in a different box. The number of complexions corresponding to some particular distribution of quantum states $\{n_i\}$ is then

$$W\{n_i\} = N!/ \Pi_i n_i! \qquad (A2.5)$$

The most probable set of quantum states $\{n_i\}$ maximizes $W\{n_i\}$ subject to the constraints already described. Thus,

$$\delta \ln W\{n_i\} = 0 \qquad (A2.6)$$

$$\delta \, (NE) = \delta \, \Sigma_i n_i E_i = 0 \qquad (A2.7)$$

$$\delta \, N = \delta \, \Sigma_i n_i = 0 \qquad (A2.8)$$

where E_i is the energy of a system of the ensemble in the ith state, NE is the total energy of the ensemble and E is the average energy over the energies of the systems in the ensemble. Using Lagrange's method of undetermined multipliers we multiply the latter two equations by $-\beta$ and $-\alpha$, respectively and add them to the first equation to obtain after suitable use of Stirling's approximation for the factorial of a large number (i.e. $N! = N\ln N - N$)

$$\Sigma_i \, \delta n_i(\alpha + \beta E_i + \ln[n_i/N]) = 0 \qquad (A2.9)$$

But, use of the undetermined multipliers can make only two of the terms in the parentheses ( ), say the jth and the kth ones equal to zero. The δn_i are completely independent of each other. Hence, the latter equation can onlybe satisfied if each term in the parentheses ( ) is equal to zero. Thus,

$$p_i = n_i/N = \exp[-\alpha - \beta E_i] \qquad (A2.10)$$

is the canonical distribution function and the fraction of the systems of the ensemble in quantum state i.

We now set about to determine the values of the undetermined multipliers by comparing functions we shall derive to thermodynamic functions. We sum p_i over all states using the normalization condition

$$\Sigma_i p_i = 1 = [\exp(-\alpha)]\Sigma_i \exp(-\beta E_i) \qquad (A2.11)$$

Solving for $[\exp(-\alpha)]$ and substituting we obtain

$$p_i = \exp(-\beta E_i)/\Sigma_i \exp(-\beta E_i) \qquad (A2.12)$$

The sum in the denominator is called the partition function and is denoted by Z.
The average energy of the ensemble is given by

$$E = \Sigma_i p_i E_i \qquad (A2.13)$$

We rewrite an equation obtained above into the following form

$$\Sigma_i p_i E_i = -\beta^{-1}\ln[\Sigma_i \exp(-\beta E_i)] - \beta^{-1}\Sigma_i p_i \ln p_i \qquad (A2.14)$$

The left hand side equals the energy E. We now draw attention to the thermodynamic relation

$$E = F - T(\partial F/\partial T)_V \qquad (A2.15)$$

Let $L = \beta^{-1} = kT$ and then we can write the above relation as

$$E = F - L(\partial F/\partial L)_V \qquad (A2.16)$$

If we now equate F to $-L\ln\Sigma_i \exp(-E_i/L)$ we will note that on taking the derivative of F with respect to L at constant volume V, which is equivalent to taking it with respect to E_i since holding the latter constant corresponds to holding the volume constant, we obtain

$$(\partial F/\partial L)_{Ei} = -[\ln\Sigma_i \exp(-E_i/L) + (1/L)\Sigma_i E_i \exp(-E_i/L)/\Sigma_i \exp(-E_i/L)]$$
$$= [F - E]/L \qquad (A2.17)$$

In arriving at the latter result we have made use of the fact that the average energy E is given by

$$E = \Sigma_i E_i \exp(-E_i/L)/\Sigma_i \exp(-E_i/L) \qquad (A2.18)$$

We thus see that our statistical relation for the Helmholz free energy F satisfies the thermodynamic relation between E and F given above providing that $\beta^{-1} = kT$. Further, since the entropy S satisfies

$$E = F + TS \qquad (A2.19)$$

and we have already identified the first two terms in the statistical equation (A2.10) as E and F. Hence, the last term must equal TS. Thus,

$$S = -k\Sigma_i p_i \ln p_i \qquad (A2.20)$$

APPENDIX 3.

Consider a polymer having N total mers constrained to lie in a plane and having the mers making 90° angles with each other, with the mers lying at 45° to the horizontal axis, as illustrated in Figure A3.1. Let the total x-projected length of the polymer equal **L** and the x-projected length of each mer equal **a**. If the number of mers pointed to the right exceeds that directed to the left by 2n then

$$2n = L/a$$

Thus, the number of mers pointed to the right is given by

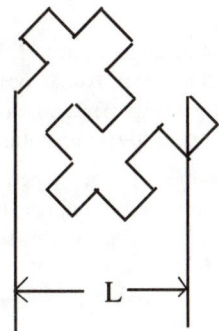

Figure A3.1 Schematic two dimensional polymer illustrating horizontal component of end-to-end length,L, as compared to total length of polymer.

$$\#\!\rightarrow\, = N/2 + n$$

while the number directed to the left is given by

$$\#\!\leftarrow\, = N/2 - n.$$

It is now possible to set down an expression for the entropy of the polymer

of x-projected length L as follows

$$S = k \ln\{2N!/[(N/2+n)!(N/2-n)!]$$

APPENDIX 4.

Specific Heat.

For a crystal containing N atoms there will be $3N$ independent normal modes of oscillation each of frequency ω_i. Einstein considered the case of $3N$ independent harmonic oscillators of the same frequency, the average energy of which is

$$3Nh\omega\,/[\exp(h\omega/kT)-1].$$

Using the definition of the specific heat $C_V=(\partial E/\partial T)_V$, then

$$C_V = 3Nkx^2e^x(e^x-1)^{-2}$$

is obtained, where $x=h\omega/kT$. For large values of T, $C_V=3Nk=3R$, in agreement with experiment. This latter result holds approximately for solid compounds also, with the proviso that the dimensions are energy per atom or energy per gm-atom. However, the variation of C_V with T, especially at low temperatures, does not agree with experiment.

Debye suggested that the solid be treated as a continuous isotropic medium, which leads to the existence of a range of frequencies from zero to some maximum ω_m, to a density of normal modes that varies as the square of the frequency up to some maximum frequency such that the total number of normal modes equals 3N. He then found that

$$E = [9NkT/(x_m)^3]\int_0^{X_m}\{x^3/(e^x-1)\}dx$$

from which the specific heat is found to be

$$C_V= [9Nk/\{2x_m^3\}]\int_0^{X_m} [x^4/\{\cosh x - 1\}]dx$$

This relation reduces to 3Nk at high temperature and at low temperature to

$$12\pi^4Nk/[5x_m^3] = CT^3$$

both of which are in agreement with experiment. Since the Debye temperature $\theta=h\omega_m/k$, it is possible to use the experimental data for the specific heat at low temperature to evaluate a value for θ. The latter temperature may also be obtained from the dependence of C_V at higher temperature (i.e. 0.5θ to 0.75θ), from elastic constant data and from thermal expansion data. The relation of the Debye temperature to the elastic constants is given by

$$\theta = (hc/2\pi k)[6\pi^2N/V]^{1/3}$$

where V is the volume of the crystal and the average elastic wave velocity for an isotropic crystal is given by

$$(c)^3 = 3/[2/(c_t)^3 + 1/(c_l)^3]$$

and c_t and c_l are the transverse and longitudinal wave velocities, respectively, i.e. $c_t=(C_t/r)^{1/2}$ and $c_l=(C_l/r)^{1/2}$, where r is the density, C_t is the transverse modulus and C_l is the longitudinal modulus.

Since the entropy and energy are related to the specific heat via the relations

Appendix 4

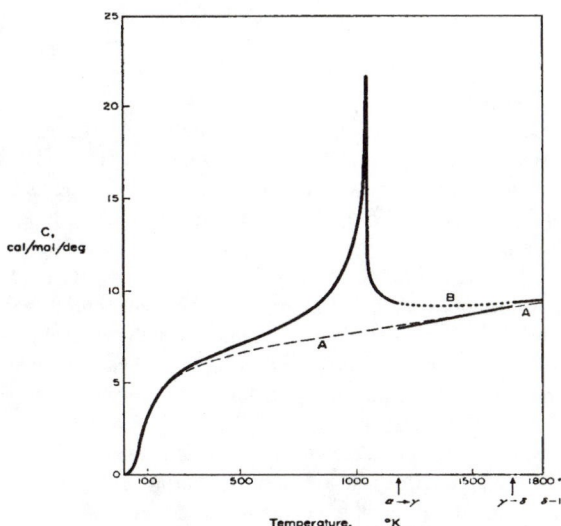

Figure A4.1. The measured specific heat of iron from T = 0°K to 1808°K (solid curves). Curve A is the estimated non-magnetic specific heat of b.c.c. iron and curve B is the estimated total specific heat of b.c.c. iron in the range over which f.c.c iron is stable,1183°K to 1673°K. The discontinuities occur at the α→γ transition temperatures.

$$S= \int_0^T (C/T)dT \text{ and } E= \int_0^T CdT$$

it is possible from the Debye theory for C to derive the related values of S and E.

The Debye model is itself an approximation for the specific heat of thermal origin. Real crystals do not have the density of normal modes depend on the square of the frequency. The assumption that solids are continuous isotropic media is too simple. The atomic nature of the solid must be taken into account. This subject forms the modern field of investigation of phonons in solids.

In the above we have not considered any contribution to the specific heat other than that due to the vibrations of the atoms. There are of course other contributions, such as electronic and magnetic ones.

At this point it is of interest to note that in second order transformations, such as the ferromagnetic-paramagnetic one in iron, the specific heat exhibits a discontinuity at the transformation temperature, and an anomalous dependence on temperature, such as is shown in Figure A4.1. This dependence is due to the disordering of the atomic magnetic moments. The theory of such disordering is known as the theory of the Ising problem. An exact solution to this problem for 3 dimensions has not yet been obtained, although there are very good approximate solutions. An analogous problem is that of disordering of a long-range-ordered superlattice.

APPENDIX 5.

Let us treat the liquid as a single system. Thus, the partition function for the whole liquid will be of the form $\{z(T)\}^N/N!$, where $z(T)$ denotes the partition function of a single molecule and division by N! accounts for the indistinguishability of the molecules. But each molecule can move anywhere in the liquid as long as it does not overlap with another molecule. If the molecular volume, Ω, is defined in terms of a sphere having a diameter defined by the close packed spacing in a solid made up of these molecules at the absolute zero of temperature, then the free volume per mole in the liquid equals $(V - N\Omega)$ or Nv. Thus, the partition function of the whole liquid will contain the factor $(Nv)^N/N!$ or $(ev)^N$. In the crystal, on the other hand, each molecule has its own equilibrium position, and the molecules are localized systems. Consequently, the partition function of the whole crystal is simply $\{z(T)\}^N$, where $z(T)$ now contains only a volume factor nearly equal to v(i.e. the volume in which the molecule *oscillates* in the liquid in its potential well is nearly equal to that for *oscillation* of the molecule in the solid state, however, in the liquid state each molecule will experience the volume Nv.) Thus, the partition function of the whole crystal contains the factor v^N, whereas that of the liquid has the factor $(ev)^N$. This leads to an excess molecular entropy in the liquid of k log e = k, where k is Boltzmann's constant.

REFERENCES.

1. P. Villars, J. Less Common Met.92,215(1983);99,33(1984);102,199(1984);109, 93 (1985).

2. R. Roy, J. Am. Cer. Soc.60, 350 (1977).

3. J.D.Fast, ENTROPY , Philips Technical Library 1962, pp.147, 258.

4. J.A. Van Vechten, Phys.Rev.7B, 1479(1973).

5. ibid p.112, Also. R.J. Weiss, SOLID STATE PHYSICS FOR METALLURGISTS, Pergamon Press 1963, p.276.

6. C. Zener, Trans. A.I.M.M.E.203, 619(1955).

7. C. Zener, Phys. Rev. 71, 846(1947).

BIBLIOGRAPHY.

Statistical Thermodynamics:

1. J. Willard Gibbs, ELEMENTARY PRINCIPLES IN STATISTICAL MECHANICS, Ox Bow Press,Woodbridge, Conn.,1981.

2. E.Schrodinger, STATISTICAL THERMODYNAMICS, Cambridge University Press, 1946.

3. L.D. Landau and E.M. Lifshitz, STATISTICAL PHYSICS, 3rd ed.,Pergamon Press,

New York, 1980.

4. R.H. Fowler and E.A. Guggenheim, STATISTICAL THERMODYNAMICS, MacMillan, Cambridge, 1939.

5. C.Kittel, ELEMENTARY STATISTICAL PHYSICS, J. Wiley & Sons, New York, 1958.

Bonding Theories:

1. D.Pettifor in PHYSICAL METALLURGY eds. R.Cahn and P.Haasen, North-Holland Publishing Co. 1985.

2. E.S.Machlin in ENCYCLOPEDIA OF MATERIALS SCIENCE, ed.M.Bever, Pergamon Press 1985.

Interatomic Potentials:

First Principle:

1. J. Hafner, FROM HAMILTONIANS TO PHASE DIAGRAMS, Springer, Berlin, 1987.

2. J.A. Moriarty, Phys. Rev.B38, 3199(1988).

Empirical:

Metals: 1.S.M. Foiles, Phys.Rev.B32, 7685(1985). (Embedded atom method.)

2. A.M. Stoneham and R. Taylor, HANDBOOK OF INTERATOMIC POTENTIALS, AERE-Harwell, Osfordshire, 1981.

Semiconductors:1. F.H. Stillinger and T.A. Weber, Phys. Rev.B31, 5262(1985).

2. J. Tersoff, Phys.Rev.B39, 5566(1989).

Ionic Crystals: 1. G.R.A. Catlow and W.C. Mackrodt in LECTURE NOTES IN PHYSICS, vol. 166, Springer, Berlin, 1982.

Phonons:

1. M.Born and K. Huang,DYNAMICAL THEORY OF CRYSTAL LATTICES,Clarendon Press, Oxford 1954.

2. W. Cochran, THE DYNAMICS OF ATOMS IN CRYSTALS, Crane, Russak, New York 1973.

Data Banks for Thermodynamic Quantities of Solids:

1. F*A*C*T*, McGill University/Ecole Polytechnique, Montreal, Quebec H3C 3A7, Canada. Att. Prof. A.D. Pelton.

2. Manlabs, Inc., 21 Erie Street, Cambridge, Mass. 02139 (Dr. L. Kaufman.)

3. Laboratoire de Thermodynamique et Physico-Chemie Metallurgique associe au CRNS, Institut National Polytechnic, B.P. 44-38401 St. Martin d'Heres, Grenoble, France. (Prof. I. Ansara, C. Bernard.)

4. MTDS, National Physical Laboratory, Teddington, Middlesex, TW11 OLW, England.

5. Institut fur Theoretische Huttenkunde, Technische Hochschule, Aachen, Germany.

6. THERMOCALC, Royal Institute of Technology, Department of Metallurgy 10044 Stockholm 70, Sweden. (Prof. Mats Hillert.)

PROBLEMS.

1. Suppose theory suggested that the B2 phase MnOs would be a superb magnetic material, but reference to the literature showed no phase information for this binary system. Use the method of Villars[1] to determine the liklihood of obtaining a stable B2 phase of MnOs.

2. Ammonium Chloride has the CsCl structure(B2)in which the NH_4 ions occupy the body-centered sites and the Cl ions the corner sites of the unit cell. There are two possible orientations for the NH_4 ions. Above a critical temperature these orientations are randomly distributed among the NH_4 ions. Below the critical temperature there is a partial long range order to the distribution of these orientations. Use a Bragg-Williams type model to calculate the configurational entropy of a partially long range ordered crystal having an excess of one orientation over the other, i.e. in the perfectly ordered crystal there is only one orientation while in the random case there are equal numbers of the two orientations.

3. The explanation of Van Vechten for the difference in entropy of fusion between metals and covalent elements appears to imply that the entropy of metals at $0°K$ will not be zero. Provide an explanation that removes this violation of the third law of thermodynamics. (Hint. Relate the Van Vechten configurational entropy of the bonding electrons in a metal to the thermal entropy of vibrational origin.)

4. Suppose by radiation that defects can be introduced into a solid and thereby raise its Gibbs free energy. Use free-energy temperature curves to determine the effect of these defects on the melting point of the irradiated solid.

5. Given a thin film coherent with its substrate. The latter is elastically much stiffer than the film. The film will not deform plastically, but only elastically up to its melting point. Also, it is known that the elastic energy per unit volume in a thin film with principal stresses, t_1 and t_2, in the plane of the film, and $t_3(=0)$ normal to the plane of the film is given by $(1/2E)[t_1^2 + t_2^2 + t_3^2] - (v/E)[t_1t_2 + t_2t_3 + t_3t_1]$. Assume E= 200GPa, v=0.3,molar volume = 10 cm³/mol and that the difference in thermal expansion coefficient between substrate and film equals 0.00001/°C. If the film is deposited coherently at 500°C, the melting point in the absence of stress is 1500°C and the difference in free energy between liquid and solid is given by $\Delta g=\Delta s(T_M-T)$ with Δs=6 cal/mol/°C, calculate the change in the melting point of the stressed solid thin film when it is heated from its deposition temperature up to its melting point as compared to the melting point of the unstressed film.

6. Suppose that in the structure corresponding to some compound ABO_4 there are two sites (types I and II) that only one of the 4 atoms of type O per molecule can occupy, per unit cell corresponding to one molecular unit. The energy of the crystal varies linearly with site occupation varying from $-E$ when a random distribution of this O type atom exists on the I and II type sites to $-E-\Delta E$ (per mole of molecules) when only the II type sites are completely occupied by this O type atom and the I type sites are vacant in the crystal. Assume the occupation of any site is independent of the occupation of other sites. Evaluate the site occupancy as a function of temperature.

7. Does the entropy of an amorphous solid have a value closer to that for the crystalline solid or to that for the amorphous liquid of the same material? Justify your answer.

II-THERMODYNAMICS OF SOLID SOLUTIONS

INTRODUCTION

We consider the following subjects in this chapter: random solid solutions of binary systems; derivation of the free energy-composition relation for random solutions; the free energy of mixtures of solid solutions. This knowledge is then used to determine the equilibrium distribution of phases in a binary system at a given temperature. We then investigate the graphical significance of the partial molar free energy; non-random solid solutions; and various models for describing the energies and entropies of these solutions. A model that attempts to evaluate the excess free energy due to difference in atom or ion radius between components, the strain free energy, is then investigated. Factors other than atomic size that affect the enthalpy and entropy of solid solutions are briefly considered. Finally, the activity and activity coefficient of solutions are defined and some behavior of these parameters is described.

1. RANDOM SOLID SOLUTIONS IN BINARY SYSTEMS.

1.1. Free energy of a random binary solid solution.

The free energy of a solid solution is given by

$$G = E + PV - TS = H - TS \qquad (2.1)$$

We assume that the enthalpy of the random solid solution, and its first and second derivatives with respect to composition are finite and continuous functions of composition. These assumptions are in agreement with experimental measurements of the composition dependence of the enthalpy of solid solutions and with first principle calculations.

In addition to the entropy contributions that exist in the pure components there will be a configurational contribution to the entropy of a random binary solid solution. The latter contribution stems from the fact that the number of different ways that the two types of atoms can be arranged on the N lattice sites differs from unity and is given by

$$W = N!/[(NX_1)!(NX_2)!]\qquad(2.2)$$

for a random distribution, where X_i is the atom fraction of component i. We make use of the Boltzmann relation between the entropy of configuration, S_{conf}, and W, which is

$$S_{conf} = k\ln W\qquad(2.3)$$

and Stirling's approximation for the factorial of a large number.(Reference to equation 1.2 of Chapter I yields the fact that W corresponds to the number of microstates g and that we have implicitly assumed in the above that each microstate is as probable as any other. This expression for W can be understood as follows. Assume that there are N distinguishable atoms that are to be arranged on N lattice sites. There are then N different ways of choosing atoms to fill the first lattice site, N-1 different ways of choosing atoms to fill the second lattice site, and so on, so that the number of different ways of placing N distinguishable atoms on N lattice sites is then N!. But, there are only two types of distinguishable atoms, 1 and 2. Thus, to arrive at the number of different ways of distributing these two types of atoms on the N sites we must divide N! by $(NX_1)!$ and $(NX_2)!$, which are the number of different ways of arranging NX_1 distinguishable atoms on NX_1 sites and NX_2 distinguishable atoms on NX_2 sites, respectively.)

The configurational entropy or entropy of mixing for a random solid solution is then

$$S_{conf} = -Nk[X_1\ln X_1 + X_2\ln X_2]\qquad(2.4)$$

We can now obtain the following relation for the free energy of a solid solution.

$$G = Nh(X_2) + NkT[X_1\ln X_1 + X_2\ln X_2] - NTs_T(X_2) \qquad (2.5)$$

(In the above the units of h and s_T, the enthalpy and non-configurational entropy, respectively, are per atom, and N is the total number of atoms in the solid solution.)

Figure 2.1 shows graphically how G depends on composition. It is important to grasp the fact that the value of the first derivative of the enthalpy, dH/dX, is finite for all values of composition, while that for the entropy term, $-T(dS/dX)$, approaches negative or positive infinity at $X_2=0$ and $X_2=1$, respectively, due to the contribution of the configurational entropy! The aforesaid fact governs the shapes assumed by the free energy-composition curves! As shown, adjacent to the terminal compositions, the free energy-composition curves must always have positive curvature values. The sign of the curvature at intermediate values of the composition may be either positive or negative depending upon whether the enthalpy of mixing,

$$N(h - X_1 h_{1o} - X_2 h_{2o}),$$

has a negative or positive value, respectively. The significance of these facts will be demonstrated after the next section in which the free energy

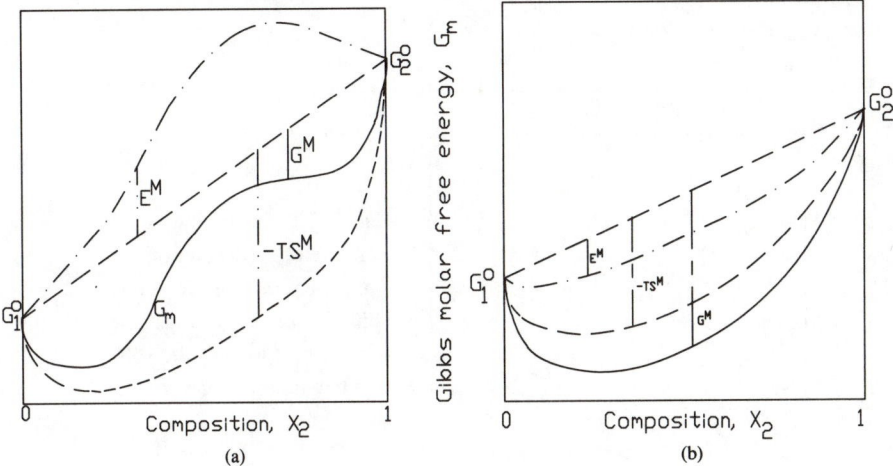

Figure 2.1. Molar free energy versus composition diagrams illustrating various molar and relative integral molar quantities.

of a mixture of two solid solutions is derived analytically.

Summarizing, the free energy of a binary solid solution must be represented as a function of composition by a single continuous curve having either of the shapes described by the full lines in Figures 2.1a and 2.1b.

1.2. Free energy of a mixture of solid solutions

We wish to derive the free energy of a mixture of two binary solutions. One of these solutions has the composition defined by X'_2 and the other has the composition defined by X''_2. These solutions have the corresponding free energies per atom g' and g", respectively. The points representing such solutions on a free energy composition map are shown in Figure 2.2. Let us further suppose that the relative amounts of these two

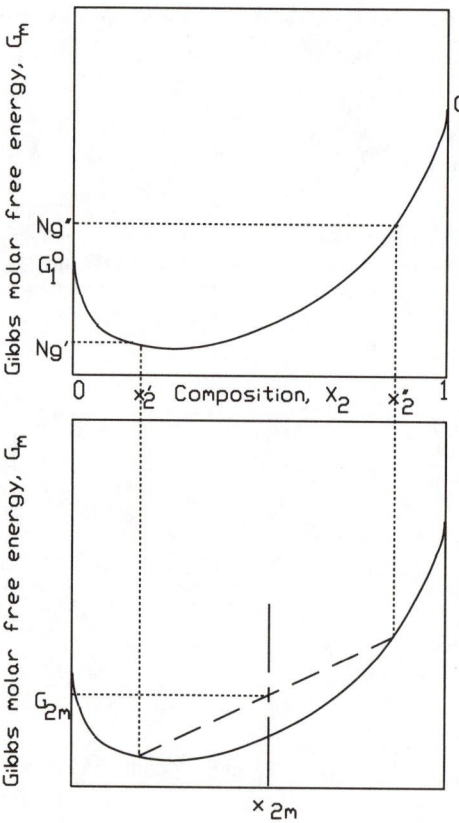

Figure 2.2. Illustrating the molar free energies of two solid solutions corresponding to the compositions, X'_2 and X''_2.

Figure 2.3. Illustrating that the mixture of solid solutions ' and " having the average composition X_{2m} is not stable with respect to the solid solu tion having this same compo sition.

solutions are such as to yield the average composition for the mixture given by X_{2m}.

According to the tie-line principle derived from conservation of atoms of a given kind, the total number of atoms of the " solution present is $N"=N(X_{2m}-X'_2)/(X"_2-X'_2)$. Similarly, the total number of the ' solution is given by $N'=N(X"_2-X_{2m})/(X"_2-X'_2)$. Hence, the free energy of the mixture is given by

$$G_{mixt} = g'N' + g"N"$$

Substituting for N' and N" yields

$$G_{mixt} = N(g'X"_2-g"X'_2)/(X"_2-X'_2)+N(g"-g')X_{2m}/(X"_2-X'_2).$$

This relation is identical to the equation for the straight line passing through the points (Ng',X'_2) and $(Ng",X"_2)$! Hence, in a free energy-composition map, points representing the free energy of mixtures of solutions lie along the line joining the points corresponding to the component solutions. From this result, it follows that for the case where the free energy-composition curve for a phase has a positive curvature that the free energy of any mixture of two compositions of this phase is not stable with respect to the solid solution having the average composition of the mixture. This situation is illustrated in Figure 2.3. In this figure the intersection of the vertical line with the dashed line yields the free energy of a mixture of solid solutions.The compositions of these component solutions are defined by the intersections of the dashed line with the full curve. The average composition of these solutions is that of the vertical line, X_{2m}.

It follows from the above analysis that when the free energy-composition curve of a solution has a negative curvature then a mixture of solutions can be stable relative to the solution corresponding to the average composition of the mixture. This fact is illustrated in Figure 2.4, where the dashed line corresponds to the mixture that has the lowest free energy in the composition range extending from X'_2 to $X"_2$. Further, our analysis has shown that the negative curvature of the free energy-composition curve is a consequence of a positive value of the enthalpy of mixing,i.e.

$$H^M = H_m - X_1 H_{1mo} + X_2 H_{2mo} > 0$$

Here, H_m is the molar enthalpy(enthalpy of one mole) of the solid solution,

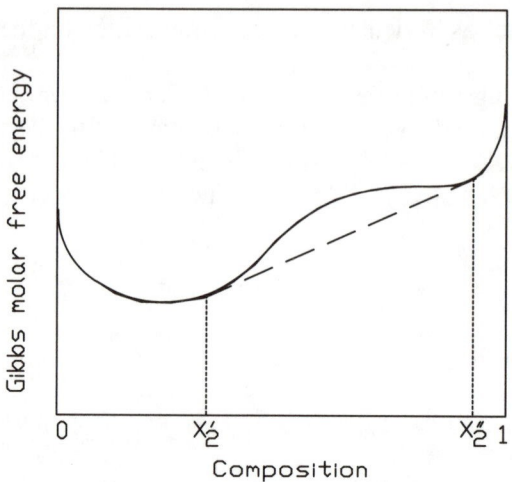

Figure 2.4 The dashed line represents the free energy of mixtures of solutions that are stable over the range of average compositions from X_2' to X_2'', relative to the solid solution having the average composition. The composition of the solutions in the mixture are X_2' and X_2''.

and subscripts io denote pure i. The enthalpy of mixing, H^M, is also known as the relative integral molar enthalpy.

1.3. Phase Boundary Compositions Corresponding to Miscibility Gap Compositions of a Random Binary Solid Solution.

When the enthalpy of mixing is positive, as noted in the previous section, there will be a miscibility gap consisting of a mixture of solid solutions that represents the equilibrium situation at low temperatures. The phase boundary compositions can be derived from two considerations: graphical or analytic. According to Figure 2.4 the following relation must be valid at equilibrium,

$$dG_m/dX_2 \Big|_{X_2=X_2'} = dG_m/dX_2 \Big|_{X_2=X_2''} = (G_m''-G_m')/(X_2''-X_2'). \quad (2.6)$$

(Cautionary note. See Problem 2.1 at end of chapter.)

Or by substitution of

$$G^M = G_m - (X_1 G_{1mo} + X_2 G_{2mo}) \qquad (2.7)$$

the relations defining the phase boundaries become

$$\left. dG^M/dX_2 \right|_{X_2 = X'_2} + G_{2mo} - G_{1mo} = (G''_m - G'_m)/(X''_2 - X'_2) \qquad (2.8)$$

and

$$\left. dG^M/dX_2 \right|_{X_2 = X''_2} + G_{2mo} - G_{1mo} = (G''_m - G'_m)/(X''_2 - X'_2). \qquad (2.9)$$

Another way of defining the compositions of the coexisting phases in the miscibility gap at equilibrium is to use the fact that then the partial molar free energies of a given component are equal in the coexisting phases, i.e.

$$\overline{G}'_i = \overline{G}''_i \quad (i=1,2) \qquad (2.10)$$

Consider the molar free energy G_m. It is related to the partial molar free energies by

$$G_m = X_1 \overline{G}_1 + X_2 \overline{G}_2 \qquad (2.11)$$

We now take the differential of equation 2.11 to obtain

$$dG_m = X_1 d\overline{G}_1 + X_2 d\overline{G}_2 + \overline{G}_1 dX_1 + \overline{G}_2 dX_2 \qquad (2.12)$$

By the Gibbs-Duhem relation, $X_1 d\overline{G}_1 + X_2 d\overline{G}_2 = 0$. Using this result and then multiplying the resulting equation by X_1/dX_2 under the constraint that $X_1 + X_2 = 1$ we obtain

$$X_1 dG_m/dX_2 = -X_1 \overline{G}_1 + X_1 \overline{G}_2$$

$$= \overline{G}_2 - (X_1 \overline{G}_1 + X_2 \overline{G}_2)$$

$$= \overline{G}_2 - G_m \qquad (2.13)$$

We rearrange (2.13) to the form

$$\overline{G}_2 = G_m + X_1 dG_m/dX_2 \qquad (2.14)$$

Similarly it can be shown that

$$\overline{G}_1 = G_m + X_2 dG_m/dX_1 \qquad (2.15)$$

Equations 2.14 and 2.15 are interpreted graphically in Figure 2.5.

The relations developed in the foregoing are independent of any model of the solid solution. If either equations 2.8 and 2.9 or 2.14 and 2.15 are to be used for the prediction of phase boundaries then it is apparent that it is necessary to know the molar free energy of the solid solution as a function of composition at the temperatures of interest. It is useful at this point to make some simplifying assumptions governing the solid solution for pedagogical reason.

1.4. Regular Solid Solutions.

Consider a regular solution in which the entropy of configuration is the same as for the random solution and the enthalpy of mixing can be expressed as a symmetric function of the composition for a binary system, i.e. $H^M = bX_1X_2$. In this case where both H^M and S^M are symmetric functions of the composition then the relative integral free energy must also be a symmetric function of the composition, as shown in Figure 2.6. The symmetric function shown in Figure 2.6, corresponding to b>0, is the only type that will yield a stable miscibility gap between the compositions X' and X". As shown in this figure at the phase boundary compositions the value of dG^M/dX_2 is zero. Consequently, by equations 2.8 and 2.9 then $G_{2mo} - G_{1mo}$ must equal the quantity $(G''_m - G'_m)/(X''_2 - X'_2)$ as well. But, for the regular solution the value of dG^M/dX_2 is given by $b(1-2X_2) + RTln(X_2/X_1)$, where b>0, for the case described in Figure 2.6. Hence, the phase boundary compositions of the miscibility gap are given by

$$X_2 = X_1 exp[-b(1-2X_2)/RT] \qquad (2.16)$$

for both the coexisting compositions.

Obviously, there must be a temperature above which the curvature of the free energy-composition relation is everywheres positive. The temperature at which the curvature first changes from positive to negative at some composition on decreasing temperature is called the critical temperature. This temperature can be derived by differentiating equation 2.16 with respect to composition, X_2, then setting $dT/dX_2=0$ and $X_2=0.5$. The result is $T_c=b/2R$. Of course, this result is applicable only to a regular

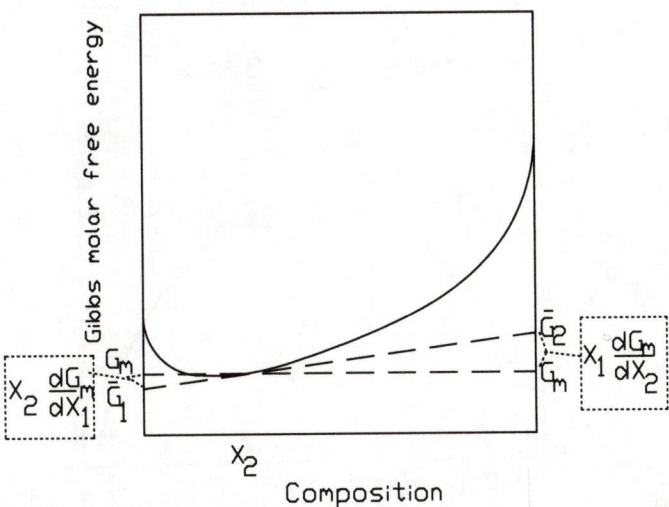

Figure 2.5. Illustrating the graphical significance of the partial molar free energy and its relation to the molar free energy. Refer to equations (2.14) and (2.15) of the text.

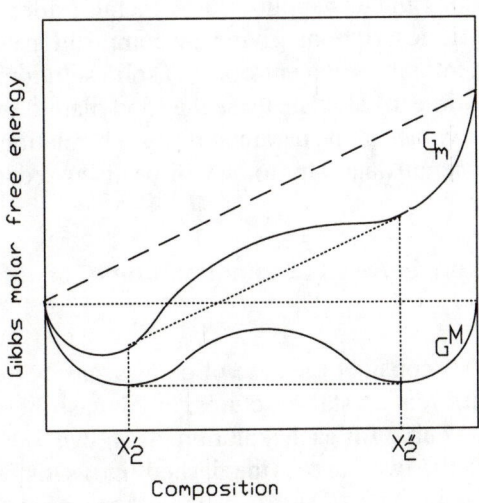

Figure 2.6. Illustrating how the relative integral molar free energy becomes a symmetric function of composition for a regular solution.

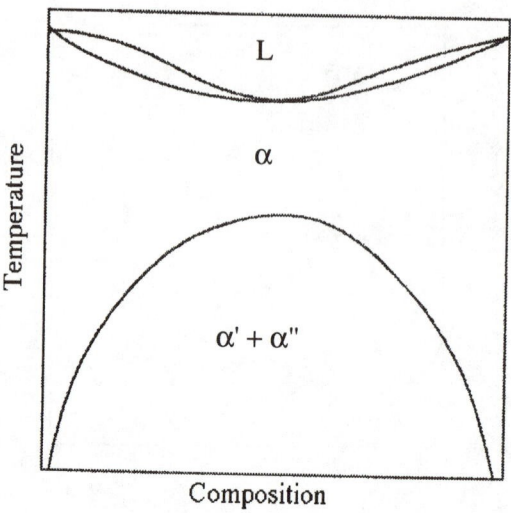

Figure 2.7. Illustrating a miscibility gap type phase diagram.

solution. A miscibility gap and the critical temperature is illustrated in Figure 2.7.

The regular solution approximation is applicable to many semiconductor solid solutions and less applicable to metallic alloys. There are, of course, other analytic descriptions giving the composition and temperature dependence of the enthalpies and entropies of solid solutions. Because it is not possible at this time to calculate these thermodynamic properties from first principles, the values of the parameters in such relations are obtained by fitting to experimental data. This topic will be discussed in greater detail in the next chapter.

1.5. Equilibrium between terminal solutions of different crystal structure.

Finally, let us consider the case of equilibrium between two solid solutions having different crystal structures. In this case there will be a free energy-composition curve for each solution, as shown in Figure 2.8. The common tangent to the two curves (the dashed line) satisfies the requirement described by equation 2.10 and thus yields the compositions of the coexisting phases in the stable two phase region (i.e. the vertical dotted lines.)

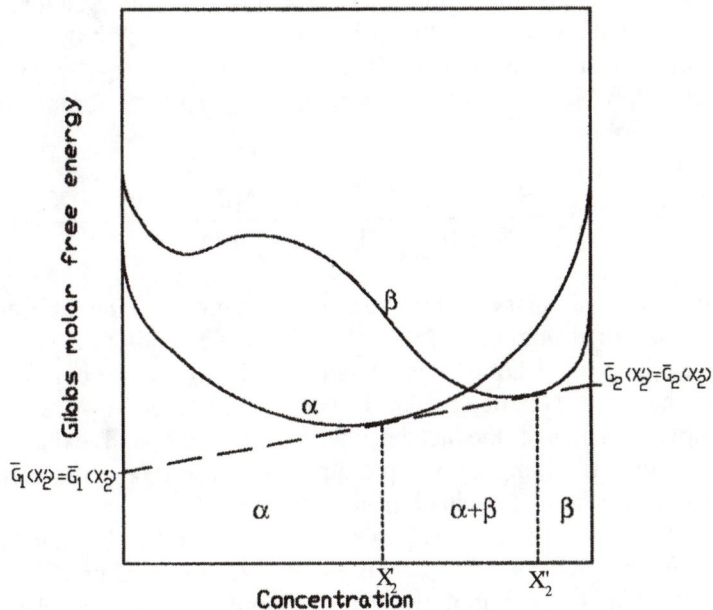

Figure 2.8. Illustrating the equilibrium developed between two different structures α and β.

2. NON-RANDOM BINARY SOLID SOLUTIONS.

The configurational entropy of non-random solutions cannot be described by equation 2.4. Also, a rigorous analytic description of the configurational entropy of a three dimensional non-random solution does not yet exist. Consequently, to obtain a thermodynamic description of a non-random solution it is necessary to resort to approximations of the configurational entropy. There are various approximations that have been used. The simplest and least exact are the "quasi-chemical" and mean-field approximations.[1] More exact approximations are based on the the use of Monte Carlo simulations and the cluster variation method.[2] We shall briefly consider these approximations.

2.1. Quasi-chemical approximation.

In this model, the short-range order is expressed in terms of the number of unlike bonds relative to the number that would exist for a random distribution of the atoms on the lattice sites. At the onset, it should be noted that this model yields poor results for the configurational entropy. In the present approximation, the configurational entropy for a binary system is given by $S=k\ln W$ or

$$S = k\ln\{\frac{(zN/2)!}{(N_{11}!N_{22}![(N_{12}/2)!]^2)} \frac{[(zN-N)X_1]! \ [(zN-N)X_2]!}{(zN-N)!}\} \quad (2.17)$$

In this expression, the first fraction in W represents the different number of ways of arranging the N_{11}, N_{22} and N_{12} type bonds in a lattice of $zN/2$ bonds. This bond lattice has zN lattice points (two per bond). The real lattice has only N lattice points. Hence, the remainder in W represents an attempt to account for the fact that on the average $z/2$ bonds exist per lattice point. This correction is only approximate because it neglects correlation, i.e. 1–1 bonds cannot emanate from 2 type atoms.

Consider the binary solution. There are only N_{11}, N_{22} and N_{12} types of nearest-neighbor bonds. Because only 1-2 and 2-2 type bonds can emanate from a 2 type atom we have the bond conservation relation:

$$2N_{22}+N_{12} = zNX_2 \quad (2.18)$$

Similarly, $2N_{11}+N_{12}=zNX_1$ represents the total number of bonds from 1 type atoms. Thus, having two independent equations between three variables, only one of the variables is independent. Choose this variable to be N_{12}.

The order dependent part of the internal energy is given by

$$E_{conf} = N_{11}E_{11}+N_{22}E_{22}+N_{12}E_{12} \quad (2.19)$$

Hence, setting $G=E-T(S+NX_1{}^\circ S_1+NX_2{}^\circ S_2)$ and minimizing G with respect to N_{12} yields

$$(N_{12}/2)^2/(N_{11}N_{22}) = \exp(-2v/(kT)) \quad (2.20a)$$

where

$$v = E_{12}-0.5(E_{11}+E_{22}). \quad (2.20b)$$

The similarity of this relation to the mass action law provides this approximation with its name-"quasi-chemical". According to the relation, $v<0$ implies $(N_{12}/2)^2>N_{11}N_{22}$. By substitution and rearrangement of terms it can be shown that

$$\frac{N_{12}}{(zNX_1X_2)} = \frac{zN\exp(-2v/(kT))}{\{N_{12}(1-\exp(-2v/(kT)))+zN\exp(-2v/(kT))\}} \quad (2.21)$$

Hence, for $v<0$, $N_{12}/(zNX_1X_2)>1$, i.e. N_{12} is larger than that corresponding to a random solution and vice versa for $v>0$. Thus, the quasi-chemical approximation relates short-range order $(v<0)$ or clustering $(v>0)$ to the sign of v.

On the assumption that the total internal energy is expressable as a sum over nearest-neighbor bond energies that are composition independent then the expression for the relative integral molar energy is particularly simple in the quasi-chemical approximation. It is

$$E^M = N_{12}v. \quad (2.22)$$

For a random distribution of the atoms $N_{12} = zNX_1X_2$. Thus, with increasing temperature, the quasi-chemical value of the relative integral molar energy approaches that of a regular solution. This cannot be interpreted to mean that regular solution behavior implies that the relative integral molar energy equals a sum over the bond energies between unlike atoms. There are other possible physical explanations for the regular solution dependence of this energy on composition. For example, in alloys the integral molar energy(the cohesive energy) is a sum of two terms. One is the configurational energy (equation 2.19) and the other is a term independent of the configuration and dependent upon the molar volume. Further, the latter term is much greater than the former in absolute value and also is composition dependent, because the average molar volume is composition dependent. Thus, the composition dependence of the relative integral molar energy is more likely to be determined by the configuration independent energy, at least for the case of alloys. Thus, $N_{12}v$ should be considered to be the quasi-chemical configuration dependent contribution to the relative integral molar energy, rather than the complete expression for the latter.

2.2. Cluster variation method

The configurational entropy in the cluster variation model depends upon the size of the maximum clusters used and the crystal structure of the solid solution. For example, the configurational entropy per lattice point, using as a maximum cluster the four-point tetrahedron, in the disordered fcc structure is

$$S_{conf} = -k[2\Sigma L(z_{ijkl}) - 6\Sigma L(y_{ij}) + 5\Sigma L(x_i)] \qquad (2.23)$$

where $L(u) = u \ln u$, z_{ijkl} is the cluster probability of a tetrahedron with configuration ijkl, y_{ij} is the probability of finding an ij pair, and x_i is the atomic concentration of component i. The relations between these parameters are

$$x_i = \Sigma_{jkl} z_{ijkl} \qquad (2.24)$$

$$y_{ij} = \Sigma_{kl} z_{ijkl} \qquad (2.25)$$

The cluster probabilities z_{ijkl} are considered to be independent of each other. Hence, the equilibrium state of short-range order is determined by minimization of the free energy with respect to each of the cluster probabilities. The configuration dependent energy term in the free energy must be approximated. The simplest approximation is to take the configurational energy to be given by the sum of pairwise interactions as in equation 2.19. These interactions need not be limited to nearest-neighbors but can also include further neighbor interactions. Also, if necessary, many-body terms can be introduced. In all these cases, the configurational energy becomes a function of the cluster probabilities and composition through relations 2.24 and 2.25 and similar relations. Hence, although the configurational entropy is determined quantitatively, the configurational energy is given in terms of fitting parameters, such as the bond interaction energy, v, described in the quasi-chemical model.

The reader is referred to the literature for detailed applications of the CVM model. The configurational entropy given by the CVM model appears to be adequate to yield agreement with parameters that are sensitive to it, such as the critical temperature, x-ray short-range order parameters and phase diagrams.

As noted above, in all models of solid solutions the internal energy cannot be adequately described by the model and recourse must be made to empirical description to obtain a good analytic description of the free energy of any solid solution phase.

2.3. Monte Carlo method

This is a method to use a computer to compute equilibrium values of properties. The method is as follows. Choose NX_1 type 1 particles and NX_2 type 2 particles. (N depends here on the interaction distance involved in evaluation of the energy of the assembly of particles.) Choose the primitive lattice of the material's crystal structure such that $\mathbf{a}_1, \mathbf{a}_2, \mathbf{a}_3$ are unit displacements along the three axes, respectively. In this way, the displacement $(p-h)\mathbf{a}_1, (q-k)\mathbf{a}_2, (r-l)\mathbf{a}_3$, where h,k,l and p,q,r, are integers, will move a particle from a lattice position $h\mathbf{a}_1, k\mathbf{a}_2, l\mathbf{a}_3$ to the lattice position at $p\mathbf{a}_1, q\mathbf{a}_2, r\mathbf{a}_3$. Use is made of periodic boundary conditions, such that the motion of a particle to a site outside the block containing N lattice sites will make the particle re-enter from the opposite side. An initial distribution of the N particles on the N sites is chosen and the energy E° for this distribution is calculated. The computer records the coordinates of all the particles. The computer then generates random values of p, q, and r. If the particle at $(p-h)\mathbf{a}_1, (q-k)\mathbf{a}_2, (r-l)\mathbf{a}_3$ differs from that at $h\mathbf{a}_1, k\mathbf{a}_2, l\mathbf{a}_3$, then the computer evaluates the new energy, E'. If $E'-E^\circ$ is negative, the move is accepted. If $E'-E^\circ$ is positive then the acceptance of the move obeys the following conditions. A new random number, z, is generated, where $0 \leq z \leq 1$. If $z < \exp[-(E'-E^\circ)/kT]$ the move is accepted, whereas for the reverse inequality the particle is put back to its original position (i.e. $h\mathbf{a}_1, k\mathbf{a}_2, l\mathbf{a}_3$.) The process is repeated for each particle and the average value of the energy $<E>$ is calculated. After M moves

$$<E> = (1/M) \Sigma_{i=1}^{M} E_i$$

The average $<E>$ so calculated will fluctuate at the start, but will stabilize after sufficient $M = M^*$. For $M > M^*$, the system will be at equilibrium. Thereafter, the computer is programmed to evaluate average values of thermodynamic quantities that can be compared with experiment or that can be used theoretically.

As for the other methods of evaluating the non-random distribution of particles the Monte Carlo method is hindered by the lack of a quantitative expression for the energy of the system of particles.One possible scheme to solve this problem is to model this energy and to evaluate the parameters by an iterative process.This process involves assuming values for them, then calculating via the Monte Carlo procedure average values for various thermodynamic properties, at equilibrium, and then changing the initial values of the parameters in the direction that minimizes the difference between the calculated average values of the thermodynamic properties and their experimental values, recalculating these average values, and so on until these deviations are acceptable. As may be imagined, a high speed computer is required to arrive at acceptable results. Progress in attaining quantitative descriptions of the energy of systems of particles, the cohesive energy problem, will remove this barrier to the prolific prediction of phase equilibria.

Another way of using both the Monte Carlo and CV methods is to assume various energy interaction schemes and then to calculate the regimes of equilbria for the various ground state structures that are consistent with the assumed interactions. For example, if it is assumed that the configurational energy is given by a sum over first nearest neighbor interactions then only certain ground state ordered configurations may be stable. For the fcc disordered structure in this example the possible ordered configurations are the $L1_0$ and the $L1_2$ structures. Figure 2.9 compares calculated phase diagrams based on two types of clusters used in the CV method (the

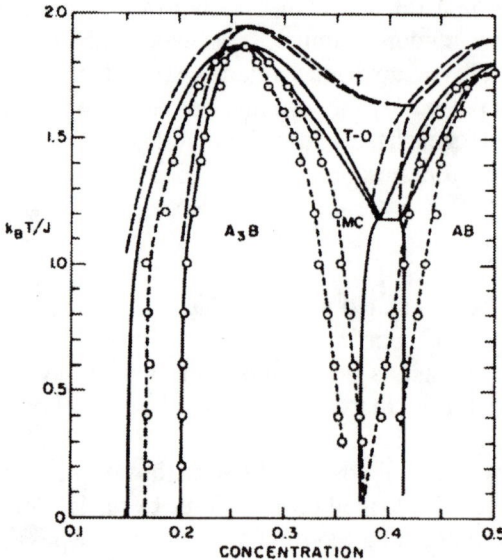

Figure 2.9. After Mohri et al(Acta Met.33,1171(1985). Comparison of phase diagram for first nearest-neighbor pair approximation. TO-CVM(full line), T-CVM(long dashes) and Monte Carlo simulation(short dashes).

tetrahedron and tetrahedron-octahedron clusters) with that deduced by the Monte Carlo method. (The parameter J in the ordinate of Figure 2.9 equals $v/2$, where v is defined by equation 2.20b.) The subjects of ground state structures, coherent phase diagrams, CVM and Monte Carlo methods, as applied to phase equilibria and Ising type problems are being rapidly developed and already the literature can yield a book on these related subjects alone.

3. CONCEPT OF STRAIN ENERGY IN SOLID SOLUTIONS.

The lattice parameter of a pure crystalline element defines the equilibrium spacing between the atoms corresponding to the absence of strain energy in the lattice. If the distance between these atoms is changed from that of the equilibrium spacing then the lattice is said to be strained and the increment in energy associated with this change in lattice parameter is termed strain energy. When solute atoms are introduced to form a dilute solid solution, x-ray evidence demonstrates that the distance between solvent atoms has altered from the value corresponding to the equilibrium spacing in the pure crystalline element. One view of the change in energy associated with the change in solvent interatomic distances is that it corresponds to the increment in energy associated with strain of the solvent-solvent interatomic bonds. This concept has certain predictive value of use in the interpretation and prediction of phase diagrams and thus a more detailed investigation of it is justified. However, it must not be inferred that the only contribution to the change in lattice parameter upon the introduction of a solute atom is due solely to strain. There are additional effects on the lattice parameter associated with a change in configuration of the solute atom's outer electrons and those of its neighbors that are not included in the strain energy concept. Such effects are minimized when the difference in electronegativity and dsp character of the outer electrons between solute and host are negligible in the case of alloy solid solutions and when the solute ions have the same valence as that of the host for the case of ionic solid solutions. Thus, the analysis of strain energy given in this section is limited to solid solutions where these other effects are minimal,e.g. Ag-Cu, (K,Na)Cl solid solutions.

The original evaluation of the strain energy due to the introduction of a solute atom into a pure solid solvent host is due to Friedel[3]. Eshelby[4] then provided a result that did not make approximations for certain elastic parameters. We shall make use of the results of the latter's analysis. The assumption is made that the introduction of a solute atom into a vacant site in the interior of the host lattice exerts a pressure on the surrounding atoms. This pressure acts to distort the host lattice. The distortion produced is assumed to be the same as that calculated from the application of elasticity theory to the equivalent continuum problem. Thus, if the atomic distance between host atoms in the pure host is $2°R_1$ then the radius of the hole into which the solute atom is introduced and at which the latter exerts its pressure on the host lattice is $°R_1$. It is assumed that this pressure acts to make the hole dilate until the outward pressure due to the solute atom is equal to the inward pressure due to the distorted host on the solute atom. Hence, the stress-free solute atom is considered to be a compressible solid with a radius equal to $°R_2$. The interesting results of this model are:

1. The state of stress in the host about the solute atom is pure shear.

2. The total strain free energy due to the introduction of one solute atom into the host is

$$G_{st} = 2\mu_1 C_6 (°V_2 - °V_1)^2 / 3V_2 \qquad (2.26)$$

where $C_6 = 3K_2/(3K_2 + 4\mu_1)$, K is the bulk modulus, μ is the shear modulus, V is the atomic volume, the superscript $°$ denotes the pure state. The strain free energy due to the introduction of solute atoms sufficient to produce the composition X_2 is then

$$G^{M,st} = NG_{st}X_2\{1 - K_2 X_2 / K_1\} \qquad (2.27)$$

for $X_2 \ll 1$ and where $(1 - K_2 X_2 / K_1)$ is a consequence of the change in the strain energy due to the interaction of the introduced solute atom with all the other solute atoms in the host. When there is no other contribution to the relative integral molar free energy, other than the entropy of mixing, then

$$G^M = G^{M,st} + NkT[X_1 \ln X_1 + X_2 \ln X_2]. \qquad (2.28)$$

3. The strain entropy, S_{st}, is given by $-dG_{st}/dT$, which upon the

substitution of representative values is closely approximated by the value of the first term in the derivative, dG_{st}/dT, namely $[2C_6(^\circ V_2-^\circ V_1)^2/3V_2](d\mu_1/dT)$. Since, the sign of the first [ ] bracket is positive, then the sign of S_{st} depends upon the sign of the derivative in the second bracket, which is negative. Hence, the sign of S_{st} is positive, i.e. the strain entropy is a positive quantity. Also, it is an excess quantity in that

$$S_m = X_1 {}^\circ S_1 + X_2 {}^\circ S_2 - Nk[X_1 \ln X_1 + X_2 \ln X_2] + S^{M,st}.$$

The strain entropy term prevents the solution from belonging to the regular solution category, because for the latter the relative integral molar entropy is equal solely to the entropy of mixing, i.e. the excess molar entropy is zero for a regular solution. (An excess quantity is given by the difference between the relative integral molar term and that for an ideal solution.)

In the absence of chemical effects between unlike atoms, as may be induced by a difference in electronegativity or d band centers of gravity if the components are transition elements or hybridization between overlapping unlike bands (i.e. d band of one component with p band of other component), the above model provides a close description of the solid solution thermodynamic quantities of alloys. For example, Figure 2.10 shows a comparison between experiment and the results of using the above model to calculate the phase boundary compositions of the solid state miscibility gap in the copper-silver binary system.(The calculation is based on setting the parameter b in equation 2.16 equal to $NG_{st}(1-K_2X_2/K_1)/(1-X_2)$ and solving for X_2 in 2.16).

One conclusion of the above discussion of strain energy is that a difference in atomic volumes between components contributes a positive term to the free energy and to the enthalpy. Since a positive value of the enthalpy of mixing leads to the formation of a miscibility gap, and, hence, limited solid solubility, it follows that a sufficient difference in atomic volumes between components will restrict the mutual solid solubility of the components. The larger is the positive enthalpy of mixing the less is the solubility(composition of the solute component at the phase boundary). This result is the explanation for the empirical rule of Hume-Rothery[5] according to which the solid solubility of a component in a host is very limited when the difference in atomic radii between host and solute exceeds 15%.

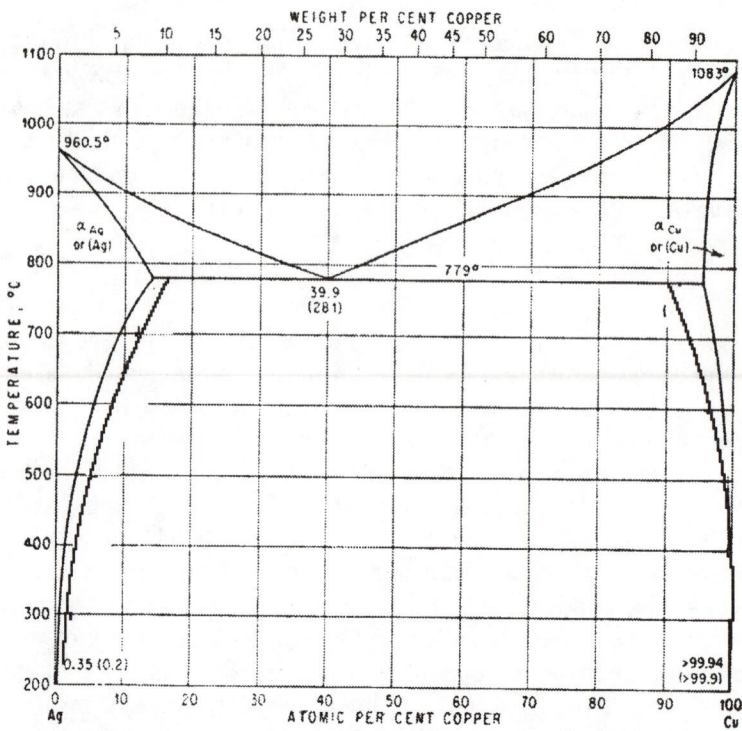

Figure 2.10. Jagged lines represent the calculated phase boundaries of the miscibility gap. The smooth lines represent the experimental phase diagram which is taken from M. Hansen, *Constitution of Binary Alloys*, McGraw-Hill Book Co., N.Y., 1958.

4. FACTORS OTHER THAN ATOMIC SIZE DEVIATION THAT AFFECT THE ENTHALPY AND ENTROPY OF SOLID SOLUTIONS

It must be remembered that factors other than size affect the solid solubility as suggested above, i.e. chemical interactions between the unlike atoms. An illustration of the latter fact is given by the Darken-Gurry plot, Figure 2.11, which describes the limit of solid solubility in a plot of atom radius difference against the Pauling electronegativity difference, for the case of alloy solid solutions. As shown in this figure the solid solubility is extremely limited when the electronegativity difference exceeds about 0.4, even if the difference in atomic size is negligible. It may be recalled that the electronegativity, according to Pauling, is a measure of an atom to attract electrons. There are various measures of the electronegativity, most of which are derived from the heats of formation of molecules. However, there are more modern measures of the electronegativity for solids. One such measure for covalently bonded solids is the dielectric based ionicity.[6]

The effect of an electronegativity difference on the solid solubility can be understood using tetrahedrally coordinated $A^N B^{8-N}$ covalently bonded semiconductors to illustrate the relation between the heat of formation of the semiconductor from its elemental components and the AB bond ionicity.[6] This ionicity is roughly proportional to the difference in Pauling electronegativity between components.The greater the bond ionicity the

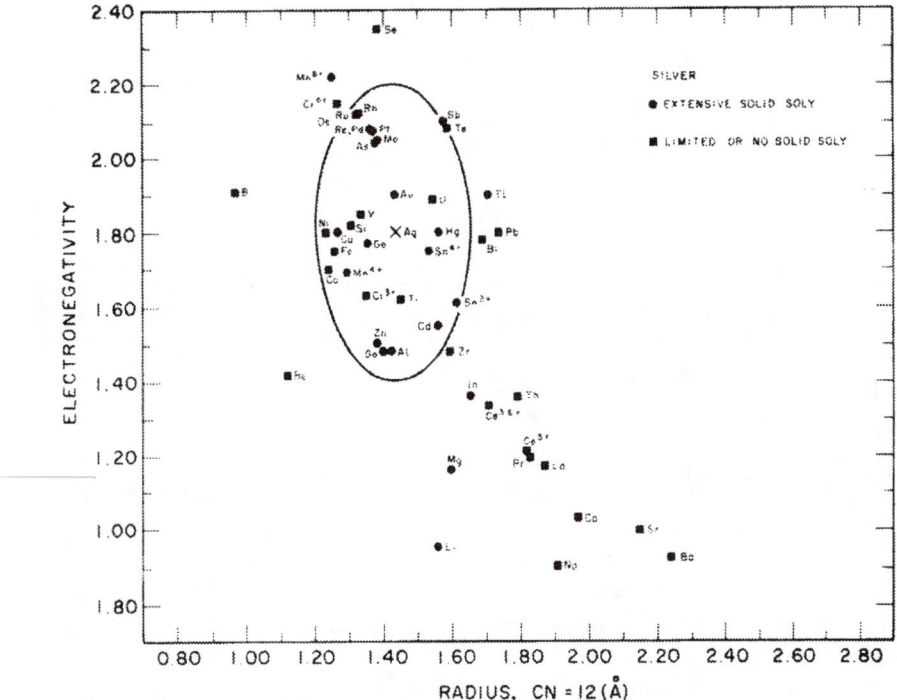

Figure 2.11. Darken-Gurry plot for various solutes dissolved in silver. Circles indicate alloys in which extensive solid solutions are found; squares indicate alloys in which limited or no solid solubility is found. After Pearson, *The Crystal Chemistry and Physics of Metals and Alloys*, Wiley-Interscience, N.Y., 1972.

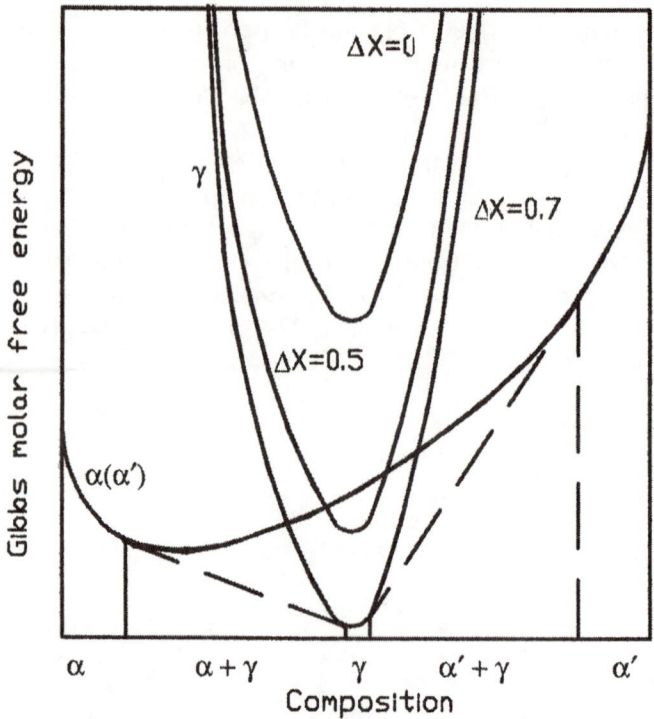

Figure 2.12. Illustrating the effect of an intermediate phase on the phase equilibria in a binary system. Changes in the stability of the intermediate phase will result in changes in the compositions defining the phase boundaries. The ΔX values represent the difference in electronegativity between the pure components.

more negative is the energy of formation of the AB phase. The same considerations apply to other covalently bonded structures. Hence, if in an AB binary system it is possible for covalently bonded intermediate phases to appear, the greater the ionicity of the AB bond (the greater the electronegativity difference) the more stable will be the intermediate phase. The relative stability of the intermediate and terminal phases will then affect the solid solubility limits (phase boundaries) of the terminal phases, as illustrated in Figure 2.12.

The same results apply when the intermediate phases are not covalently bonded semiconductors in that the electronegativity difference is roughly proportional to the parameter that, for the solid in question, determines the energy of formation of the intermediate phase. For example, the energy of formation of transition metal alloys is governed mainly by the difference in centers of gravity of the d bands of the component transition metals.[7] The latter difference is roughly proportional to the electronegativity difference. Thus, in general, the greater is the electronegativity difference the more stable is the intermediate phase relative to the terminal phases and the less is the limit of solubility of the terminal phases for the solute component.

This effect of electronegativity difference on the solid solubility is not a consequence of an effect on the thermodynamic properties of the terminal phase, but rather of an effect on the thermodynamic properties of the intermediate phase in equilibrium with the terminal phase. This role differs from that of the atomic size difference on the solid solubility in that the latter is the result of an effect on the thermodynamic properties of the terminal phase itself.

The thermodynamic properties of the terminal phase may be affected by other than atomic size difference. Empirically, Hume-Rothery showed that for alloys based on copper, silver and gold that the phase diagrams could almost be superimposed using outer electron concentration in place of atom composition as the abscissa. The effect of valence of the solute was originally explained qualitatively by Jones and Mott[8] and then given a quantitative explanation in the work of Blandin et al[9]. Other factors are known to affect the thermodynamic properties of metallic solid solutions, such as the overlapping of the d and p levels of the transition and simple metal components, respectively, in their alloys. However, a more thorough discussion of this matter would take us far outside the scope of this book.

Strain energy, ionicity (difference in electronegativity) also affect the equilibrium solubility of donor and acceptor species in semiconductor solid solutions. However, the problem of describing the thermodynamic values of solutes in semiconductor compounds and ionic crystals is much more complicated in that a given solute species can exist simultaneously as a charged or neutral species in substitutional or interstitial sites and sometimes in two oxidation states. Thus, in these solutions each distinguishable species of the same element must be considered a different solute species.

5. ACTIVITY AND ACTIVITY COEFFICIENT OF SOLUTIONS

The partial molar free energy of a solute component, $\overline{G}_i$, is related to the activity of that component, a_i, by the relation

$$\overline{G}_i - G_i^o = RT \ln a_i.$$

where G_i^o is the free energy of one mole of component i in its standard state. Since the standard state is arbitrary, the activity is arbitrary. Usually, for solid solutions, the standard state is chosen to be the pure component. Also, the activity coefficient is defined by $\gamma_i = a_i/X_i$. According to Figure 2.5 and the corresponding section in this chapter, the partial molar free energy of a solution at some composition X_i in the binary system is given graphically by the intercept of the line that is tangent to the molar free energy of the solution at this composition with the coordinate axis representing pure i. Thus, for compositions where the molar free energy versus composition curve has a positive curvature the partial molar free energy of the solute component will increase with increase in the composition of the solute component while that for the solvent component will decrease. A transition from positive curvature to negative curvature of the molar free energy versus composition curve implies that the term $d\ln\gamma_i/dX_i$ must be negative and exceed in magnitude beyond some composition X_i the value of the positive quantity $d\ln X_i/dX_i = 1/X_i$. Any value of the activity coefficient differing from unity implies that the solution is no longer ideal. When the activity coefficient of the solute equals a constant the solution obeys Henry's law. Strain energy and chemical interactions contribute non-zero terms to the solute dependence of the activity coefficient, but of opposite sign.

In general, the results we have derived concerning the thermodynamics of solutions apply, in principle, not only to metallic solid solutions, but to ionic solid solutions, to covalently bonded solid solutions and to polymer solutions. We shall discuss the case of polymer solutions in the next section.

6. POLYMER ALLOYS.

At one time it was thought that polymers did not form solutions. This deduction was based on two facts and one assumption. One is that the relative integral free energy for the solution must be a negative quantity if it is to form. The other fact is that the entropy of mixing for a high molecular weight polymer solution is very small. The assumption was that the enthalpy of mixing was necessarily a positive quantity. That the entropy of mixing of a high molecular weight polymer solution is small can be obtained from (2.4), on the assumption that the two polymer components of the solution have the same molecular weight and shape. N in (2.4) is the number of molecules. Thus, the ratio of the configurational entropy for the polymer solution to that for a metallic solution is the reciprocal of the number of mers in the polymer molecule. For a normal high molecular weight polymer molecule the configurational entropy is a negligible quantity. For components that have different numbers of mers per molecule the relation for the configurational entropy per mer in the solution becomes

$$s_{conf} = -k[(\Phi_1/r_1)\ln \Phi_1 + (\Phi_2/r_2)\ln \Phi_2]$$

where r_i = number of mers in strand type i and Φ_i = site fraction occupied by component i.

That polymer solutions do form is now well recognized. Robeson[10] has reported the existence of more than 160 miscible pairs of polymers. It is apparent that the assumption concerning the sign of the enthalpy of mixing for polymer solutions is wrong and that the latter can be a negative quantity. It is interesting to consider the possible origins of a negative enthalpy of mixing in polymer solutions. Polymers have satisfied primary bonds. Thus, the only bonding contribution to the enthalpy of mixing must involve secondary bonds, such as dipole-dipole, hydrogen bonding and van der Waal type bonding. As an example, when the radicals on the component polymers are conducive to hydrogen bonding, as in DNA or PVC/Polyester mixtures, it is possible that the enthalpy of mixing of such component polymers will be negative. Because many useful products can be made from polymer alloys, and because the concept is relatively new, this field is experiencing a concentration of activity(see bibliography).

7. EQUILIBRIUM IN A STRESSED SOLID.

In stressed solids the criteria defining equilibrium that have been considered in this chapter may no longer be applicable. In particular, equilibrium may not be described by equations (2.8)-(2.10). Indeed, the concept of a chemical potential of a species in the solid may become meaningless. Gibbs showed that a homogeneously stressed solid containing a component that does not diffuse, when equilibrated with a fluid, develops different chemical potentials of the non-diffusing species in the saturated fluid dependent on the orientation of its surface in contact with the fluid. On the other hand, species that do diffuse in the Gibbs type solid have defined chemical potentials that are constant throughout all phases at equilibrium, even if some of them are stressed. An example of such a solid containing non-diffusing components and diffusing components is Si (as the non-diffusing component) containing transition element solute, such as Ni (as the diffusing component). The Ni atoms are in interstitial sites in the diamond cubic lattice of Si.

Larche and Cahn[11] have developed this concept based on use of a lattice network in which the same place can be located after diffusion as the equivalent of Gibbs's non-diffusing solid. In this network the thermodynamic properties are functions of the elastic strain and local composition as defined by the network. (A Bravais solid where lattice sites are occupied by substitutional atoms is one example of such a network solid.) Equilibrium may exist in the presence of a non-homogeneous distribution of stress and composition in a network solid. In the network solid the useful thermodynamic quantities are the diffusion potential, to be defined shortly, the Helmhotz free energy density (per unit volume) and the stress tensor. At equilibrium in the stressed network solid, the diffusion potential is constant throughout the solid and the divergence of the stress tensor is zero to achieve mechanical equilibrium.

The diffusion potential, M_{IK} is defined by

$$M_{IK} = (1/\rho'_o)(\partial f'/\partial C_{IK})_{\theta, E_{ij}}$$

where ρ'_o is the molar density(number per unit volume) of lattice sites in the reference state relative to which the strain is defined, f' is the specific Helmholtz free energy(energy per unit volume) in the reference state θ is the absolute temperature. E_{ij} is the strain defined by $E_{ij} = 0.5(u_{ij} + u_{ji})$, where $u_{ij} = \partial u_i/\partial x$, u_i is the displacement in the i direction and $(\partial/\partial C_{IK})_{C_{j \neq I,K}}$ is the unit composition increase in species I, and equal decrease in species K, holding the composition of all other substitutional species on that type site fixed. For binary solutions, $C = C_1$ and $(\partial/\partial C_{12}) = (\partial/\partial C)$. In the case of equilibrium of a network solid with a fluid

$$M_{IK} = \mu_I^L - \mu_K^L$$

where μ_i^L is the chemical potential of i in the fluid.

A fluid cannot resist shear stress and has a constant hydrostatic pressure throughout at equilibrium. Chemical potentials of species in a fluid are meaningful and the concepts in the other sections of this chapter apply to fluids. Dislocations can generate and move readily in some solids above about one half the absolute melting point. These solids are effectively fluids and again chemical potentials of species are meaningful in them.

Larche and Cahn[11] have considered many applications of these concepts. A complete review here is outside the scope of this book. However, we shall undertake to describe some results of their analysis in later chapters in this book at their appropriate places.

8. SUMMARY

Summarizing, we have investigated the forms that the free energy-composition curves can take in binary systems. We have derived the relation for the free energies of mixtures of two solutions and thereby have been able to describe when such a mixture will be stable and when the solution will be stable relative to the mixture of solutions. Also, we have graphically interpreted the significance of the partial molar free energy. Further, we have discussed the concepts of short-range order, strain energy and chemical interactions between the components of a homogeneous solid solution. We have introduced the concepts of polymer alloys and of equilibrium in stressed network solids. In the latter, chemical potentials may not have meaning, but diffusion potentials replace them as the quantity that is constant throughout coexistant phases in equilibrium.

APPENDIX

Definitions of Thermodynamic Quantities

Partial Molar Quantities.

$$\overline{Q} = (\partial Q/ \partial n_i)_{P,T, n.} \quad (k \neq i)$$

where n_i is the number of moles of i. (Replacing n by the number of atoms N and with

Q, the total Gibbs free energy, yields the chemical potential.)

Relative Integral Molar Quantities.

$$Q^M = Q_m - \Sigma(X_i \,{}^\circ Q_i)$$

(Superscript zero refers to the pure state, subscript i to the component i, and subscript m to the value for one mole of substance.)

Relative Partial Molar Quantities.

$$Q_i^M = \overline{Q}_i - {}^\circ Q_i$$

Excess Quantities.

$$\text{Integral:} \quad Q^{xs} = Q^M - Q^{id}$$

$$\text{Partial:} \quad \overline{Q}^{xs} = Q_i^M - Q_i^{Mid}$$

(Superscript id refers to the ideal solution value defined below.)
Types of Solutions:

$$\text{Ideal:} \quad H^M = 0; \quad S^M = -R\Sigma[X_i \ln X_i].$$

$$\text{Regular:} \quad H^M = bX_1X_2 \text{ (binary solution) with } b \neq 0; \quad S^M = S^{Mid}$$

Binary Solutions - Correlating Functions.

$$\alpha_1 = G_1^{Mxs}/(X_2)^2 = [RT\ln \gamma_1]/(X_2)^2$$

$$\alpha_2 = [RT\ln \gamma_2]/(X_1)^2 = -X_2\alpha_1/X_1 + (X_1^{-2}) \int_0^x \alpha_1 \, dX_1$$

$$\beta_1 = S_1^{M.xs}/X_2^2 = -d\alpha_1/dT$$

Relations between Quantities.

$$Q_m = X_1\overline{Q}_1 + X_2\overline{Q}_2$$

$$Q^M = X_1Q_1^M + X_2Q_2^M$$

$\overline{Q}_1 = Q_m + (1- X_1) (\partial Q_m /\partial X_1)_{X_2}$ ...(Multicomponent solution; for binary solution substitute total derivative in place of partial derivative.)

$Q_1^M = Q^M + (1- X_1)(\partial Q/\partial X_1)_{X_2}$...(Multicomponent solution; for binary solution substitute total derivative in place of partial derivative.)

$$X_1 d\overline{Q}_1 + X_2 d\overline{Q}_2 = 0 \quad \text{(Gibbs-Duhem)}$$

$$(\partial Q_m / \partial X_2)_{X_1 \ldots} = \overline{Q}_2 - \overline{Q}_1$$

$$Q_1^{Mxs} = Q_1^M - Q_1^{Mid}$$

REFERENCES

1. A.G. Khachaturyan, THEORY OF STRUCTURAL TRANSFORMATIONS IN SOLIDS, J. Wiley, NY, 1983

2. D. de Fontaine, Solid State Physics 50, 73(1979); T.Mohri, J.M. Sanchez and D.deFontaine, Acta Met.33, 1171(1985).

3. J. Friedel, Advances in Physics 3, 446(1954).

4. J.D. Eshelby, Solid State Physics 3, 79(1956).

5. W.Hume-Rothery and G.V.Raynor,THE STRUCTURE OF METALS AND ALLOYS,The Institute of Metals,London,1962.

6. J.C. Phillips, BONDS AND BANDS IN SEMICONDUCTORS, Academic Press, NY, 1973.

7. D.G. Pettifor, Calphad 1, 305(1977).

8. N.F. Mott and H.Jones, THE THEORY OF THE PROPERTIES OF METALS AND ALLOYS, Oxford University Press, Oxford, 1936.

9. A. Blandin in ALLOYING BEHAVIOR AND EFFECTS OF CONCENTRATED SOLID SOLUTIONS (ed. T.B. Massalski), Gordon and Breach, NY, 1963; in PHASE STABILITY IN METALS AND ALLOYS (ed. P.S. Rudman, J. Stringer and R.I. Jaffee, McGraw-Hill, NY, 1967.

10. L.M. Robeson, in POLYMER COMPATIBILITY AND INCOMPATIBILITY: PRINCIPLES AND PRACTICE, MMI SYMP.SER, 3, ed. K. Solc, Cooper Station, NY(1981).

11. F. Larche and J.W. Cahn Acta Met.21, 1051(1973):26, 1579(1978);30, 1835(1982).

BIBLIOGRAPHY

General:

1.R.A.Swalin,THERMODYNAMICS OF SOLIDS, 2nd ed.,J.Wiley, New York, 1972.

2. D. Gaskell,INTRODUCTION TO METALLURGICAL THERMODYNAMICS, Hemishpere Publ.Corp, N.Y., 1981.

Metal Solutions:

L.S. Darken and R.W. Gurry, PHYSICAL CHEMISTRY OF METALS, McGraw-Hill, NY, 1953.

Semiconductor Solutions:

G.B. Stringfellow, ORGANOMETALLIC VAPOR-PHASE EPITAXY, THEORY AND PRAC-TICE, Academic Press, N.Y., 1989, Chapter 3. This chapter is an excellent summary of the thermodynamic developments in semiconductor solid solutions.

Polymer Solutions:

1. J.W.Barlow and D.R. Paul, Ann.Rev.Mat.Sci.11, 299(1981).
2. L.A. Utracki, POLYMER ALLOYS AND BLENDS, Hanser, Oxford U. Press, N.Y. 1989.

PROBLEMS

1. Show why the relation $dG/dX_2|_\alpha = dG/dX_2|_\beta$ is insufficient to define the compositions of the coexisting phases α and β at thermodynamic equilibrium.

2. Why does the functional dependence of free energy on composition in a binary system exhibit either a positive curvature at all compositions, or a sequence in which the curvature alternates from positive to negative and then back to positive as the composition varies from pure solvent to pure solute?

3. Why is there a thermodynamic driving force for impurities to form a solution with an absolutely pure element?

4. What thermodynamic factor is responsible for the existence of a positive curvature to the dependence of free energy of a binary solid solution on composition?

5.Evaluate the activity coefficient corresponding to a regular solution where the factor b equals 1000 cal/gatom at $X_2 = 0.3$.

6. Show where, in a plot of molar free energy versus composition for a single solution, the points corresponding to the partial molar free energy of the solute component (2) and that of the solvent component (1) lie, for a composition of the solution equal to $X_2 = 0.75$.

7. Demonstrate in a plot of molar free energy versus composition that the partial molar free energies in two different phases are not equal at compositions differing from the solvus compositions for both phases, whereas show that they are equal at the latter compositions.

8. Use the appropriate relation in the Appendix to show that a positive partial excess molar enthalpy of any component will yield a miscibility gap in a binary phase diagram.

9. The Ge-Si binary phase diagram reveals complete mutual solid solubility of the components. However, the tetrahedral covalent radii are 1.225 and 1.173 Angstroms for Ge and Si, respectively. Calculate the critical temperature corresponding to the expected miscibility gap based on the assumption that strain energy provides the only contribution to the excess enthalpy of solution.

10. All pseudo-binary solid solutions of semiconductor binary compounds reveal positive values for the energy of mixing, whereas the energy of formation of the binary compounds from their elements is negative. Also, despite the positive energies of mixing in the pseudo-binary solutions, in many cases ordered arrangements of the components are found in these solutions. Provide reasonable explanations for these facts.

III-FREE ENERGY AND PHASE DIAGRAMS

INTRODUCTION

In the previous chapter the bases for the development of an understanding of the relationship between free energy-composition diagrams and phase diagrams were explored. We shall in Section A attempt to provide an understanding of the factors responsible for the various temperature dependences exhibited by solidus and liquidus phase boundary compositions; the influence of intermediate phases on phase diagrams; the concept of metastable phases; the development of phase diagrams from the temperature dependence of the free energy-composition curves; and, finally, methods for the prediction of ternary phase diagrams. In Section B we shall present first the basic principles regarding heterogeneous chemical equilibria and then we shall relate these principles to ternary phase diagrams and equilibrium between contacting compounds.

A. FREE ENERGY AND PHASE DIAGRAMS-BINARY SYSTEMS

1. SOLID-LIQUID EQUILIBRIA.

1.1. Phase boundary compositions and their distribution coefficient.

From the condition governing the equilibrium between phases the relations for the solid-liquid equilibrium in a binary alloy can be set down as follows:

$$\overline{G}_1^S = \overline{G}_1^L \quad ; \quad \overline{G}_2^S = \overline{G}_2^L \tag{3.1}$$

But, $\overline{G}_i = {}^{xs}G_i + {}^oG_i + RT\ln X_i$. (Actually, the latter relation acts to define the excess molar free energy of component i, ${}^{xs}G_i$, in terms of the partial molar free energy of component i and the molar free energy of component i in its pure state, ${}^oG_i$.) Substitution into equations 3.1 and solution of the resulting equations yield

$$^SX_2/^LX_2 = \exp\{[(^{oL}G_2 - {}^{oS}G_2) + (^{xsL}G_2 - {}^{xsS}G_2)]/RT\} \qquad (3.2a)$$

$$^SX_1/^LX_1 = \exp\{[(^{oL}G_1 - {}^{oS}G_1) + (^{xsL}G_1 - {}^{xsS}G_1)]/RT\} \qquad (3.2b)$$

Let us postpone consideration of the significance of these equations for the distribution coefficient (i.e. ratio of solidus to liquidus compositions) until we have considered another concept that will enable us to understand which factors are responsible for the change in "average melting point" on addition of solute to a host solvent.

1.2. Composition dependence of 'average' of solidus and liquidus temperatures.

Figure 3.1 shows the free energy-composition curves for a solid phase and a liquid phase. Attention is directed to the composition X^* at which the two phases have both the same composition and same values of free energy, i.e. the composition at which the two curves intersect. From our previous discussion, we are able to conclude that X^* lies in the stable two phase solid plus liquid region between the solidus composition ^SX and the liquidus composition ^LX for the temperature corresponding to the free energy curves. Thus, $X^*(T)$ may be considered to be an average of the solidus and liquidus compositions at the temperature T and hence the inverse function $T(X^*)$ may be considered to be an average of the solidus and liquidus temperatures corresponding to the composition X^*. For the sake of brevity we shall call $T(X^*)$ the 'average melting point'. An attempt will now be made to derive the temperature dependence of the composition X^*.

The function $X^*(T)$ is defined by

$$G_m^L(X^*) = G_m^S(X^*) \qquad (3.3)$$

Equation 3.3 can also be written as follows

$$^*X_1\overline{G}_1^L + {}^*X_2\overline{G}_2^L = {}^*X_1\overline{G}_1^S + {}^*X_2\overline{G}_2^S \qquad (3.4)$$

Let $\Delta G_i = {}^L G_i - {}^S G_i$. Also by the definition for the excess free energy, i.e.

$$\overline{G}_i = {}^\circ G_i + {}^{xs}G_i + RT\ln X_i$$

Substitution into 3.4 yields

$$^*X_1(\Delta^\circ G_1 + \Delta^{xs}G_1) + {}^*X_2(\Delta^\circ G_2 + \Delta^{xs}G_2) = 0 \tag{3.5}$$

Taking the derivative of 3.5 with respect to T yields

$$0 = {}^*X_1(\Delta^\circ S_1 - \Delta^{xs}S_1) + {}^*X_2(\Delta^\circ S_2 - \Delta^{xs}S_2) + (\Delta^\circ G_2 - \Delta^\circ G_1 + \Delta^{xs}G_2 - \Delta^{xs}G_1)[d^*X_2/dT]$$

Solving yields

$$[dT/d^*X_2] = \frac{\{\Delta^\circ G_2 - \Delta^\circ G_1 + \Delta^{xs}G_2 - \Delta^{xs}G_1\}}{({}^*X_1[\Delta^\circ S_1 + \Delta^{xs}S_1] + {}^*X_2[\Delta^\circ S_2 + \Delta^{xs}S_2])} \tag{3.6}$$

To simplify, we consider the case where

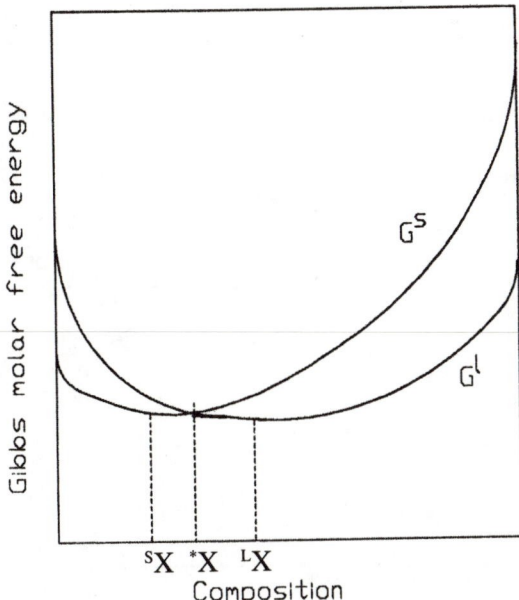

Figure 3.1. Illustrating definition of composition where $G^S = G^L$. Since free energy curves correspond to a particular temperature T then $T(^*X)$ is also defined.

$$\Delta S = \Delta^{\circ}S_1 = \Delta^{\circ}S_2$$

(Thus, the following applies only to those binary systems for which the approximation that the entropy of fusion is the same for both components is valid.) It is also reasonable to assume that $\Delta^{xs}S << \Delta^{\circ}S$. We will further make use of the approximation that near the melting point, ^{M}T

$$\Delta^{\circ}G_i = (^{M}T_i - T)\Delta^{\circ}S_i$$

With these substitutions in 3.6 we obtain

$$[dT/d^*X_2] = (^{M}T_2 - {}^{M}T_1) + \{(\Delta^{xs}G_2 - \Delta^{xs}G_1)/\Delta^{\circ}S\} \qquad (3.7)$$

Thus, the first term on the right hand side of 3.7 is the slope of the line joining the melting points of the two pure components in the phase diagram. See Figure 3.2. The second term on the right hand side of 3.7 equals zero for an ideal solution. Hence, for an ideal solution, the liquidus and solidus curves must lie on either side of the line joining the melting points of the two

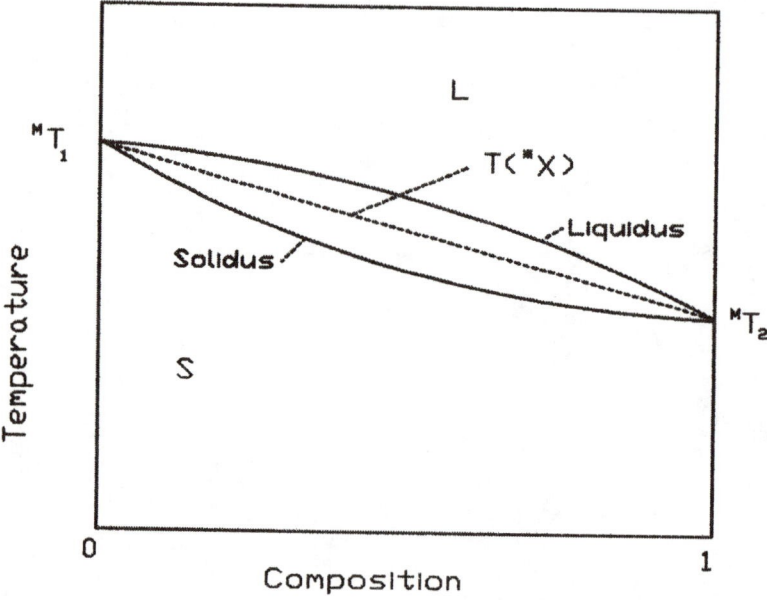

Figure 3.2. Schematic phase diagram for ideal liquid and solid solutions.

components, as shown in Figure 3.2!

If the binary system is not ideal then we may use equation 3.7 to determine the behavior of the 'average melting point'. Let component 1 denote the solvent. For $X_2 \ll X_1$ the assumption that the absolute value of the excess free energy of component 1 is less than that of component 2 is a good approximation. Thus, the sign of the deviation of the slope of the 'average melting point' from the line joining the melting points of the two components is given by the sign of $\Delta^{xs}G_2$. Generalizing, at a given composition, if the excess free energy of any component in its solid phase is more positive than that in its liquid phase, then for increasing composition of that component there will be a negative deviation of the 'average melting point' relative to that for the ideal solution, as shown in Figure 3.3. With this relationship between these excess free energies reversed the 'average melting point' will exceed that for the ideal solution, as shown in Figure 3.4.

In Chapter IV we will make use of equations 3.2 and 3.7 to provide a basis for the prediction of the Gibbs adsorption of solute to interfaces on the assumption that the local atomic or molecular environments at interfaces approximates to those in liquids. Also, we shall show in the next

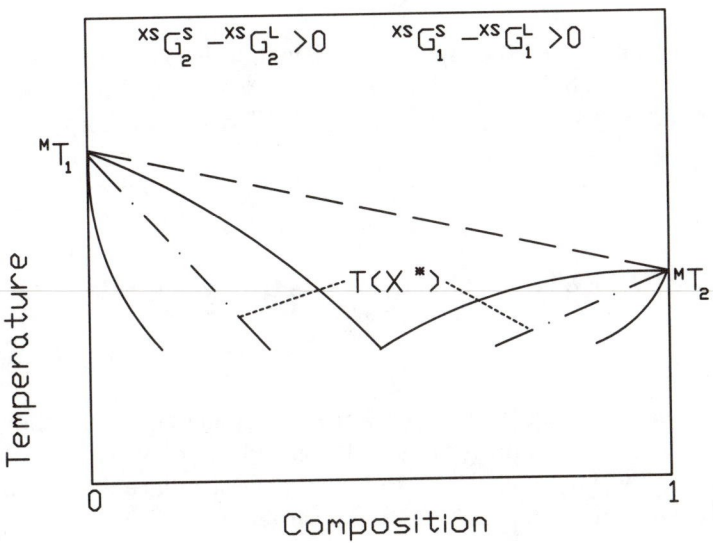

Figure 3.3. Schematic section of a phase diagram illustrating the negative deviation of $T(X^*)$ when the partial molar excess free energies for the solid exceed those for the liquid.

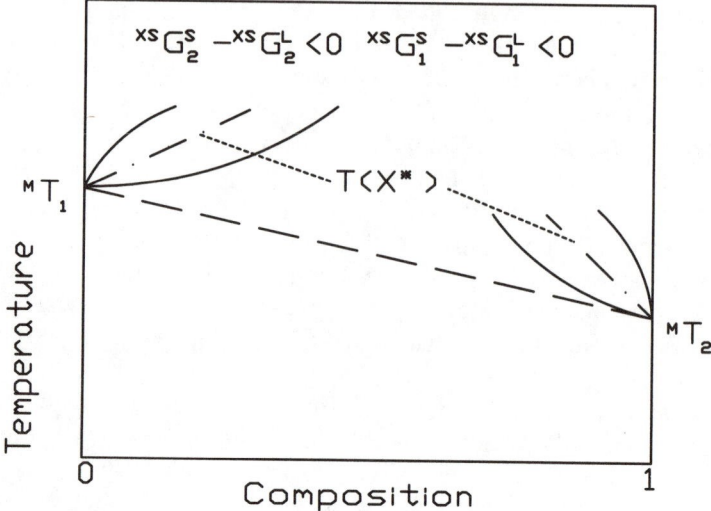

Figure 3.4. Schematic section of a phase diagram illustrating the positive deviation of $T(X^*)$ when the partial molar excess free energies for the solid are less than those for the liquid.

section why sometimes the sign of the deviation between the "average melting point" and the weighted mean pure component melting point will correlate to the existence of phase separation or order in isomorphic phase diagrams.

2. ORIGIN OF EUTECTIC PHASE DIAGRAM IN ISOMORPHIC SYSTEMS.

In Section 1.2 we showed that the necessary and sufficient condition to produce phase separation or a miscibility gap in an isomorphic system is that the enthalpy of mixing, H^M (the relative integral molar enthalpy), have a positive value. It is also true that if the partial excess molar enthalpy is positive for $0 < X_2 < 1$ then there must be a miscibility gap in the system. (The partial molar excess enthalpy for the solute component 2 corresponding to some composition X_2^* is given by the intercept with the ordinate at $X_2 = 1$ of the tangent to the molar enthalpy curve at that composition for $0 < X_2^* < 1$.)

Thus, the conditions required to produce an eutectic type phase diagram are that $^{xs}H_2^s > 0$ and that $^{xs}H_2^s > {}^{xs}H_2^l$. These conditions are satisfied by the case where the <u>only contribution to the partial molar excess enthalpies is that due to atomic(ionic) size induced strain energy</u>. This situation is found to occur unambiguously in phase diagrams involving ionically bonded solids in which the cation is isovalent. For example, Figure 3.5 shows the phase diagrams between first column fluorides and CsF, where the deviation in ionic radius between Cs and the other first column elements increases in the sequence: Rb, K, Na, Li. These phase diagrams qualitatively obey equations 3.6 and 2.27 in that as the deviation in ionic radius between Cs and the other cations increases the negative deviation of the "average melting point" from the weighted mean melting points of the pure components increases. Also, the solid solubility of one component in the other decreases in this sequence. The lack of first neighbor interactions

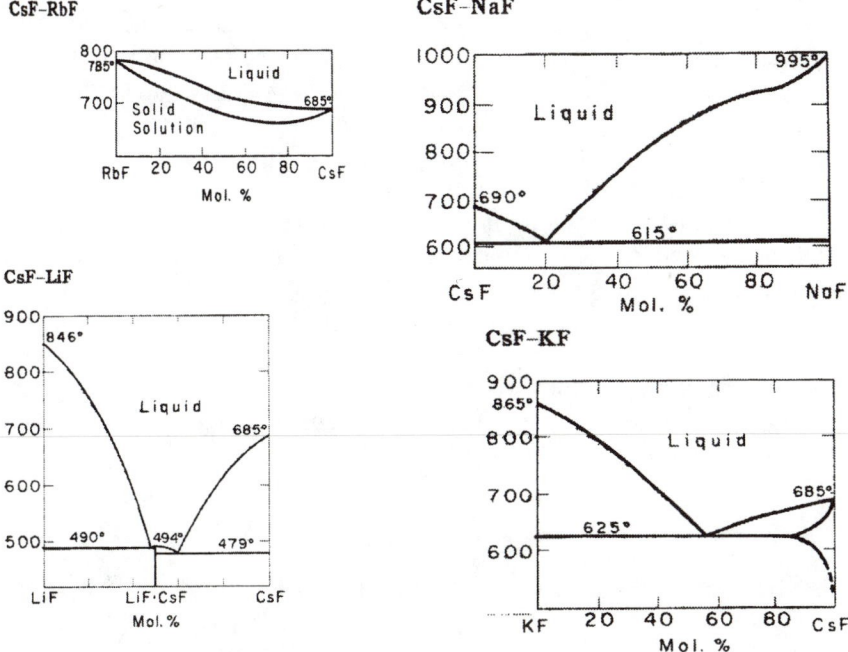

Figure 3.5. Phase diagrams of CsF with the other isovalent fluorides. The deviation in ionic size increases in the sequence Rb, K, Na, Li. Phase diagrams are from E.M. Levin, C.R. Robbins and H.F. McMurdie,PHASE DIAGRAMS FOR CERAMISTS, The American Ceramic Society, Columbus, 1964.

between cations in ionically bonded solids eliminates many interactions that occur in alloy systems. A consequence of this fact is that it is possible to illustrate the effect of the strain energy in phase diagrams of ionic solids by limiting the cation to an isovalent sequence.

In ionically bonded solids, solutes differing in valence from that of the solvent will have corresponding excess partial molar enthalpies that are positive, because the introduction of such solutes into the lattice requires the concomitant production of defects to maintain electroneutrality. Further, this enthalpy will be less positive in the liquid state where the excess charge of the solute can be easily screened by motion of the anions relative to it. Thus, except for the cases where long-range interactions, excepting the electrostatic interaction, can exist between cations (e.g. magnetic interactions), it is expected that the phase diagrams of ionically bonded systems will not exhibit peritectic behavior or positive deviations of the "average melting point" from the weighted mean pure component melting point. Examination of ceramic phase diagrams involving ionically bonded solids reveals that this expectation is obeyed (i.e. most phase diagrams exhibit eutectic type behavior of the solidus and liquidus phase boundaries.)

In alloy and covalently bonded systems nearest-neighbor or many-body interactions are prevalent. Hence, it is much more common to observe positive deviations of the "average melting point" from that of the ideal solution in such phase diagrams than in ionic systems. One source of this behavior stems from bonding interactions that are characteristic of the solid state and are absent in the liquid state. In multistructural systems the first term in equation 3.6 provides a negative deviation of the "average melting point" from the weighted mean melting point of the pure components. For these systems $\{(G_2^{o,s} - G_2^{o,l}) - (G_1^{o,s} - G_1^{o,l})\}$ is always negative when 2 is the solute. The solute in multistructural (non-isomorphous) phase diagrams always has a crystal structure that is metastable with respect to the stable structure of the solvent (component 1.)

3. SOLID-SOLID EQUILIBRIA EQUIVALENT TO THE SOLID-LIQUID CASE.

The relationships derived above can be applied to the equilibria between polymorphic phases that extend into the composition coordinate merely by substituting for the subscript S that corresponding to the phase that is stable at low temperature in the pure host, for the subscript L that corresponding to the phase stable at high temperature and interpret $^{o}T_{M}$ as the equilibrium transformation temperature between these two structures. It is usually difficult to predict how solute affects the relative stability of the two competing polymorphs. However, if one of these competing polymorphs has the same crystal structure as the stable structure of the solute, then the effect of solute is to stabilize this structure. Examples illustrating this effect are given in Figures 3.6 and 3.7. The stable structure of Ni is fcc and, as shown, Ni as a solute stabilizes the fcc structure of Fe. The stable structure of Mo is bcc and Mo as a solute stabilizes the bcc structure of Fe. This particular role of solute is exerted through the term $\Delta^{o}G_{2}$ in equation (3.6).

The term $\Delta^{xs}G_{2}$ in equation (3.6) also can exert significant effects on the relative stability of competing polymorphs. For example, the effect of solute on the magnetic interaction between Fe atoms is manifested by means of this term.[1] Also, the difference in strain energies induced in the competing polymorphs by interstitial solutes, as described quantitatively in equation (2.27), enters into this term. For example, interstitial solutes, such as carbon and nitrogen are larger than the cavities into which they fit in either bcc or fcc iron. However, the misfit is greater for the bcc structure. As a

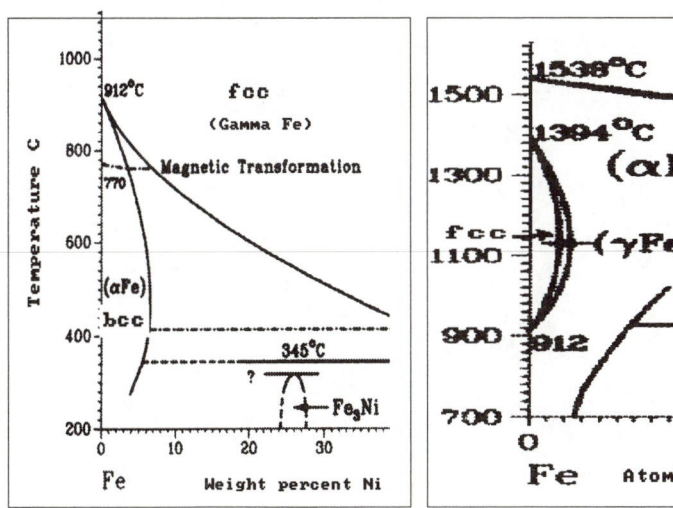
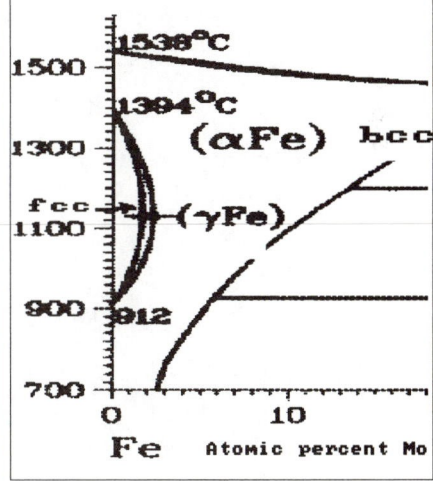

Figure 3.6. Fe-Ni Phase Diagram   Figure 3.7. Fe-Mo Phase Diagram
Extracted from BINARY ALLOY PHASE DIAGRAMS,1, ASM,1986.

consequence, carbon or nitrogen as a solute in iron stabilizes the fcc structure, as shown in Figure 3.8. Similar effects can be shown to exist in ceramic phase diagrams and in phase diagrams for covalently bonded systems.

The solid state transformations differ from the melting transition in monostructural systems in one important aspect. The entropy of the transformation between two given crystal structures can differ markedly between elements. Thus, equations 3.2 and 3.6 are appropriate descriptions, whereas equation 3.7 is not, of the dependence on solute of the transformation temperature. For example, although for most normal metals $S_{bcc} - S_{fcc} > 0$, as suggested in Chapter 1, the reverse is valid for the Periodic Groups V and VI transition elements and for the low temperature transition in iron. This effect of the entropy of transformation is also likely to be valid for solid state transformations in ionic and covalently bonded solids.

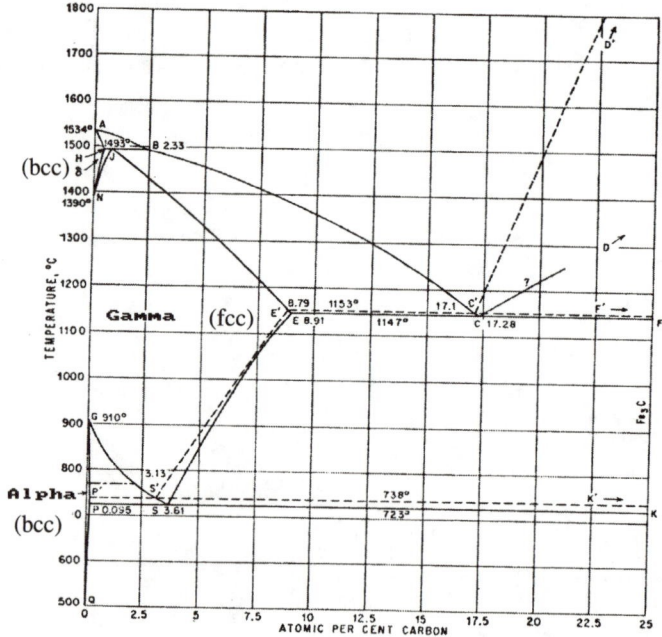

Figure 3.8. Iron-carbon phase diagram. Note both stable and metastable diagrams. After M. Hansen, CONSTITUTION OF BINARY ALLOYS, McGraw-Hill, NY, 1958.

4. INTERMEDIATE PHASES.

Intermediate phases that tend to be centered about stoichiometric compositions will also have associated free energy-composition curves that tend to have minima at the corresponding stoichiometric composition as shown in Figure 2.12 Normally, the curvature in the vicinity of the minimum free energy will tend to be large, although this is not a necessary condition. If any part of the free energy-composition curve for the intermediate phase falls at a lower free energy value than that corresponding to the most stable arrangement for the terminal phases, at any composition, then the intermediate phase becomes stable relative to the partially stable situation for the terminal phases , as shown in Figure 2.12.

Examination of Figure 2.12 leads to the conclusion that any intermediate phase that is stable must have a range of composition over which this phase is stable. The larger is the curvature in the vicinity of the minimum in the free energy-composition curve for the intermediate phase the smaller is the composition range of stability of this phase. There have been attempts in the past to categorize intermediate phases into two groups: one exhibiting a large range of compositional stability and the other exhibiting a negligible one. It is questionable that the population distribution frequency as a function of the compositional range of stability is bi-modal. At least, this author has not seen any evidence concerning the modality of this distribution. In any case, it should be apparent that the limits of composition over which the intermediate phase is stable are functions also of equilibrium with the adjoining phases (i.e. of where the common tangents lie) and not only of the curvature of the free energy-composition curve for the intermediate phase at its minimum.

In general, in alloy systems, intermediate phases tend to have more unlike nearest-neighbor bonds than do solid solutions at the same composition. Hence, if there is a tendency for the unlike nearest-neighbor bonds to be more stable than the average for the like atom bonds then there is a tendency for the intermediate phase to be stabilized. This tendency to stabilize unlike atom bonds can be related to several factors, depending upon the nature of the components. Alloy theory is concerned with the identification and description of these factors. There are intermediate phases, however, which depend upon other factors than those related to the maximization of the number of unlike bonds for their stability. Indeed, in the phase diagrams of ceramic systems we encounter intermediate phases the stability of which have different origins than in alloy systems. For

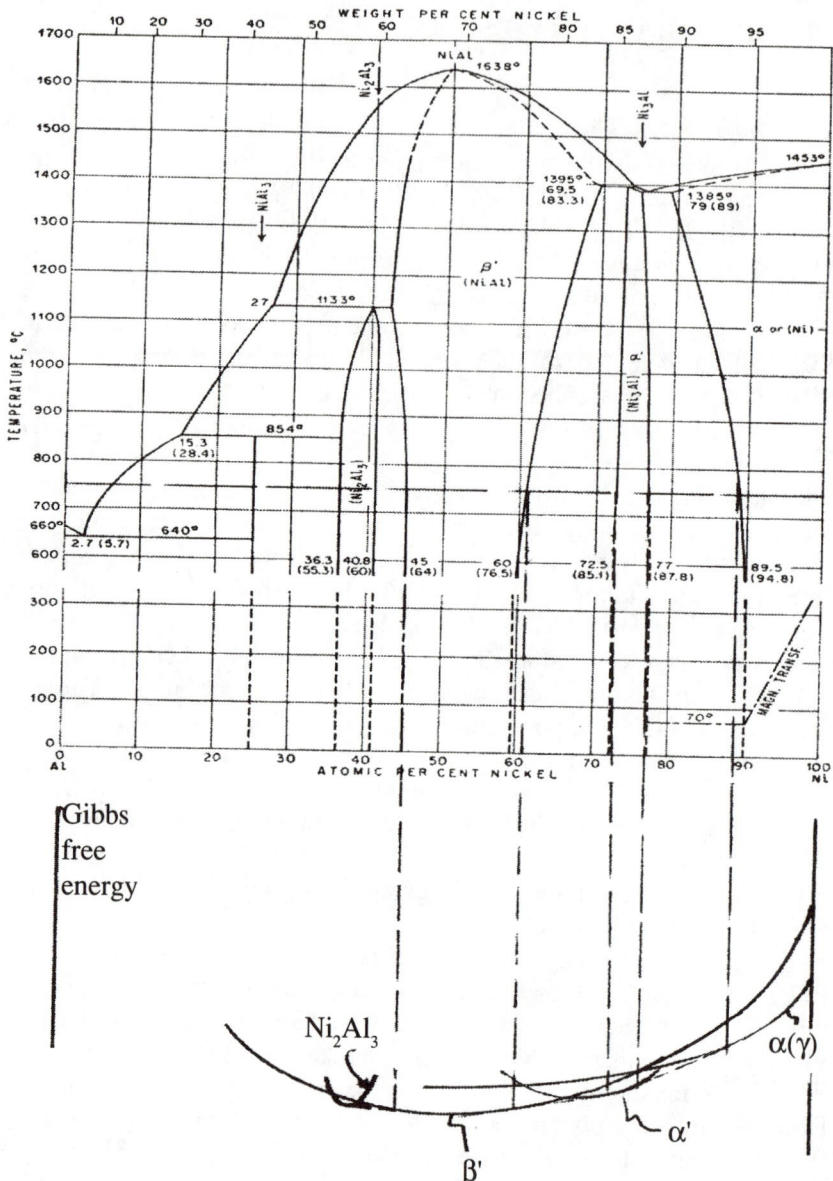

Figure 3.9. Ni-Al phase diagram showing relation to free energy-composition curves for the various phases. Wide composition ranges of phase stability must coincide with small deviations in free energy between coexisting phases.

example, the spinel phase $MgAl_2O_4$ is an intermediate phase in the MgO-Al_2O_3 binary system and does not depend for its stability on Mg-Al bonds, which do not exist in the spinel structure. An exploration of the origins of the stability of intermediate phases would be of interest, but is extraneous to the present objective. Here, we are concerned only with the effect of stable intermediate phases upon the phase diagram.

If the intermediate phase is only just slightly more stable than the most stable mixture of its bounding phases, as occurs for example in the Ni-Al system where the intermediate phase Ni_3Al has the terminal Ni base solid solution and another intermediate phase NiAl as its bounding phases (see Figure 3.9), then it is possible that there will be relatively large compositional ranges of stability for all three phases. On the other hand, the more general situation is that the intermediate phase is sufficiently more stable than the most stable mixture of its bounding phases so that these compositional ranges of stability are limited. If one of the bounding phases is a terminal solid solution then the solvus line corresponding to the limit of solubility of this phase will fall close to the terminal composition, i.e. it is said in this case that the solubility of the terminal solid solution is limited.

5. METASTABILITY.

The concept of the metastability of a phase in a binary system and some consequences of this metastability can now be discussed. Consider the free energy-composition diagram shown in Figure 3.10. As shown, the intermediate phase is not stable relative to the mixture of terminal phases alpha and beta. However, suppose that the kinetics of formation of the beta phase are so slow relative to that for the intermediate phase, gamma, as to allow the gamma phase to form. In this case, it is possible to attain a metastable equilibrium between the alpha and gamma phases described by the common tangent to the free energy-composition curves for these two phases. One consequence of the fact that the gamma phase is unstable relative to the mixture of the alpha and beta phases is that the solvus line for the alpha phase in <u>metastable equilibrium with the gamma phase</u> must be contained within the solvus lines corresponding to the <u>stable equilibrium</u> between the alpha and beta phases. A more general rule of phase diagram construction follows from this property of metastability—**the solvus lines separating single phase regions from two phase regions in a binary**

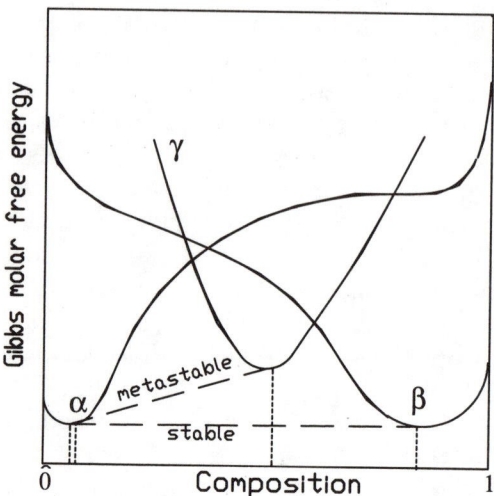

Figure 3.10. Illustrating the partial equilibrium possible between a metastable phase and a stable phase.

system must extrapolate into two phase regions. The student is urged to construct the free energy-composition curves corresponding to the two solid phases and liquid phase for an eutectic phase diagram and for two temperatures-one above and one below the eutectic temperature as a means of appreciating the basis for this rule.

6. TEMPERATURE DEPENDENCE OF FREE ENERGY COMPOSITION CURVES AND PHASE DIAGRAMS

It should be apparent by now that the phase boundary compositions at a given temperature in some system of binary components are determined by the points of tangency of the lowest common tangents to the free energy-composition curves of the possible phases in the system. Also, whether a single phase or mixture of phases will be stable at some composition and temperature in the phase diagram is determined by the line in the free energy-composition diagram that represents the lowest free energy for these conditions. Obviously, a variation in temperature involves changes in the free energy-composition curves of the possible phases and hence in the identity of the envelope that represents the lowest possible free energy in the system, as is illustrated in Figure 3.11. Consequently, derivation of a phase diagram from free energy-composition curves for possible phases requires that the latter curves be known as a function of temperature. It is easier to grasp how phase diagrams develop from free energy-composition curves using a dynamic mode of illustrating this relationship. Such a dynamic mode is available in the form of computer programs.(See Bibliography.)

The development of phase diagrams based on the assumption that the solutions obey regular solution criteria is edifying and demonstrated in Figure 3.12. The parameter Ω shown both on the ordinate and abscissa(top) corresponds to the parameter b in equation 2.16.

It is only recently that a concerted attempt has been made to derive phase diagrams from the bases of G(T,X) relations for the possible phases in the system. This attempt is described briefly in the next section.

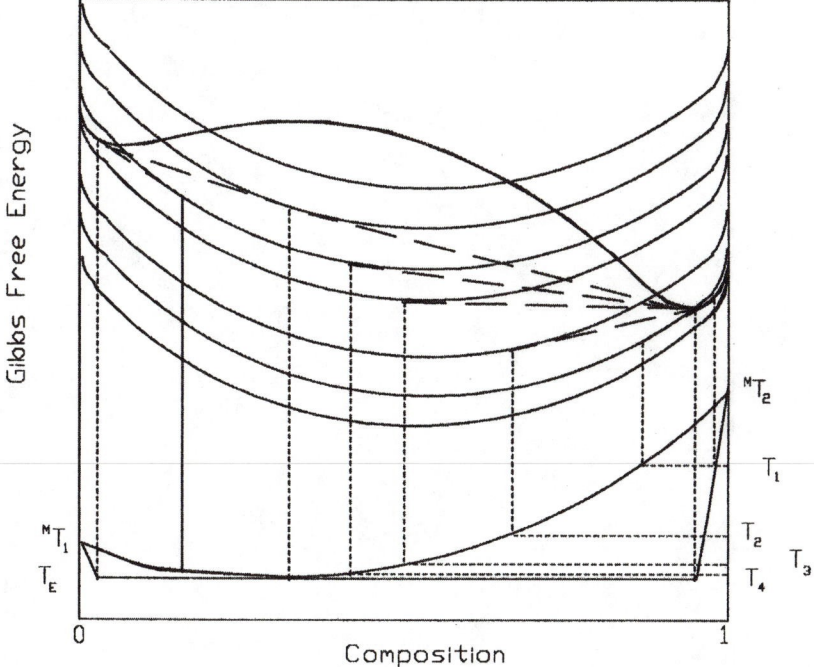

Figure 3.11. Freeenergy-composition curves at various temperatures for a liquid phase relative to a solid phase(held stationary for convenience in drawing only). The relation between the common tangent points and the solidus and liquidus phase boundaries is also shown. The eutectic temperature T_E and its related significance in the free energy curves is also apparent.

7. PREDICTION OF PHASE DIAGRAMS.

It is apparent from the above that given a knowledge of the composition and temperature dependence of the free energy of phases it should then be possible to derive the corresponding phase diagram. Although there exist some thermodynamic data for binary systems, these data are far from sufficient to allow such calculations to be made for most of the

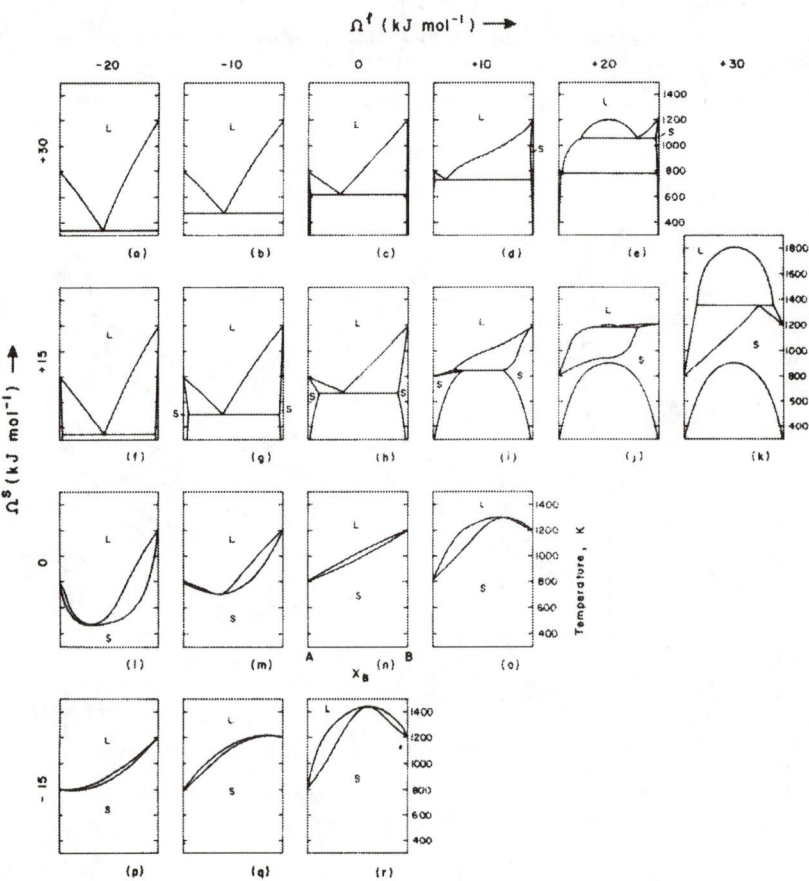

Figure 3.12. Sequence of phase diagrams corresponding to changes in regular solution parameters($\Omega\equiv b$) for liquid and solid solutions. After Pelton and Thompson, Progress Solid State Chemistry **10**, part 3, 119(1976).

binary systems. Hence, a different point of view has arisen concerning a strategy of prediction of phase diagrams. This strategy is based on the existence of a large body of binary phase diagram data and is as follows.

From the knowledge that the free energy is both composition and temperature dependent, functions have been assumed to give this dependence analytic form. The composition dependence of several of these functions for the excess free energy is described in Table 3.1. The Gibbs free energy for one phase in a binary system is usually written in the form

$$G = \sum X_i {}^oG_i + RT\ X_i \ln X_i + {}^{xs}G \qquad (i = 1,2)$$

Since equilibrium is determined by the equality of partial molar free energies it is more instructive to consider the equivalent expression for the latter quantity for a phase as follows.

$$\overline{G}_i = {}^oG_i + RT\ln X_i + {}^{xs}\overline{G}_i \qquad (3.8)$$

where

$$ {}^{xs}\overline{G}_i = {}^{xs}G + (1 - X_i) d^{xs}G/dX_i$$

as can be ascertained by substituting equation 3.8 into equation 2.16. It is now apparent that by equating the partial molar free energies for a given component in the two phases being equilibrated the parameters characteristic of the elements in the two phases appear as differences in such equations. This difference $({}^oG_i(alpha) - {}^oG_i(beta))$ is called the lattice stability of component i for the two phases alpha and beta and is the quantity already listed in Appendix 1 of Chapter I.

The unknown terms in the resulting equations for the compositions of the solvus lines for the two phases in equilibrium are the lattice stabilities and the partial molar excess free energies. A power series is assumed for the temperature dependence of each of the lattice stabilities and for the parameter, such as C_p in the Redlich equation in Table 3.1. Thus, for example, if we consider two phases involved in a two phase equilibrium, and if we consider terms only up to the quadratic, then there will be 14 adjustable parameters to be fitted to the two solvus lines in the phase diagram representing this two phase equilibrium. (From the following equation for the dependence of the molar free energy of a pure component as a function of temperature

$$ {}^oG = m_1 - m_2 T + m_3 (T - \ln T) - m_4 T^2 / 2 + \dots$$

TABLE 3.1

$$\text{Margules:} \quad {}^{xs}G = X_1 X_2 \sum_{p=0}^{n} A_p X_2^p$$

$$\text{Borelius:} \quad {}^{xs}G = X_1 X_2 \sum_{p=0}^{n} B_p X_2^{n-p} X_2^p$$

$$\text{Redlich \& Kister:} \quad {}^{xs}G = X_1 X_2 \sum_{p=0}^{n} C_p (X_1 - X_2)^p$$

$$\text{Bale \& Pelton:} \quad {}^{xs}G = X_1 X_2 \sum_{p=0}^{n} D_p (P_p(X_1 - X_2))$$

$P_p(X_1 - X_2)$ = Legendre polynomial of degree p of argument $(X_1 - X_2)$

we deduce that there are 4 parameters for each component in the lattice stability contribution to the free energy and three for each phase in the excess free energy contribution. Thus, there will be a total of 4*2+3*2=14 adjustable parameters.) The equations necessary to evaluate these parameters are obtained both from known thermodynamic data and from the phase diagram using the equality of partial molar free energies at the solvus compositions. If there are not sufficient independent data then the temperature dependence is limited to fewer terms in the expansion.

In order to minimize the computational problem, it has been customary to assume that there is no solubility range for the intermediate phases and to treat them as compounds. It would be useful to have experimental values for the free energy of formation of such compounds. Various procedures have developed for estimating values of the energy of formation of such compounds and these are outlined in the Appendix to this chapter. Using such estimated values it is then possible by a trial and error procedure to fit the parameters required to describe quantitatively any binary phase diagram, consistent with the assumption that the intermediate phases are compounds.

Using procedures similar to those described above, various centers around the world have developed data banks containing values for the adjustable parameters that are required to yield values for the various lattice stabilities of elements and the excess free energy of binary systems. These data now provide a basis for the prediction of multicomponent phase diagrams using the following procedure.

The free energy of a phase in the multicomponent system is assumed to be given by

$$G = \sum_{i=1}^{n} X_i \,^{\circ}G_i + RT \sum_{i=1}^{n} X_i \ln X_i + \sum_{k=2}^{n} \sum_{j=1}^{\left|\substack{n \\ k}\right|} \,^{xs}G_{j,k}$$

where $\left|\substack{n \\ k}\right|$ is the binomial coefficient $= n!/[(n-k)!k!]$

Knowing the terms for the binary interactions for each phase in a two phase equilibrium it is then possible to evaluate the equilibrium either by calculating the minimum energy for the mixture of phases or by equating the appropriate partial molar free energies. At first, it is assumed that there are no ternary excess free energy terms in the first attempt to predict a ternary phase diagram and only if there are sufficient data to allow for a correction to the predicted phase diagram are the ternary terms fitted to these data.

The CALPHAD procedure in evaluating the free energy of a phase assumes that any deviation from a random mixture of the components of a phase may be described by the excess free energy of mixing. Although this is a viable procedure for the large majority of phases, it tends to lead to complicated expressions for the latter when a phase involves either partial long range order or significant short range order for the distribution of the components on the lattice sites. In the latter case, it is better to attempt to obtain a more accurate description of the entropy of configuration and a simpler expression for the excess free energy relative to this modified entropy. The latter is the procedure followed by the group using the CV method.[2]

Summarizing, we have determined the conditions governing the effect of solute addition upon the solidus and liquidus temperatures and found that a more positive excess free energy for the solid phase relative to the liquid phase results in a decrease in the "average melting temperature". Further, the relations governing the distribution coefficient for solidus to liquidus compositions have been derived. One origin of the eutectic phase diagram has been shown to be the existence of a positive enthalpy of mixing in the solid state. The origin of intermediate phases in phase diagrams has been explored, as well as the effect of metastability on solvus composition. The integral relation between free energy-composition curves for possible phases as a function of temperature and the phase diagram has been demonstrated. Finally, a strategy for the prediction of phase diagrams has been outlined.

B. HETEROGENEOUS CHEMICAL EQUILIBRIA AND PHASE DIAGRAMS.

8. THERMODYNAMICS OF HETEROGENEOUS CHEMICAL REACTIONS.

It is worthwhile summarizing here the thermodynamic relations governing heterogeneous chemical reactions because a knowledge of this discipline is often of use in materials science, ceramics and metallurgy. We note first that the vapor pressures of substances should be exponentially related to the reciprocal temperature, as can be derived using the Clapeyron relation as follows. We note first that in the transformation from a solid or a liquid (a condensed phase) to a vapor phase the change in molar volume is to a good approximation equal to that for the vapor phase itself, i.e. $V_v - V_s = V_v$. Also, for an ideal gas, the molar volume of the vapor is given by

$$V_v = RT/p$$

where p represents the partial pressure of the vapor. Substitution into the Clapeyron equation and rearranging terms yields

$$dlnp/d(1/T) = -\Delta H/R$$

Integration then yields that

$$p = p_o e^{-\Delta H/RT}$$

Thus, as the temperature increases the vapor pressure of a substance in equilibrium with it increases greatly.

Let us now consider reactions between two different substances, e.g. the reaction between aluminum and oxygen which forms the oxide Al_2O_3. The reaction can be written as follows

$$(4/3)Al + O_2 = (2/3)Al_2O_3 \qquad (3.A)$$

The equilibrium constant for this reaction $K=1/p(O_2)$ and this corresponds to a certain value of the standard free energy for this reaction and temperature, where the standard free energy of reaction refers to all components being in their standard reference states at the temperature and at <u>one atmosphere pressure</u>. Values for the standard free energy of reaction ΔG° are given in various books (see bibliography). A convenient method of providing such information is via Ellingham diagrams, which presents the standard free energy of reactions with a given component, such as O_2, as a function of temperature.(See Figure 3.13).

There may be some question in the reader's mind as to the origin of the equilibrium partial pressures of O_2 and partial pressure ratios in the Ellingham diagram has certain other features which are useful and which we shall demonstrate. Consider the reaction given above. Suppose that we are interested in the values corresponding to 300 °C. According to the Ellingham diagram, the standard free energy for the above reaction at this temperature is-996 kJoules/mol($-237,000$ cal/mol). If now we draw a straight line through the coordinate $(0,0^{..}K)$ in the upper left hand region of the diagram and the point defined by $\Delta G^\circ=-996$ and T=300 °C, it will be found to intercept the outermost line to the right and down at the value of $p(O_2)=10^{-90}$ atmospheres. The significance of this result is that the latter is the value of the partial pressure of O_2 in equilibrium with Al and Al_2O_3. A partial pressure of O_2 less than this value will reduce the oxide and a partial pressure of O_2 above this value will oxidize the Al. For example, since even ultra high vacuum chambers contain a partial pressure of O_2 more than this value it is obvious that the surface of aluminum conducting stripes formed in vacuo for integrated circuits is covered with an oxide.

There may be some question in the reader's mind as to the origin of the equilibrium partial pressures of O_2 and partial pressure ratios in the Ellingham diagram. The basis for this relation is simple. It must be recalled that the standard free energy of a reaction corresponds to vapor phases having a partial pressure of one atmosphere. The standard free energy change for the reaction given by (3.A) is $-237,000$ cal/mol as noted above. At equilibrium, the change in free energy for this reaction would of course equal zero. But, the standard state of the O_2 gas in this reaction, one atmosphere pressure does not correspond to the partial pressure in equilibrium with this reaction, i.e. the partial pressure that would make the change in free energy equal to zero. Thus, to (3.A) must be added the reaction

$$O_2(p_{eq}) = O_2(p^\circ=1) \qquad (3.9)$$

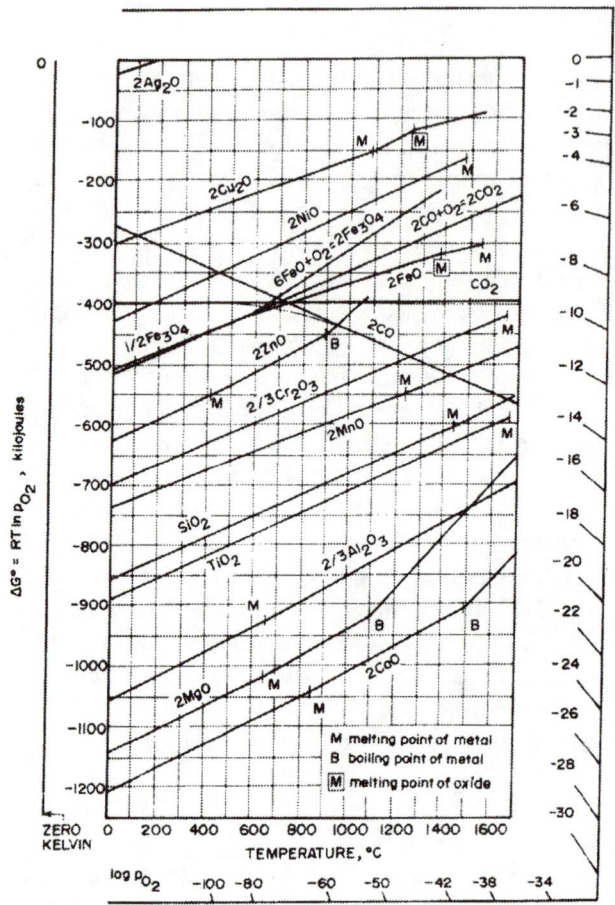

Figure 3.13. Ellingham diagram for several oxides. After Gaskell in PHYSICAL METALLURGY, eds. R.W. Cahn and P. Haasen, North-Holland Physics Publ. NY,1983.

for which the change in free energy, considering that O_2 acts as an ideal gas, is

$$\Delta G' = RT\ln(p^\circ/p_{eq}) \qquad (3.10)$$

Thus, for equilibrium, $\Delta G^\circ + \Delta G' = 0$ or in this case

$$RT\ln(p^\circ/p_{eq}) - 237,000 = 0$$

Rearranging the terms and using the fact that $p^o=1$ then

$$p_{eq} = e^{(-237,000/RT)}$$

Substituting 573 °K for T and solving yields $p_{eq}=1.5*10^{-90}$ atm., the result we found previously using the Ellingham diagram.

In general, if a reaction involves m_i moles for the ith reactant for a total of n gases of which j are reactants (to be denoted by the superscript r)and n−j are products (to be denoted by the superscript pr) then we would have

$$\prod_{i=j+1}^{n-j}(p^{pr}_{eq,i})^m / \prod_{i=1}^{j}(p^r_{eq,i})^m = e^{(-\Delta G/RT)} = K \qquad (3.11)$$

where K is the equilibrium constant.

For the case where one or more of the elemental constituents taking place in a reaction is a solute in a solid solution then the reaction must be corrected to account for the free energy of solution of this solute in its solvent. The correction is analogous to that already made for the change in state of a gas from the standard state of one atmosphere to the equilibrium pressure. In the case that the solute is a reactant rather than a product, we must add to the reaction equation the reaction

$$[A] = A^o$$

for which the free energy of the reaction is $-RT\ln a_A$, where a_A is the activity of component A in its solvent. (The activity of A in its pure standard state is equal to 1.) Since, at equilibrium, the total free energy of the reaction is zero, we would then obtain that

$$RT\ln a_{A,eq} = \Delta G^o.$$

If more than one solute is involved or there are also constituents in gaseous states then the corresponding equations would have to be added to obtain the relation between their partial pressures, the activities of the dissolved constituents and the standard free energy of reaction for the equilibrium situation.

The Ellingham diagram is also useful in indicating which metal will reduce which oxide. Any metal corresponding to a reaction having more negative free energy of formation values than those for the formation of some other oxide will reduce the latter oxide. For example, Ti will reduce Cr_2O_3.

9. TERNARY PHASE DIAGRAMS AND HETEROGENEOUS CHEMICAL EQUILIBRIA.

For the sake of simplicity let us limit our consideration of contacting species to three elemental components. A typical problem involving just three elemental components is the contact between a metal and a compound diffusion barrier, e.g. Al/TiN. It is desired to know what should the actual contacting phases be to have a thermodynamically stable contact. We refer back to the definition of thermodynamic stability, according to which the partial molar free energy of each component must be equal in the contacting phases. This situation can occur in a ternary system in a three phase region or along the tie lines bounding the three phase region. Suppose that the ternary phase diagram for the system is not known, but that the stable phases in the binary systems are known. We use the Ti-Al-N ternary to illustrate how to determine the stable phases in this ternary and the effect on the TiN/Al contact.

The phases that are stable in the three binary systems are shown in Figure 3.14. The phases TiN and Al are joined in this figure by a dashed line. All the possible tie lines between phases that intersect this dashed line are shown. Let us consider the reactants to be TiN and Al while the possible products are AlN and each phase along the bottom abscissa to which it is joined by a line. We need to determine whether any of the product combinations are stable with respect to the reactants. To do this we calculate the free energy for each reaction corresponding to the temperature, 600 C. The possible reactions between Al and TiN and the approximate free energies for these reactions are as listed below.

$$TiN + 4Al = TiAl_3 + AlN \qquad -29,800$$

$$TiN + 2Al = TiAl + AlN \qquad -13,600$$

$$5TiN + 7Al = Ti_5Al_2 + 5AlN \qquad +2850$$

$$TiN + Al = Ti + AlN \qquad +5400$$

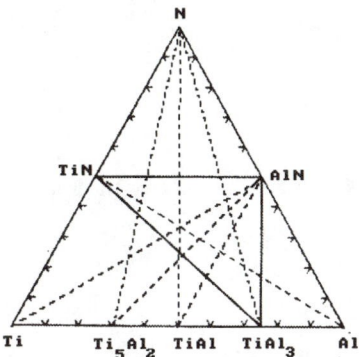

Figure 3.14. Ternary phase diagram for the Ti-N-Al system. T=600°C.

(The free energies shown are for the reactions in units of cal.) Thus, only the first two reactions yield stable products and of these the most stable reaction yields AlN + TiAl$_3$.

We continue this process considering N and each of the intermetallic compounds as reactants with two possible sets of products: TiN + TiAl$_3$ and TiN + AlN. In these reactions we find that the products are stable. Also, we consider TiN and TiAl$_3$ as reactants and as products AlN plus a titanium aluminide (and Ti), which have tie lines that cross the reactant tie line, as the products. We find that the reactants are stable with respect to the possible products. Hence, the tie-lines between AlN and TiAl$_3$, between TiN and AlN, and between TiAl$_3$ and TiN represent stable tie-lines. Thus, the region bounded by the solid lines in Figure 3.14 is a stable three phase region on the assumption that there is no stable intermediate phase in this ternary system. Also, there is another stable three phase region consisting of Al +TiAl$_3$ + AlN.

How may this phase diagram information be used in the contact or junction problem? If a mixture of AlN and TiAl$_3$ is deposited onto Al and then TiN is deposited on this intermediate layer then the contacting layers will be in equilibrium. (We neglect the small concentrations of the coexisting phases in solution in each of the layers at the corresponding phase boundaries required to achieve the true equilibrium. Also, we neglect the real possibility that intermediate ternary phases may be stable and that interface energies may change the free energy of the thin film system so as to stabilize what would be in bulk a metastable phase.)

Summarizing this section, we have provided a review of important aspects of heterogeneous chemical equilibria and have illustrated the use of these concepts in problems met in materials science.

APPENDIX-Methods of Predicting Values for the Energy of Formation of Intermediate Phases

Intermetallic compounds.

One empirical method, developed by Miedema[A1] and his collaborators, has had some success in the prediction of energies of formation for intermediate phases. Another[A2], for equiatomic intermediate phases formed between transition elements, is based on a microscopic theory of bonding due originally to Friedel and Pettifor which assumes a rectangular d band density of states. Others[A3,A4], using more exact approximations have obtained values for the energies of formation for all compositions of alloys formed between transition metals. Finally, Williams and others[A5] have developed computer programs to yield first-principle and 'ab-initio' energy of formation values for any solid binary intermediate phase having a cubic crystal structure.

Covalent compounds

Phillips[A6] describes a method for predicting the heat of formation of tetrahedrally bonded covalent compounds. Miedema's method may also be used for this class of materials.

Ionic compounds

Harrison[A7] provides a clear description of how the energy of ionic, as well as covalent compounds may be calculated to a good first approximation.

REFERENCES TO APPENDIX.

A1. A.R. Miedema, R. Boom and F.R. de Boer, J.Less Com. Metals 41,263(1975);A.R. Miedema, J.Less Com. Metals 46,67(1976); A.K. Niessen, F.R. de Boer, R. Boom, P.F. de Chatel, W.C.M. Mattens and A.R. Miedema, CALPHAD 7,51(1983).
A2. R.E. Watson and L.H. Bennett, Phys.Rev.Lett.43,1130(1979).
A3. A.Pasturel, C.Colinet and P. Hicter, CALPHAD 9,349(1985).
A4. J.van der Rest, F. Gautier and F. Brouers, J. Phys F5, 2283(1975).
A5. C.D.Gelatt.Jr., A.R. Williams and V.L. Moruzzi, Phys.Rev.B27, 2005(1983).

A6. J.C. Phillips, BONDS AND BANDS IN SEMICONDUCTORS, Academic Press, NY, 1973.
A7. W.A. Harrison, ELECTRONIC STRUCTURE AND THE PROPERTIES OF SOLIDS, Freeman, San Francisco, 1980.

REFERENCES

1. M.Hillert, T.Wada and H.Wada, J. Iron Steel Inst.205, 539(1967); H.Harvig, G.Kirchner and M. Hillert, Met.Trans.3, 329(1972).
2. See Bibliography, Chapter II.

BIBLIOGRAPHY.

1. A.D.Pelton in PHYSICAL METALLURGY, R.W. Cahn and P.Haasen, eds.,3rd edition,North Holland Physics Publishing, New York,1983.
2. A.M. Alper, ed., PHASE DIAGRAMS-MATERIALS SCIENCE AND TECHNOLOGY, vol. 1-5, Academic Press, New York, 1970-1978.
3. P.Gordon,PRINCIPLES OF PHASE DIAGRAMS IN MATERIALS SYSTEMS, McGraw-Hill, New York, 1968.
4. FREE ENERGY CURVES, Code 435, Institute of Metals, Old Post Road, Brookfield, Vt 05036.
5. O. Kubaschewski and C.B. Alcock, METALLURGICAL THERMOCHEMISTRY, 5th ed., Pergamon Press, NY, 1979.

PROBLEMS

1. Describe the behaviour of the liquidus and solidus phase boundaries for the following three conditions:ideal liquid and solid solutions;excess free energy of solute component for the solid exceeds that for the liquid and vice versa.
2. If the excess free energy required to insert one solute atom in the solvent is 0.216 ev/ atom what is the approximate limit to the solid solubility at 500°K?
3. Draw a series of phase diagrams corresponding to more positive values of the difference $^{xs,l}G_{solute} - {}^{xs,s}G_{solute}$. In this case, do you expect the onset of a miscibility gap in the solid state or the formation of an intermetallic compound or ordered phase?
4. Demonstrate that the strain energy due to the insertion of carbon atoms into the interstitial sites at 0,0,1/2 in both the bcc and fcc crystal structures is greater for the bcc structure.

5. Using a free energy-composition diagram demonstrate how it is possible in some cases for the stable composition range of an intermediate phase **not** to include the stoichiometric composition.

6. How does the free energy-composition curve for a superlattice differ from that for an intermetallic compound?

7. If there is a large electronegativity difference between the components of a binary system is it likely or unlikely that there will be intermediate phases in this system?

8. Why is the solvus composition of a terminal phase in equilibrium with a metastable phase always richer in solute than that corresponding to the stable equilibrium? (Hint. Use free energy-composition diagrams to help provide the answer to the question.)

9. Draw the free energy-composition curves for the phases that appear in Figure 3.6 corresponding to 600 and 1400 °C.

10. Figure 3.5 reveals that an intermediate phase is stable in the LiF-CsF system. Speculate as to whether the Li and Cs ions occupy points on the same sub-lattice in a random fashion or are arranged on different sub-lattices in an ordered array? Provide a justification for your answer.

11. If atomic jumps are effectively frozen-in at absolute temperatures below half the absolute melting point, and if you assume Nature chooses the polymorph that is most stable how would you predict the range of stability of amorphous solids in eutectic type phase diagram systems?

12. In an atmosphere consisting of a partial pressure of O_2 equal to 10 atmospheres, above what temperature will SiO_2 decompose to Si and O_2?

IV-THERMODYNAMICS OF INTERFACES

INTRODUCTION

For the materials scientist one of the more important set of properties is that associated with surfaces and interfaces. This chapter treats their thermodynamic aspects. The excess free energy associated with interfaces provides a driving force for a variety of kinetic processes in the form of an effect of particle size. These relations are developed quantitatively. Two contrasting approaches to the concept of the surface or interface, those of Gibbs and Guggenheim, are described, and applied to the phenomenon of adsorption in this chapter. Also, the various types of interfaces are investigated briefly and expressions governing their local equilibria are developed. Finally, applications are made of these concepts to some important problems in materials science.

1. CONCEPT OF SURFACE QUANTITIES.

Let us perform a thought experiment as follows. Separate a solid crystal into two halves along some plane (say plane A), parallel to a crystal plane. The separation plane A is chosen so as not to contain lattice points nor equilibrium positions of atoms. In the approximation where the cohesive energy is given by a sum over bond energies, the energy difference, energy of original solid minus energy of resulting halves, is then equal to the sum of the energies of the bonds that intersect the separation plane A in the original solid. The specific surface energy, e_s, per unit area of surface, in this approximation is then one-half of the negative of the bond energy sum over the bonds intercepting unit area of the separation plane A. This follows from Gibb's definition of a specific surface quantity as being the excess per unit area of surface of the object having the surface over that given by the product of the specific quantity per unit volume by the volume of the object. Thus, the specific surface energy is given by

$$e_S = (2E_{1/2} - E^o)/(2A) \qquad (4.1)$$

where A is the area of plane A, $E_{1/2}$ is the energy of one of the product halves and E^o is the energy of the original solid.

2. AN APPROXIMATE MODEL FOR EVALUATING THE SURFACE ENERGY.

The specific surface energy has been calculated by Friedel et al[1], and many others subsequently, using a nearest-neighbor bond approximation for the cohesive energy and the assumption that the specific surface energy corresponds to one-half the sum of the energies of nearest-neighbor bonds that have been broken in the formation of the unit area of surface in the process of separating a crystal into two halves along some crystal plane with indices h,k,l, with the result that

$$e_S = -2(2h+l)(h^2+k^2+l^2)^{-1/2}e_b/a^2 \quad \text{(fcc)} \qquad (4.2)$$

where $h \geq k \geq l$, e_b=energy/bond=cohesive energy/(Nz/2), z = coordination number and a = lattice parameter.

Values of specific surface energy deduced using the above relation are shown below and are compared to measured values of the average specific surface free energy.

Metal	e_b	a^2	e_s, ergs/cm$_2$			
	10^{-12} ergs/bond	10^{-16} cm^2	calculated (111)	(110)	exp'tal (100)	
Ag	−0.792	16.6	1650	1350	1905	1690
Au	−0.98	16.58	2100	1715	2425	2175
Cu	−0.94	13.02	2500	2080	2890	2260

It is apparent that the bond model yields a good approximation for the value of the specific surface energy and thus the concept that the excess energy associated with the surface is related to the defect of bonds across the surface has rough predictive value, at least for metals and covalent bonded materials. A calculation of the work required to separate an ionic solid into two halves has been performed[2] using a sinusoidal function to approximate the force resisting the separation with the result that $e_s \approx Ya/4\pi^2$, where Y is Young's modulus and a is the interatomic spacing, with a similar match to experimental values.

2.1. Wulff plot of surface energy.

Herring[3] has shown for any monatomic crystal with a center of symmetry that a) if the total energy of the crystal can be expressed as a sum over bond energies and b) if the surface atoms are at perfect periodic lattice points then the plot of surface energy versus orientation must consist of portions of spheres that pass through the origin when extended. Figure 4.1 illustrates such a polar plot of surface energy versus orientation for the case of surface planes having a common [001] rotation axis.

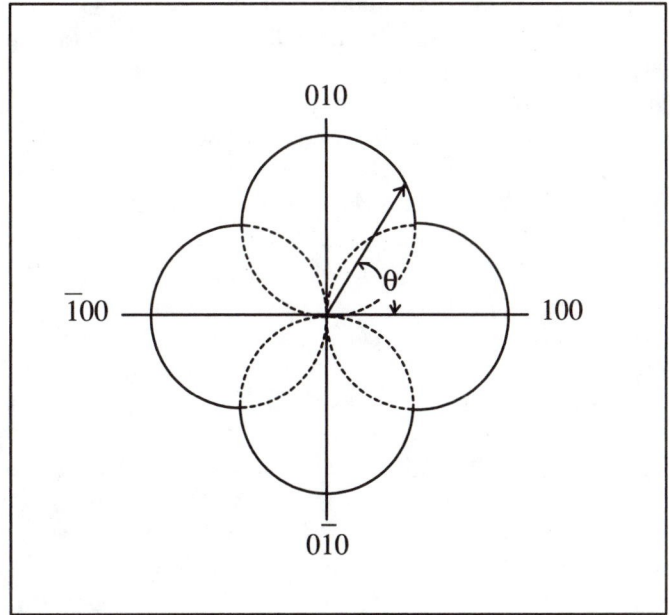

Figure 4.1. Wulff plot showing surface energy, σ, as a function of polar orientation, θ.

Mackenzie[3] has calculated specific surface energies using this concept. The result, as schematically illustrated in Fig 4.1, is that at certain low index planes, the specific surface energy has a minimum value relative to orientations just displaced from the low index orientations. Thus, the specific surface energy-orientation plot consists of such singular cusps and regions where the specific surface energy, and its first and second derivatives with respect to orientation are continuous functions. These polar plots of specific surface energy versus orientation are called Wulff plots.

The difference in surface energy between various crystal planes can exert important effects. For example, consider a thin polycrystalline film of silicon deposited on an inert substrate. If the average grain diameter is larger than the film thickness, then the difference in free energy between the various silicon grains in the thin film is primarily the difference in the excess free surface energy. For small diameter grains and a thin film this excess free energy can change the melting points of the various grains significantly. This effect is illustrated qualitatively in Figure 4.2. As shown, the specific free energy of the small grains in the thin film will vary according to the orientation of the surface plane. (The molten silicon-thin film interface energy shows a much smaller relative change in value with variation in orientation of the crystal plane parallel to the interface.) A consequence of this variation is that the melting point of differently oriented grains will differ. The orientation exhibiting the lowest surface energy will have the highest melting point, as shown. In the case of silicon, this plane is the (100) plane. Hence, it should be possible to retain selectively the (100) oriented grain in a molten thin film on slowly raising the temperature. Fortunately, the shapes of such grains are anisotropic, so that they can be aligned along ridges. By slowly moving a temperature gradient transversely along the thin molten film and parallel to the ridges, it should then be possible to grow out a single crystal thin film having a (100) orientation. Indeed, this objective has sometimes been achieved in SOI (Silicon on Insulator) technology via the use of a traveling molten zone and is a phenomenon belonging to the class called "Artificial Epitaxy" or "Graphoepitaxy".

Also, as suggested by the above analysis, the difference in surface energies can produce a driving force for the solid-state growth of grains in thin films having the lowest surface energy orientation at the expense of neighboring grains having surface orientations of higher specific surface energy. An example where this effect operates is in the heat treatment used to produce (100) oriented Fe-Si alloy transformer sheet. It may also be useful in attempts to produce desired epitaxial single crystal films.

3. SURFACE RECONSTRUCTION.

The implication of the description of crystal surfaces given in the previous sections is that the atoms at the surface are at lattice sites of the bulk crystal. In general, this concept is not valid. Although, in many cases the atoms at the surface are at lattice sites of the bulk crystal structure, the lattice parameters of this crystal structure are distorted. Normally, the lattice planes parallel to the surface are closer to each other than in the bulk crystal, although oscillations in this spacing have also been observed on passing from the surface layer into the bulk. However, in many other cases, the surface atoms are in a different crystallographic arrangement as compared to that described by the bulk crystal structure. In the latter case, the surface is said to have undergone surface reconstruction. In both cases, the driving force for these deviations from the sites defined by the bulk crystal structure is the tendency to decrease the free energy of the material. If there are

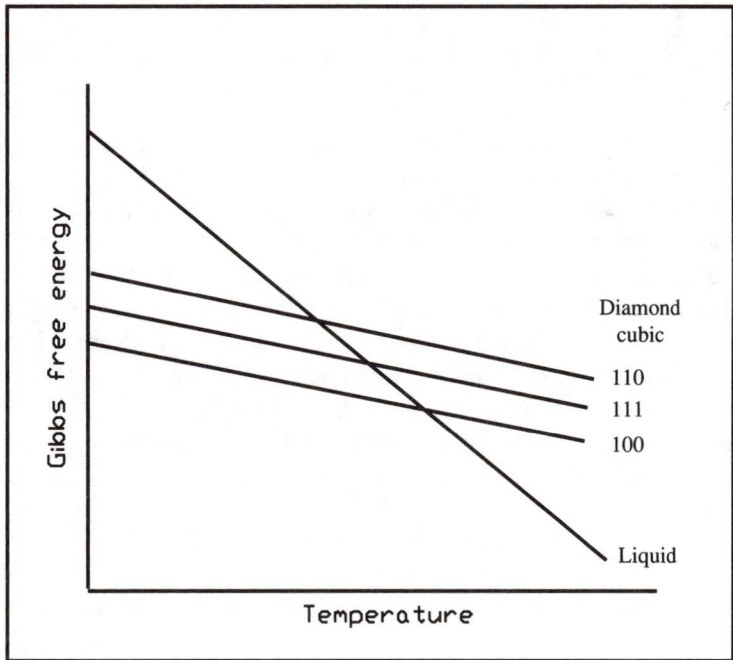

Figure 4.2. Showing the melting points of crystals having different surface facets in the diamond cubic structure.

reliable interatomic potentials for a pure material then it should be possible to calculate the energies of various surface arrangements and thereby determine the most stable one of these arrangements. Unfortunately, such reliable interatomic potentials exist only for a few elements. Hence, at this writing, little can be predicted concerning the relaxations leading to the stable configuration of the atoms adjoining the surface. The energy change associated with these relaxations is small compared to the increment in energy that comprises the surface energy itself. Thus, it is not surprising that the rough bond model used in the previous sections yields a reasonable estimate for the magnitude of the surface energy even if it neglects surface relaxations of interatomic distances or reconstruction of unit cells into others. The interested reader is referred to the monograph by Somorjai (see Bibliography) for additional material on this subject.

4. SOME PARTICLE SIZE EFFECTS.

4.1. Effect of particle size on difference in pressure between a small spherical isotropic solid and an external fluid in local equilibrium.

Let the system consist of one component and let δn molecules be transferred from a fluid phase to a spherical, isotropic solid phase in such a way that the change in Helmholtz free energy for the total system is zero, that the temperature is constant and the specific Helmholtz free energy in each phase is essentially constant (i.e. the latter is not strictly true as will be discussed later.) Thus, there is an increase in volume of the solid phase, which also involves an increase in the solid/fluid interface area. There are two main contributions to the change in free energy of the system: a volume contribution $-(p_b-p_f)\delta v$ and a surface contribution $\sigma\delta A$. (For a single component system, the surface tension σ equals the specific surface excess Helmholtz free energy f^s.)* Summing these contributions yields at equilibrium that

$$0 = -(p_b-p_f)\delta v + \sigma\delta A$$

or $\quad (p_b-p_f) = \sigma\delta A/\delta v = 2\sigma/r$ (for the spherical particle).   (4.3)

(For r small, $2\sigma/r \gg p_f$, i.e. for $\sigma = 1000$ ergs/cm^2 and $r = 10^{-6}$ cm, then $p_b = 1000$ atm.) In the above we have neglected second order terms such as

---

* The surface tension is used as a name for the surface free energy σ in the literature, although Gibbs used it for the work to stretch a surface. We shall use the former sense of σ and discuss the work to stretch a surface in the section on surface stress.

$V_b \delta f_b \approx (4\pi r^3/3)\kappa\, p_b \delta p_b \approx (8\pi r/3)\sigma\, p_b \kappa \delta r$. Substituting reasonable values into the ratio of $V_b \delta f_b/\sigma \delta A$ we obtain a value of about 0.001, which substantiates the neglect of the change in free energy of the solid due to the change in pressure.

4.2. Effect of particle size on difference in pressure between a crystal particle that develops surface facets and external fluid in local equilibrium.

We shall not derive the result used in this section, but make use of it. For a derivation the reader is referred to Defay et al(see Bibliography). If the corners of the facets on a single crystal particle are joined by lines to a point O in the crystal and the normal distances to these facets from this point are denoted by $h^{(i)}$, then it can be shown that at equilibrium the point O exists such that

$$\frac{2\sigma^{(1)}}{h^{(1)}} = \frac{2\sigma^{(2)}}{h^{(2)}} = \ldots \frac{2\sigma^{(N)}}{h^{(N)}} = p_b - p_f \qquad (4.4)$$

These relations are known are Wulff's relations.

Equation 4.4 states that the pressure inside a particle is a function of its size and not the radius of curvature as is implied by equation 4.3. Relations 4.3 and 4.4, which were derived for single component systems hold equally as well for multicomponent systems.

4.3. Dependence of equilibrium vapor pressure on particle size:

Consider particles all having the same size and crystal facets or isotropic spherical particles having the same radius in equilibrium with a vapor phase. The particles and vapor may have many components. For each component at equilibrium, the chemical potentials in the coexisting phases are equal, i.e.

$$\mu_{ib}(T,p_b) = \mu_{iv}(T,p_v) \qquad (4.5)$$

But,

$$p_b = p_v + 2\sigma/r \ (= p_v + 2\sigma^{(\gamma)}/h^{(\gamma)})$$

Hence, expanding the left hand term in 4.5 yields

$$\mu_{ib}(T,p_b) = \mu_{ib}(T,p_v) + 2v_{ib}\sigma/r, \qquad (4.6)$$

where v_{ib} is the derivative of the chemical potential with respect to pressure. Hence, the chemical potential in a small particle is also a function of the size of the particle.

Now, assuming the vapor acts like an ideal gas

$$\mu_{iv}(T,p_v) = \mu_{iv}^* + RT\ln p_{iv} \qquad (4.7)$$

where the superscript $*$ denotes the standard state dependent on temperature only. Consider that there is an equilibrium displacement such that the temperature and composition of the solid phase are maintained constant. For this displacement

$$\delta\mu_{ib}(T,p_b) = v_{ib}\delta p_b \qquad (4.8)$$

But, by equation 4.5

$$\delta\mu_{ib}(T,p_b) = \delta\mu_{iv}(T,p_v)$$

Hence,

$$\delta p_b = \delta\mu_{iv}(T,p_v)/\ v_{ib} = (RT/v_{ib}\ )\delta(\ln p_{iv}\ ) \qquad (4.9)$$

But,

$$\delta p_b = \delta p_v + \delta(2\sigma/r)$$

Hence,

$$\delta(2\sigma/r) = (RT/v_{ib}\ )\delta(\ln p_{iv}\ ) - \delta p_v \qquad (4.10)$$

Integrating from $1/r=0$ to $1/r$ and assuming that v_{ib} is constant during this integration yields

$$2\sigma/r = (RT/v_{ib}\ )\ln(p_{iv}/p_{iv}^\circ\ ) - (p_v-p_v^\circ) \qquad (4.11)$$

For small particles, $(p_v-p_v^\circ) \ll 2\sigma/r$ so that a good approximation is

$$\ln(p_{iv}/p_{iv}^\circ\ ) = (2\sigma/r)(v_{ib}/RT) \qquad (4.12)$$

This relation also holds for a single component system.

Thus, the vapor pressure in equilibrium with particles of radius of curvature r increases as the radius r decreases. This provides the driving force for vapor transport of atoms from small particles to large particles, or from above surfaces having small radius of curvature to surfaces having large radius of curvature. Further, the dependence of chemical potential on curvature provides the driving force for bulk or surface diffusion from regions of small radius of curvature (or negative radius of curvature) to regions of large radius of curvature, as occurs during the process of sintering.

There will always be a driving force for the establishment of equilibrium between a particle and its vapor. If there are particles of different size, each attempting to establish local equilibrium with its vapor, it is obvious that the vapor pressure cannot be uniform everywhere. Rather, the local equilibrium vapor pressures will set up gradients of vapor pressure that bring about transport of atoms from the high vapor pressure regions to the low vapor pressure regions. This transport will distort the vapor pressures corresponding to local equilibrium and the ever present tendency to maintain local equilibrium results in the evaporation and consequent disappearance of small particles and the growth of large particles. Such growth of large particles at the expense of small particles occurs also when the particles (precipitates) are embedded in a host matrix and attempt to approach local equilibrium with their environment. This process is called Ostwald Ripening and is considered below.

4.4. Dependence of solvus composition on precipitate particle size

Consider spherical particles of a precipitate of radius r containing only pure solute B, which are in equilibrium with a flat solvent solid solution by means of being heated in a box so that the same vapor pressure of B is in equilibrium with both the particles and the solid solution. Thus,

$$\mu_B|_{solution} = \mu_B|_{vapor} = \mu_B|_{particles} \qquad (4.13)$$

Now, the change in chemical potential of B in the solid solution with change in concentration of B is given by

$$(\mu_B - \mu_{Bo})|_{solution} = RT\ln(C/C_o) = RT(C - C_o)/C_o$$

(for small deviations of C from C_o, where the latter composition corresponds to that in equilibrium with flat particles of B.) But,

$$\mu_B|_{particles} = \mu_{Bo}|_{particles} + 2\sigma v_B/r.$$

Thus,

$$RT(C-C_o)/C_o = 2\sigma v_B /r. \tag{4.14}$$

This result states that the solubility limit of a solid solution in equilibrium with particles of pure solute depend on the radius of the latter particles. A graphical illustration of this result is shown in Figure 4.3. In the derivation of phase equilibria we normally neglect the effect of surface energy. As shown above this neglect is not justified if any of the phases is in the form of small particles.

The result just derived can be shown to apply to the case of any precipitate phase that is in equilibrium with a host phase, even if the precipitate is itself a solution or compound phase.[5] That is, the phase

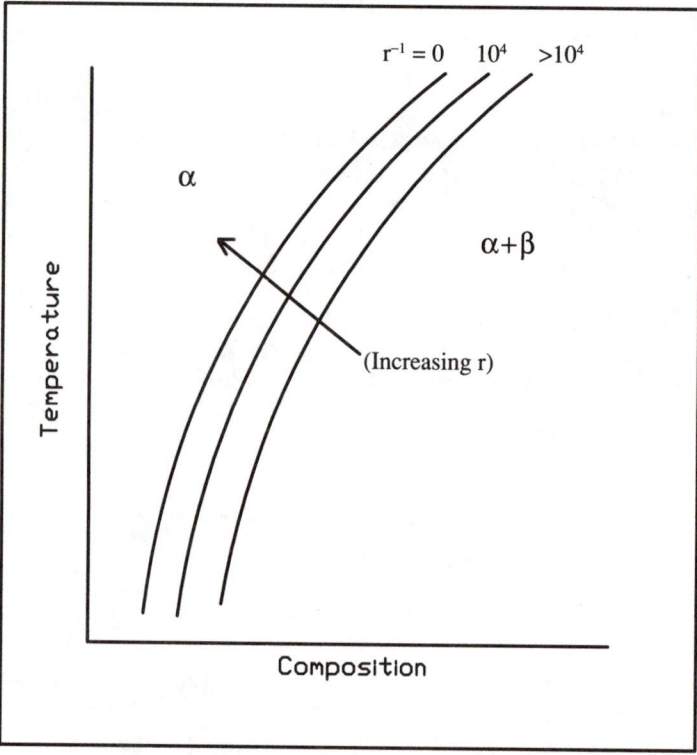

Figure 4.3. Showing the schematic effect of curvature on the solvus composition is equilibrium with particles that vary in size.

boundary compositions of a phase in equilibrium with another solid phase are dependent on the size of the latter particles. The latter result can be demonstrated by yet another approach, as follows. It can be shown[5] that the difference in pressure across an interface is given by $P'-P''= K\sigma$, where K is the mean curvature (i.e.2/r, for a spherical particle). If the interface is between a host phase and a small included particle of an intermediate compound phase, it is usually assumed that the pressure in the host phase is that in equilibrium with its environment. Thus, the increase in pressure relative to the host phase is sensed by the particle itself. The effect of this increase in pressure can be demonstrated graphically. Figure 4.4 shows the free energy-composition curves corresponding to the pressure of the envi-

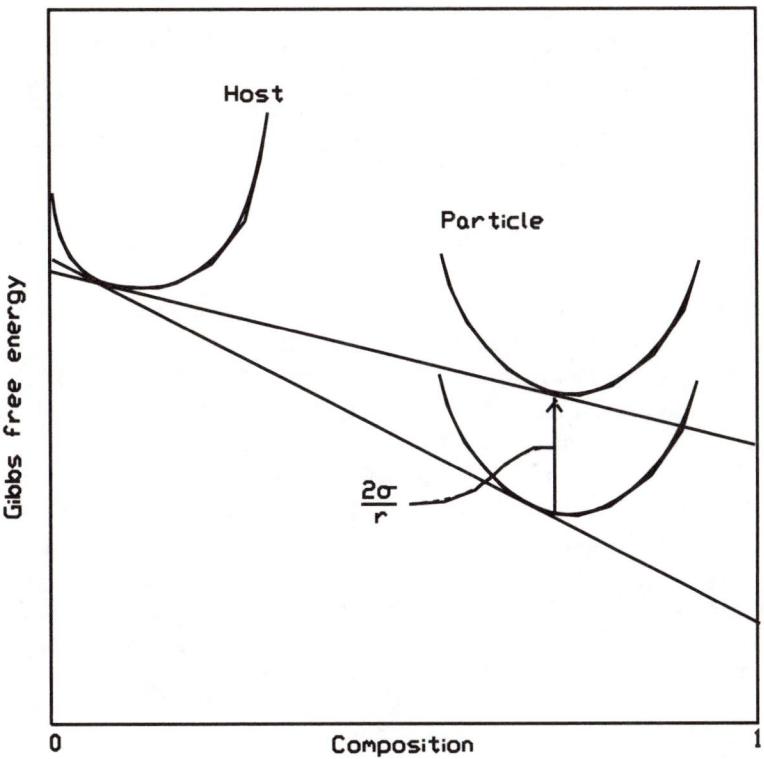

Figure 4.4. Schematic illustration of the effect of particle size on the free energy-composition curve for particles of an intermediate phase as precipitates in the host phase and the effect on the solvus composition.(r is the radius of the particle.)

ronment for both host and particle phases and also that for a spherical particle phase subjected to an increase in pressure relative to that of the environment corresponding to $2\sigma/r$. If common tangents are drawn, as shown, then it is apparent that the solvus composition of the host phase is dependent on the radius of the particle phase and that the partial molar energy (chemical potential) of the solute at the solvus composition increases with decrease in the particle radius.

5. ADSORPTION

5.1. Gibbs Adsorption:

Let us now consider the surface-volume equilibrium for a multi-component system. We follow the method of Gibbs in defining surface excess quantities as being equal to the excess of a quantity over that given by the sum of the volume densities of the quantity multiplied by the volume of the corresponding phase, i.e.

$$Q^s = Q - \Sigma_i q_i V_i \qquad (4.15)$$

Let there be a large block of the single phase in contact with a thermal reservoir at temperature T and n chemical reservoirs each capable of maintaining the chemical potentials within the single phase block at their specified values $\mu_1, \mu_2...\mu_n$. Let us now create new surface area in this block (by separating the block in such a way as not to affect the specific quantities q_i applicable to the block. The work dW is done to create the new surface area dA (reversibly), with the reservoirs maintaining the temperature T and the chemical potentials μ_i. We define $\sigma \equiv dW/dA$. The production of area dA will be accompanied by a transfer of dN_i moles of component i from the ith reservoir, maintaining the chemical potential at μ_i. Since μ_i in the bulk was unaffected by the creation of dA, the dN_i corresponds to a surface excess defined by $\Gamma_i = dN_i/dA$. The change in Helmholz free energy of the block is

$$dF = dW + \Sigma \mu_i dN_i = \sigma dA + \Sigma \mu_i \Gamma_i dA \qquad (4.16)$$

This change must be a surface excess by the constraints, i.e. $dF = dF^s$. Per unit area then f^s (=dF^s/dA) is then

$$f^s = \sigma + \Sigma \mu_i \Gamma_i \qquad (4.17)$$

We now allow a change dT in the temperature of the reservoir and $d\mu_i$ in the chemical potential of the reservoirs at constant surface area. The change in Helmholz free energy of the system block and surface is

$$dF = -SdT + \sum \mu_i \, dN_i \tag{4.18}$$

But, $S = S_b + S^s$ and $dN_i = dN_{ib} + d^sN_i$. Also,

$$dF - dF_b = dF^s = -S^s dT + \sum \mu_i d^sN_i \; .$$

Dividing by the area of the surface of the block then

$$df^s = dF^s/A = -s^s dT + \sum \mu_i \, d\Gamma_i \tag{4.19}$$

But, $df^s = d\sigma + \sum \mu_i d\,\Gamma_i + \sum \Gamma_i d\mu_i$ by previous equation. Hence,

$$d\,\sigma = -s^s dT - \sum \Gamma_i d\mu_i. \tag{4.20}$$

This is the Gibb's adsorption relation between the variables σ, T and μ. For T constant, we obtain

$$\left. \frac{\partial \sigma}{\partial \mu_i} \right|_{T, \mu_j} = -\Gamma_i \tag{4.21}$$

which can be interpreted as meaning that upon increasing the quantity i in the surface, then σ is decreased on raising μ_i or if by raising μ_i, the work to produce new surface, σ, is decreased, then an excess of i will be found in the surface. Perhaps, this can be more easily understood by applying the Gibb's equation to a binary system at constant T. Then,

$$(d\sigma/dX_2) = -\Gamma_1 (d\mu_1/dX_2) - \Gamma_2 (d\mu_2/dX_2) \tag{4.22}$$

But, by the Gibbs-Duhem relation $X_1 d\mu_1/dX_2 + X_2 d\mu_2/dX_2 = 0$, then

$$(d\sigma/dX_2) = (\Gamma_1 X_2/X_1 - \Gamma_2)d\mu_2/dX_2 \equiv -\Gamma d\mu_2/dX_2$$

$$= -(\Gamma_2 - \Gamma_1 X_2/X_1)d\mu_2/dX_2 \equiv -\Gamma d\mu_2/dX_2 \tag{4.23}$$

$$(\Gamma \equiv \Gamma_2 - \Gamma_1 X_2/X_1)$$

(Gibbs chose the dividing plane of the surface such that $\Gamma_1=0$.)

We have already learned in Chapter II that in the composition range where a solid solution is stable the sign of $d\mu_2/dX_2$ must be positive. Hence, in stable solid solutions there will be an excess of the solute 2 at the surface when the addition of solute acts to decrease the surface tension.

The Gibbs adsorption isotherm that has been derived for a surface applies as well to any interface, such as a grain boundary in a single phase system. Application to multiphase systems is somewhat complicated. (See Cahn in INTERFACIAL SEGREGATION.)

5.2. Guggenheim's pseudo-thermodynamic model of an interface phase.

Contrary to the conclusions we are led to make on the basis of Gibb's approach to the thermodynamics of interfaces, we now start with the hypothesis that we can describe an interface as a separate phase with homogeneous properties. This approach was originally developed by Guggenheim[6] and more extensively by others[7].

Thus, the surface phase will have a Gibb's free energy, ${}^sG_m$, per mole that depends upon composition in the normal way. However, the value of this molar energy at each terminal composition is related to the Gibb's molar free energy in the corresponding bulk phase by

$$\begin{aligned}{}^sG_{1o} - {}^bG_{1o} &= \sigma_{1o}\,{}^sA_{1o} \\ {}^sG_{2o} - {}^bG_{2o} &= \sigma_{2o}\,{}^sA_{2o}\end{aligned} \qquad (4.24)$$

(${}^sA_{1o}$ is the area per mole of pure component 1 of surface phase s.)

We are now concerned with evaluation of the distribution coefficient per solute for equilibrium between the bulk (b) and surface phases. Let the virtual transfer of atoms correspond to the exchange of a solvent atom in the boundary phase with a solute atom in the bulk phase at constant number of atoms in each of the surface and bulk phases. For this equilibrium, the change in free energy of the surface phase in the exchange must equal that for the bulk phase. Thus,

$$ {}^s\mu_2 - {}^s\mu_1 = {}^b\mu_2 - {}^b\mu_1 \qquad (4.25) $$

or

$$ {}^s\overline{G}_2 - {}^s\overline{G}_1 = {}^b\overline{G}_2 - {}^b\overline{G}_1 $$

describes this equilibrium. But,

$$\overline{G}_i = G_{io} + RT\ln X_i + ^{xs}G_i \tag{4.26}$$

Hence,

$$^s\overline{G}_2 - {}^s\overline{G}_1 = {}^sG_{2o} + RT\ln {}^sX_2 + {}^{s,xs}G_2 - {}^sG_{1o} - RT\ln {}^sX_1 - {}^{s,xs}G_1 =$$

$$^b\overline{G}_2 - {}^b\overline{G}_1 = {}^bG_{2o} + RT\ln {}^bX_2 + {}^{b,xs}G_2 - {}^bG_{1o} - RT\ln {}^bX_1 - {}^{b,xs}G_1 \tag{4.27}$$

Rearranging terms yields

$$^sX_2/{}^sX_1 =$$
$$(^bX_2/{}^bX_1)\exp\{-[(^sG_{2o} - {}^bG_{2o}) - (^sG_{1o} - {}^bG_{1o}) + (^{s,xs}G_2 - {}^{b,xs}G_2) - (^{s,xs}G_1 - {}^{b,xs}G_1)]/(RT)\}$$

$$\tag{4.28}$$

$$^sX_2/{}^bX_2 =$$
$$(^sX_1/{}^bX_1)\exp\{-[(\sigma_{2o}{}^sA_{2o} - \sigma_{1o}{}^sA_{1o}) + \Delta^{xs}G_2 - \Delta^{xs}G_1]/(RT)\}$$

Thus, we obtain again a distribution coefficient relation equivalent to that between liquid and solid phases (see equation 3.2) except that the terms have somewhat different significance. Segregation of solute 2 to the interface phase is promoted by a value of $^{b,xs}G_2 >> {}^{s,xs}G_2$ and $^{s,xs}G_1 \approx {}^{b,xs}G_1$. Note that these quantities are evaluated at the respective concentrations of the interface (^sX_i) and bulk (^bX_i) phases. If $^{b,xs}G_2$ is due to strain energy, then certainly the similarity of the liquid/solid distribution coefficient to that of the surface/bulk one is apparent. This analogy has proved to be useful in predicting the likely segregation of solute to interfaces on the basis of experimental phase diagrams for the system in question.

The approach of this section can be illustrated graphically, as follows. We draw the free energy-composition curves for the host phase and for the host surface phase as shown in Figure 4.5. We note that the equilibrium being considered involves an exchange of a solvent atom in the surface phase with a solute atom in the host phase or vice versa. In either case, the condition describing the equilibrium is given by equation (4.25) and the relation described by (4.25) is shown in Figure 4.5, where the straight lines shown have the same slope.

It must be emphasized that the relations obtained in this section are not rigorous. Further, they have been developed only for the special case of

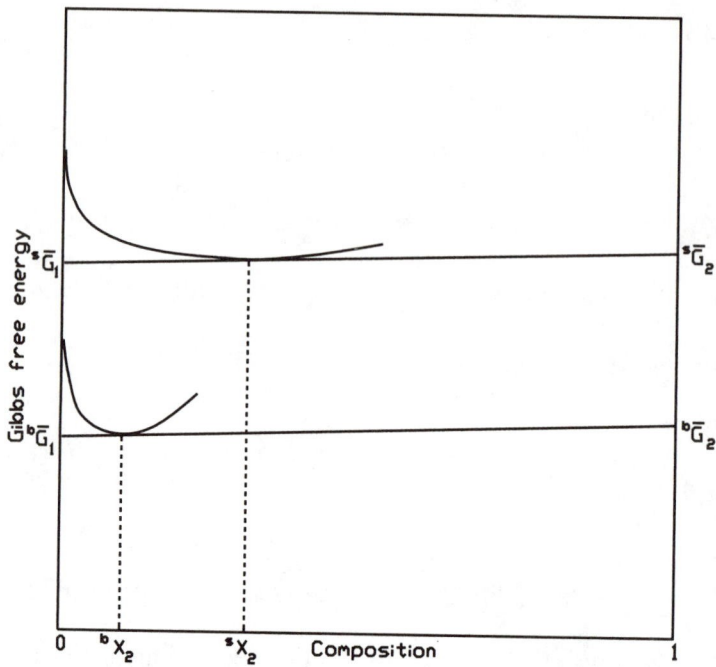

Figure 4.5. Free-energy-composition curves for a bulk and its surface phase illustrating the relation for the equilibrium between these phases.

exchange of a bulk sited atom with one in the so-called surface phase. The possibility exists that surface adsorption can occur without the necessity of such an exchange. For example, interstitial sited atoms in the bulk can move to the surface region without requiring surface atoms to move into the bulk. There are more than enough vacant interstitial sites in both regions to accommodate the transfer of interstitial sited atoms from one region to the other. Nevertheless, the relations (4.28) are useful where applicable.

A correlation between the segregation ratio, ${}^{s}X_{2}/{}^{b}X_{2}$, and the solid solubility has been found.[8] We can develop the bases for this correlation using relations already considered. In particular, we note that for a regular solution that the difference in partial excess molar free energies is given by

$$^{xs}G_{2} - {}^{xs}G_{1} = -RT\ln({}^{*}X_{2}/{}^{*}X_{1}).$$

where $^*X_i$ are the compositions at the solvus or solid solubility limit.

Substitution of this relation into 4.28 yields for the segregation ratio

$$\beta = {}^sX_2/{}^bX_2 = ({}^sX_1/{}^bX_1)({}^*X_1/{}^*X_2)\exp[-\Delta G_s/RT) \qquad (4.29)$$

where $\Delta G_s = [(\sigma_{2_0}{}^sA_{2_0}/{}^sN_{2_0} - \sigma_{1_0}{}^sA_{1_0}/{}^sN_{1_0}) + ({}^{s,xs}G_2 - {}^{s,xs}G_1)]$.

Now, if we consider only the case of dilute solutions, then (4.29) becomes

$$\beta = (1/{}^*X_2)\exp[-\Delta G_s/RT).$$

Figure 4.6 shows a plot of experimental values of the enrichment ratio for adsorption at grain boundaries versus the solid solubility in metallic hosts. The scatter of a factor of 5 in the enrichment ratio corresponds to a small scatter in the value of G_s. Incidentally, the data at a value of the solubility equal to unity (complete solid solubility) appear to obey the correlation although the theory should apply only to dilute solution. Little is known about the segregation of solutes to surfaces in ceramics, except that such segregation can be marked and have significant effects on properties. Since the above relations are general and not specific to any material, there is no reason not to expect them to operate in ceramics, as well as they do in metals.

The enrichment of a surface in solute can lead to surface phase transitions. Little is known about this phenomenon. However, it has been found that the segregation of sulfur on Pt and H_2 on Mo leads to several transitions from one surface unit cell to another. Presumeably, further studies will develop many more examples of this phenomenon.

5.3. Other adsorption isotherms.

An implicit assumption of the derivations in the previous section is that all sites at the surface are available to be occupied by the solute species. If this assumption is not met then other adsorption isotherms are possible. In particular, if the fraction of the total number of sites at the interface that can be occupied by solute equals $^sX_{2_0}$ then replacement of $^sX_1$ by $^sX_{2_0} - {}^sX_2$ in 4.28 yields the relation known as the Langmuir-McLean isotherm. The same substitution, but into 4.29, and with application to dilute solution only yields the "Truncated BET" isotherm. Distinguishing between these iso-

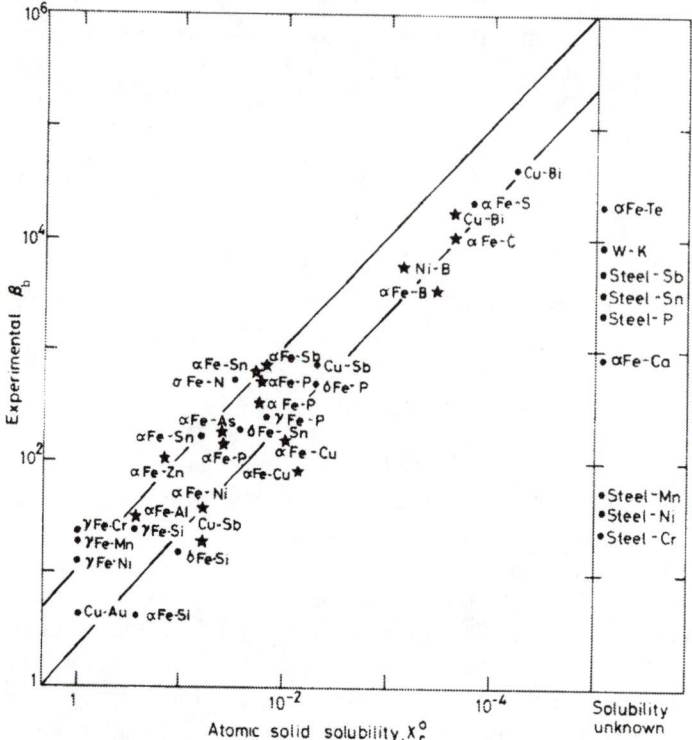

Figure 4.6. After Hondros and Seah[8]. Correlation of measured grain-boundary enrichment ratios with the atomic solid solubility.

therms requires measurements of the composition dependence of the
segregation ratio for dilute solution, which is difficult to accomplish.

6. SURFACE STRESS.

A solid phase differs from a fluid phase in that a change in surface
area is not accomplished by an exchange of atoms between the interior and
the surface, but by elastic strain. Let us consider this situation. A slab is
shown in Figure 4.7 of thickness d<<L, the width of the slab. There is a
surface stress tensor $g_{\mu\nu}$, representing the force acting on unit length of a
line in the surface of the slab, where the index μ can stand for the directions

x,y,z and v for x and y. There is a body stress acting in the bulk of the slab corresponding to $g_{\mu v}$ and induced by the latter equal to

$$p_{\mu v} = -2g_{\mu v}/d$$

At equilibrium, the free energy must be stationary with respect to any virtual distortion, such as a homogeneous strain $e_{\mu v}$. If A is the area of the face of the slab, then the change in energy in the bulk due to the strain is

$$Ad\Sigma\, p_{\mu v}e_{\mu v} = -2A\Sigma\, g_{\mu v}e_{\mu v}$$

and that in the surface is

$$2A\Sigma(\partial\sigma/\partial e_{\mu v})e_{\mu v} + 2\sigma\,\Sigma(\partial A/\partial e_{\mu v})e_{\mu v}.$$

Now for μ=x,y and v=x,y $(\partial A/\partial e_{\mu v})$=A $\delta_{\mu v}$, where $\delta_{\mu v}$ is Kronecker's

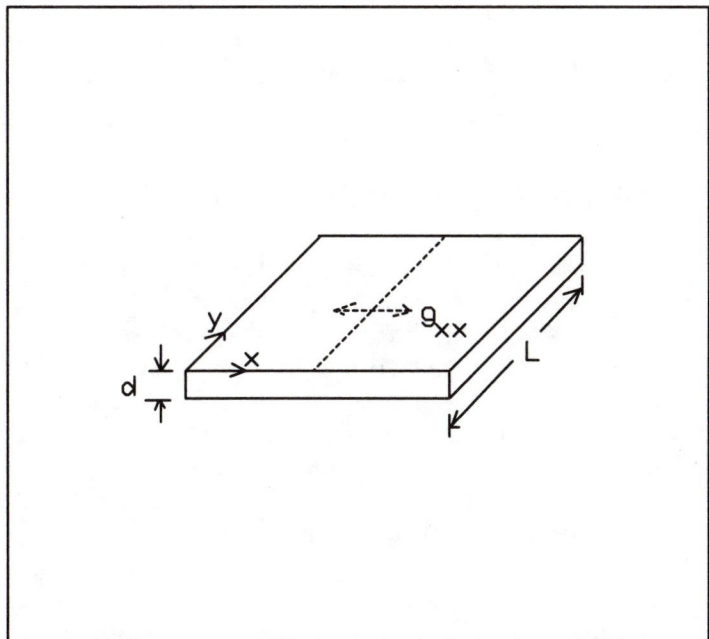

Figure 4.7. Definition of quantities for a surface slab. See text.

delta(i.e. $\delta_{\mu v}$=0 for $\mu \neq v$ and $\delta_{\mu v}$ =1 for $\mu = v$.) Adding and noting that the strains $e_{\mu v}$ are arbitrary, we must then have that the coefficients of each $e_{\mu v}$ vanish or

$$-2Ag_{\mu v} + 2A(\partial \sigma / \partial e_{\mu v}) + 2 \sigma A \; \delta_{\mu v} = 0$$

or
$$g_{\mu v} = \sigma \; \delta_{\mu v} + (\partial \sigma / \partial e_{\mu v}).$$

In words, the surface stress equals the surface tension (i.e. Gibbs terminology in which the surface tension is a specific surface free energy) plus the variation of the surface tension with respect to surface strain. For solids the latter quantity is not equal to zero for temperatures less than about one-half the absolute melting temperature. Solids act as fluids above this temperature. It is thus understandable why there exists an enhanced pressure on the concave side of an interface, when, as shown above, the surface stress is equal to the surface tension plus the variation of this tension with respect to surface strain.

7.SURFACE ENERGIES OR SOLID-GAS AND LIQUID-GAS INTERFACE ENERGIES

Table 4.1 incorporates representative values of surface energies for various liquids. A generalization from these data is that the surface free energies of liquid salts and oxides(ionic materials) are lower than those for metals. Table 4.2 presents the same comparison for solid surface energies. Again the same conclusion can be drawn. If the interphase energy between an oxide of a metal and the metal is low, then it is apparent that the oxide phase will tend to cover the metal phase to provide the surface phase. Indeed, the observation that oxides do cover their metals even after prolonged heating in a gas in equilibrium with the oxide suggests that the sum of the metal-oxide and oxide-vapor interface energies is less than that of the respective metal surface energies. This fact plays a significant role in the important technologies of brazing and soldering. The joining together of two similar or dissimilar metals by either of these methods requires a liquid metal or alloy "cement" to flow between the two solids to be joined. Given the low surface free energies of the oxides, for the "cement" to function it is necessary that either it have a lower interface energy to the

Table 4.1
Surface Free Energy of Liquid Metals

Metal	Surface Energy (ergs/cm^2)	Temperature
Al	866	600(m.p.)
Ag	895	1000
Au	1140	1063(mp)
Ba	225	720
Be	1100	1500
Ca	335	850
Cs	60	29(mp)
Cd	590	321(mp)
Cr	1590	1950
Co	1880	1495(mp)
Cb	1900	2473(mp)
Cu	1300	1083(mp)
Fe	1880	1535(mp)
Ga	720	30.3(mp)
Ge	610	960
Hf	1630	2230(mp)
Hg	475	20
In	560	156(mp)
Ir	2250	2454(mp)
K	101	64(mp)
Li	398	180(mp)
Mg	540	651(mp)
Mn	1060	1245(mp)
Mo	2250	2620(mp)
Na	191	98(mp)
Ni	1780	1455(mp)
Os	2500	3000(mp)
Pb	450	327(mp)
Pd	1500	1547(mp)
Pt	1800	1773(mp)
Rb	76	39(mp)
Re	2700	3200(mp)
Rh	2000	1966(mp)
Ru	2250	2250(mp)
Sb	380	640
Si	730	1410(mp)
Sn	550	232(mp)
Ta	2150	3020(mp)
Tc	2100	2220(mp)
Ti	1650	1730(mp)
Tl	440	318
U	1550	1132(mp)
V	1950	1710(mp)
W	2400	3410(mp)
Zn	770	420(mp)

Table 4.1 (continued)
Surface Energies of Typical Molten Salts and Inorganic Materials

Salts	Surface Energy (ergs/cm²)	Temperature °K
LiF	236	1121(mp)
NaF	186	1269(mp)
UF$_4$	196	1309(mp)
LiCl	126	883(mp)
NaCl	114	1074(mp)
CuCl	92	703(mp)
AgCl	179	728(mp)
CaBr$_2$	120	1015(mp)
Li$_2$CO$_3$	244	993(mp)
LiNO$_3$	116	527(mp)
Na$_2$SO$_4$	195	1157(mp)
NaNO$_3$	120	583(mp)
B$_2$O$_3$	80	1173
FeO	585	1693
Al$_2$O$_3$	700	2353
ZnBr$_2$	51	675(mp)
CdCl$_2$	100	841(mp)
SnCl$_2$	104	518(mp)
PbCl$_2$	138	771(mp)
GaCl$_3$	27	351(mp)
HgCl$_2$	56	550(mp)
BaBr$_2$	153	1127(mp)

oxide (than the oxide surface energy) or that the oxide be removed while the liquid "cement" is in contact both with the surfaces to be joined and the flux that removes the surface oxides.

There is a potential for application of the surface energy data described in this section that continually amazes this author. For example, the so-called C4 process of joining integrated circuit chips to their packages by a self-aligning procedure, whereby hundreds of electrical connectors separated by microscopic dimensions are made in one operation, makes clever use of these data. In this process, each molten ball of solder is prevented from spreading beyond its connector area by the presence of oxides separating the connectors, which themselves are gold plated to prevent oxidation of the connector surfaces and to wet the solder.

Table 4.2
Surface Free Energies of Solids

Metals	Surface Energy (ergs/cm^2)	Temperature °C
Al	980	450
	1100	200
Ag	1100	950
	1200	850
Au	1400	1000
Bi	500	240
Cd	670	300
Co	2420	1400
	1970	1354
Cr	2090	1700
Cu	1520	1000
	1780	925
Fe(bcc)	1930	1475
Fe(fcc)	2100	1350
Ga	770	20
In	630	140
Mo	2630	2400
Nb	2150	2225
Ni	1940	1400
	2280	1060
Pb	560	300
Pt	1950	1700
	2200	1300
Sn	670	200
	685	223
Ta	2480	2700
Ti	1940	1300
Tl	560	280
W	2800	2000
Zn	870	400

Non-metals		
NiO	1100	25
MgO	1000	25
KCl	110	25
NaCl(100)	300	25
Al_2O_3	905	1850
TiC	1190	1100
LiF(100)	340	25
CaF_2(111)	450	25
$CaCO_3$(1010)	230	25

8. SOLID-LIQUID INTERFACES.

Much has been written about the solid-liquid interface[9], perhaps because it is involved in the technologically important problem of solidification. However, little is known about it. The one conclusion that can be deduced from a comparison of the data in Table 4.3 with those for the same metals in Table 4.1 is that the solid-liquid interface energy is about an order of magnitude smaller than the surface energy and smaller than the difference between solid and liquid surface energies. Thus, we have the general rule that liquids wet their own solids.

A rough rule for the solid-liquid interface energy for metals is that the molar interface energy ($= N^{1/3}v^{2/3}\sigma_{sl}$, where N is Avogadro's number and v is the molar volume) equals about one-half the heat of fusion. For semimetals and organic compounds the proportionality constant is about 0.3.

The small value for the solid-liquid interface energy suggests that the excess volume associated with the interface must be small. This brings up an interesting point. In Gibbsian thermodynamics of surfaces, the surface excess of volume is zero by definition. In fact, it isn't. Cahn[10] has described a surface thermodynamic method in which the excess volume has meaning and for relating such excess quantities to measureable quantities. A fuller description of Cahn's method is outside the scope of this chapter because the main advantages of using Cahn's method are derived from more detailed applications to special cases, such as the case of interfaces when there are three-or-more coexisting phases.

Table 4.3
Solid/Liquid Interface Energies

Metal	Interface Energy (ergs/cm^2)	Temperature °C
Al	93	600
Au	132	1063
Bi	61	271
Cu	177	1083
Fe	204	1535
Ni	255	1455
Pt	240	1773
Sn	55	232

9. SOLID-SOLID INTERFACES.

9.1. Grain Boundaries.

There are a variety of solid-solid interfaces as listed in the Table 4.4 for copper. It should be noted that the interface energy can vary from a very small number for coherent type interfaces up to about 1/3 the surface energy for high angle grain boundaries. This is a general result for metals. In the case of ionic solids, the few measurements that have been made suggest that the ratio of the high angle grain boundary energy to the surface energy can reach a much higher value. In NiO this ratio approaches unity for [110] symmetric tilt boundaries.

The energy of very low angle tilt or twist boundaries is described quite well by dislocation theory (see any current textbook on dislocation theory). The tilt and twist small angle boundaries are illustrated in Figures 4.8 and 4.9, which show the dislocation content of such boundaries clearly. Also, implicit in these figures is the relationship of these boundaries to an axis of rotation of one grain relative to its bounding grain. In the case of the tilt boundary this axis of rotation is in the plane of the grain boundary and in the case of the twist boundary it is perpendicular to the plane of the grain boundary. One explanation for the difference in grain boundary energies between ionic materials and metals for similar boundaries is that the Burger's vector in ionic solids is about twice that in metals due to the larger unit cell.

Table 4.4 Interface Energies for Copper	
Interface	Energy (ergs/cm^2)
(High angle grain boundary)	625
(Coherent twin boundary)	24
(Non-coherent twin boundary)	498
(Intrinsic stacking-fault)	78

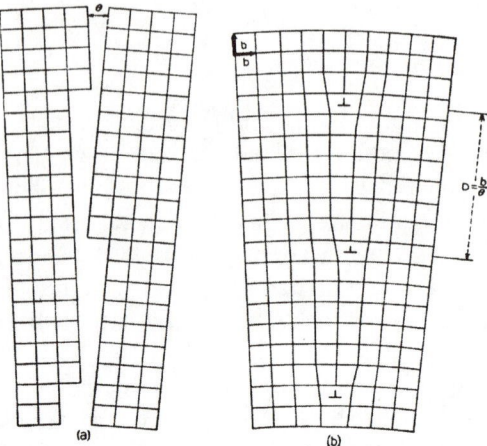

Figure 4.8. Schematic illustration of a tilt boundary formed by joining a bicrystal, each crystal of which has a common axis normal to the plane of the paper. One crystal is rotated relative to the other about this axis by the angle θ.

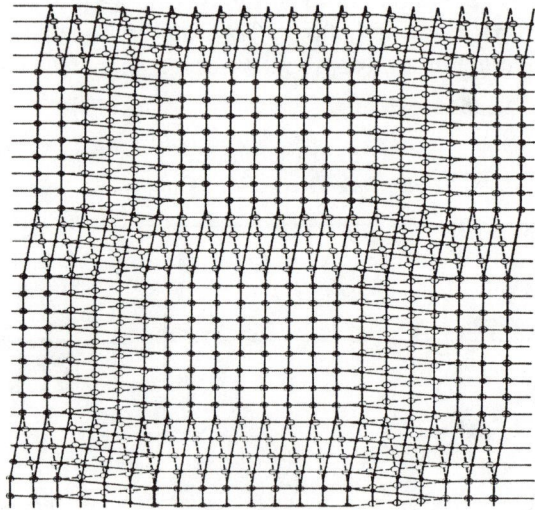

Figure 4-9. Schematic illustration of a twist boundary formed by rotating one half of a bicrystal about an axis normal to the paper and boundary relative to the other half crystal.

Computer simulation of higher angle tilt and twist grain boundaries have provided some insight concerning the structure and energies of such boundaries. This subject is still under development and the interested reader is urged to peruse the articles in the bibliography to this subject at the end of this chapter. At this time, it appears that a reasonable model of such grain boundaries yields the result that the atoms along such tilt and twist boundaries tend to configure themselves into arrangements that describe polyhedra of the type that Bernal has suggested exist in liquids. The orientations at which the boundaries consist only of one type of polyhedra are those at which the grain boundary energy exhibits minima or cusps in the plot of energy versus tilt or twist orientation.

The energy of high angle boundaries is less readily predicted by theory, although current computer simulation models of grain boundaries, using appropriate interatomic potentials and taking into account macroscopic relations that can occur due to relative translation and rotation of grains adjoining the boundary, yield fair approximations for such energies.[11] The high angle boundary is not a well defined species since it is a function of 5 variables (2 for the relative orientation of the grains adjoining the boundary and 3 to define the orientation of the grain boundary plane normal relative to the crystal axes of one of the bounding grains.) Hence, the structure of a high angle boundary can vary enormously as can its energy. Indeed, at special high angle orientations a high degree of lattice sites at the boundary can be common to both adjoining crystals. Such special boundaries will have lower energy. Consequently, one expects the existence of low energy cusps in the polar plots of the orientation dependence of grain boundary energies-that is, the grain boundary energy is expected to be orientation dependent. Indeed, experiment is consistent with this expectation as shown in Figure 4.10. This feature, as will be shown in the following section, is responsible for the existence of torques acting on boundaries tending to reorient them into lower energy orientations.

The brief discussion of grain boundary energies in the previous paragraph is applicable to all materials. However, materials do differ in the details concerning their grain boundary structure and energy. In particular, ceramics(ionic) must obey the constraint of electroneutrality, with a consequent effect on the structure of their grain boundaries, the development of a cloud of defects adjacent to the grain boundary to screen any net charge on the boundary and the sensitivity of these grain boundaries to impurities. Also, the tendency for the amorphous to crystalline transition to be sluggish in many ceramics leads to the presence of amorphous phases in ceramics that usually lie along the boundaries of crystalline grains

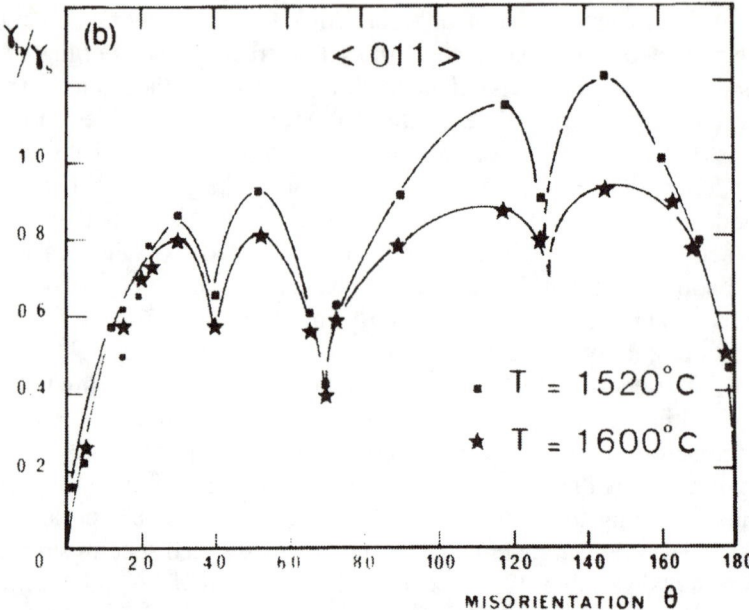

Figure 4.10. Relative <011> symmetrical tilt boundary energy in NiO as a function of the tilt angle. After G.Dhalenne et al.in ADVANCES IN CERAMICS, vol.6, 1983, 139.

possibly in a state of metastable equilibrium. We have already noted that liquid-solid interface energies in a given material are smaller in magnitude than general high angle grain boundaries and this observation may explain the microstructure of some of these glass ceramics. The state of knowledge about grain boundaries in ceramics and covalent materials is much less developed than that for metallic materials.

9.2 Interphase Interfaces.

In the case of grain boundaries, the interface is between two differently oriented crystals of the same homogeneous phase. Interfaces exist also between two different phases. These can include solid-vapor, solid-liquid and solid-solid interfaces. Solid-solid interphase interfaces occur between precipitates in solid hosts and the host phase. Such interfaces can be coherent, if there is an appropriate matching of the relative orientations of the two phases, of the crystal structures and of the lattice parameters.Table 4.5 lists some interphase interface energies.

Table 4.5
Interphase Interface Energies

Interphase system	Interface energy (ergs/cm^2)	Temperature °C
Au/Al$_2$O$_3$	1725	1000
Ag/Al$_2$O$_3$	1630	700
Cu/Al$_2$O$_3$	1925	850
Ni/Al$_2$O$_3$	2140	1000
Pt/Al$_2$O$_3$	1050	1400
Ni/ThO$_2$	2000	1200

The interphase interface energies in the table are for non-coherent interfaces, and, as shown, these values can be large. Figure 4.11a illustrates a coherent interface between a precipitate and its host phase. As shown there is a continuity of rows and planes of atoms across the interface.Figure 4.11c illustrates a non-coherent interface. It is apparent that the non-coherent interface exhibits much greater misfit than does the coherent interface.   It is no surprise therefore that the non-coherent interphase interface energies in Table 4.5 are large.

Figure 4.12 shows a lattice image electron micrograph for an interface between a silver host and a CdO precipitate, where the continuity of atomic rows across the interface is apparent. By computer simulation and comparison it is demonstrated that the outermost plane of the oxide consists of oxygen ions and not cadmium. The interface energy of an interphase

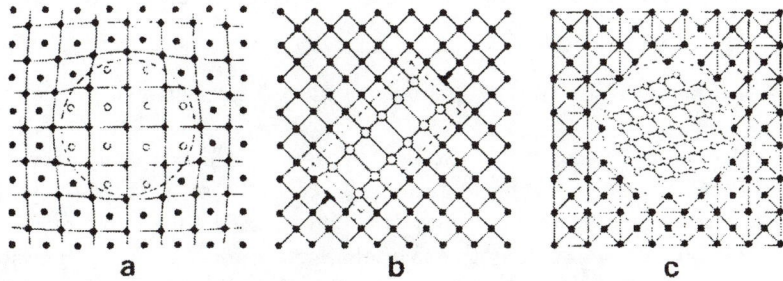

a   b   c

Figure 4.11. Schematic illustration of a precipitate: a)quasi-coherent or coherent with positive misfit; b) semi-coherent; c) non-coherent. After J.W.Martin, MICROMECHANISMS IN PARTICLE -HARDENED ALLOYS, Cambridge University Press, 1980.

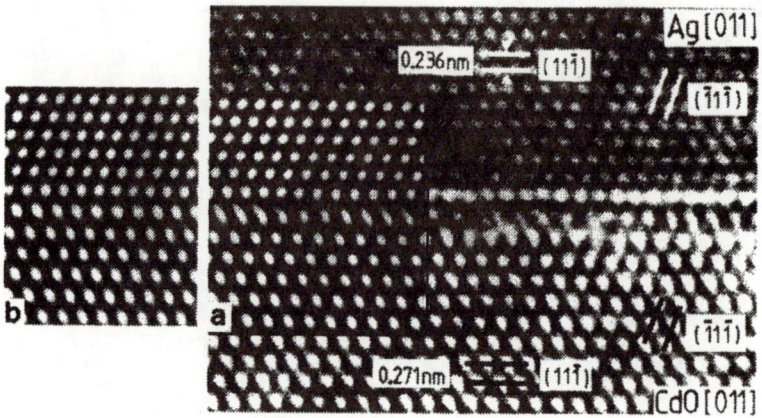

Figure 4.12. Lattice image TEM of Ag/CdO interface. Inserts a) inside box and b)show calculated contrast for a) oxygen and b) cadmium as the outermost plane of the oxide. After H.F. Fischmeister et al MAT. RES. SOC. SYMP. PROC. 122, Mat. Res. Soc., 1988.

interface can usually be partitioned into two parts. One is of chemical origin involving the change in composition that occurs across the interface and the other is geometric in the sense that it includes the energy associated with the misfit between the two lattices at the interface. It should come as no surprise therefore that the geometric portion of the interphase interface energy for a coherent interface is negligible. Very little is known about the energy of interphase interfaces at this writing although it is an active area of research, in part, because such information is vital to the solution of joining problems involving dissimilar materials. The misfit energy associated with the deviation from coherence, such as occurs in a semi-coherent boundary containing misfit dislocations, can be estimated from dislocation theory.[12] Much remains to be done to elucidate the energy and structure of interphase interfaces.

9.3. Local equilibrium at grain boundary intersections.

Consider the situation described in Figure 4.13. We shall allow, in turn, virtual motion of the grain boundaries from the position of the solid lines to those of the dotted lines and we shall calculate the change in free energy accompanying such motion. The change in interface area induces the change in free energy given by

$$(\sigma_1 - \sigma_2 \cos\varphi - \sigma_3 \cos\theta)\overline{OP}$$

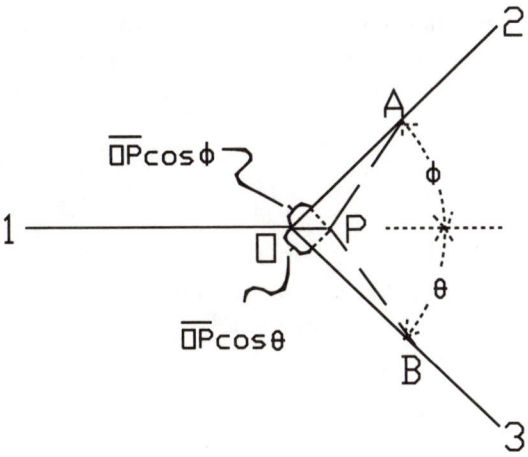

Figure 4.13. Schematic illustration of virtual displacement of grain boundaries and the changes in area and orientation that occur.

The change in free energy due to the change in grain boundary orientation is

$$\overline{AP}(\partial\sigma_2/\partial\varphi)\delta\varphi + \overline{BP}(\partial\sigma_3/\partial\theta)\delta\theta$$

$$\text{But, } \delta\varphi = (\overline{OP}\sin\varphi)/\overline{AP} \text{ and } \delta\theta = (\overline{OP}\sin\theta)/\overline{BP}$$

Setting the total free energy change equal to zero yields

$$\sigma_1 - \sigma_2\cos\varphi - \sigma_3\cos\theta + \sin\varphi(\partial\sigma_2/\partial\varphi) + \sin\theta(\partial\sigma_3/\partial\theta) = 0$$

Similar equations may be obtained for the virtual displacement of the boundaries in the direction of the vector contained in the stationary boundary and normal to the line of intersection of the three boundaries. The resulting relations can be described by the following equation.

$$\sum_{i=1}^{3} (\sigma_i \mathbf{t}_i + (\partial\sigma_i/\partial\mathbf{t}_i)) = 0$$

where $\mathbf{t}_i$ is a unit vector in the plane of the ith grain boundary, perpendicular to the line of intersection of the grain boundaries and pointing away from this line. The derivative term on the left hand side of the equation is called a "torque" term because it is a driving force tending to reorient the grain boundary. If these torque terms are zero in value then the above equation can be shown to reduce to

$$\sigma_1 / \sin\theta_1 = \sigma_2 / \sin\theta_2 = \sigma_3 / \sin\theta_3$$

where θ_i is the angle opposite the i boundary. This relation is the same as found in Mechanics for the vector resolution of forces acting at a vertex. This analogy leads to the simple relations describing the local equilibria shown in Figure 4.14.

10. DIFFUSE INTERFACES.

In the above we have implicitly assumed that the interface is a

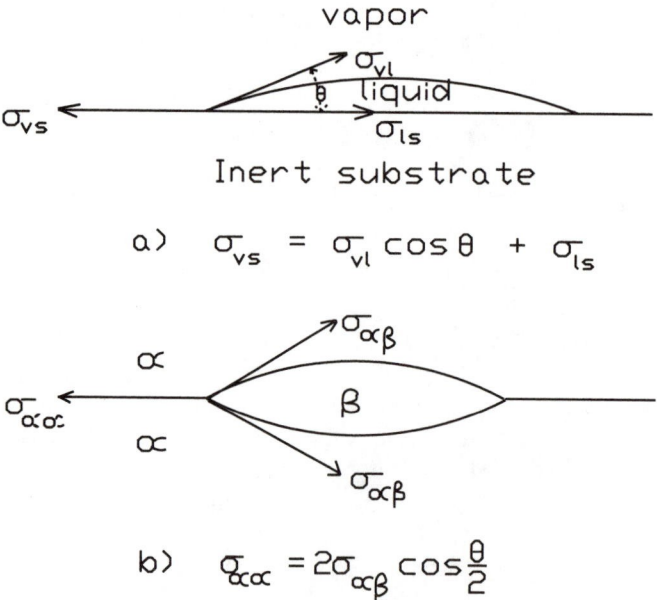

Figure 4.14. Schematic illustration of the "law of forces" representation of the surface tensions in local equilibrium at an intersection.

region of transition between its bounding phases, with the thickness of the interface on the order of atomic dimensions. In most cases of interfaces this assumption is in accord with experiment. However, there is a class of phenomena in which the interface is much thicker. For this case, a new method has arisen to describe the thermodynamics of the transition region.[13] It is no longer sufficient to express the free energy in terms of the specific free energy, but it is also necessary to include terms dependent on the gradients of the independent quantities that vary in the transition region. Here we shall merely give some of the interesting results without providing a detailed derivation of them.

Let us consider the interface between two phases in equilibrium, as defined by the common tangent to the free energy-composition curve shown in Figure 4.15. The compositions of the two phases in equilibrium are denoted by C'_e and C''_e. Suppose that the composition in the interface between the two phases in equilibrium with each other varied with distance normal to the interface, as shown in Figure 4.16. A thickness of the interface region can be defined to be

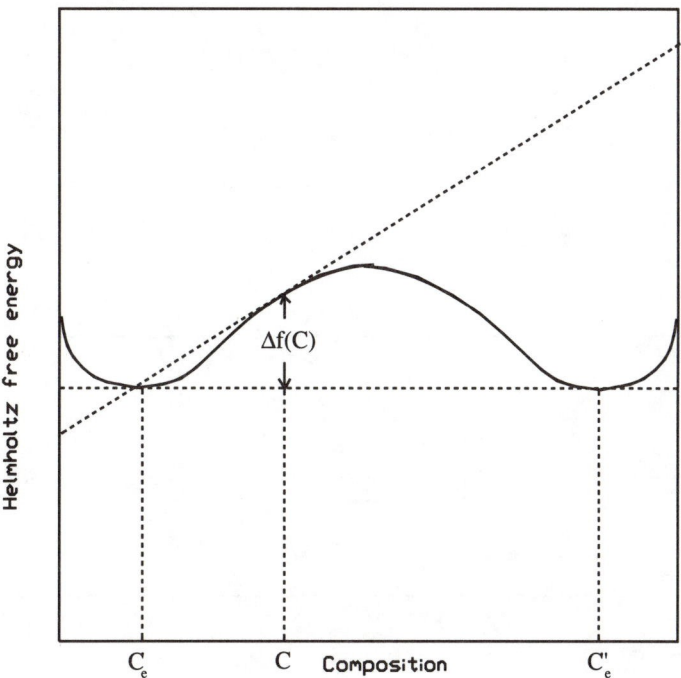

Figure 4.15. Defining compositions and free energies used in expression for thickness of diffuse interface.

$$l = (C''_e - C'_e)/(dC/dx)_{max}$$

where the slope of the composition is the maximum in the transition region. Now it can be shown that at any composition C, the slope corresponding to that composition, dC/dx, is related to the difference between the specific Helmholz free energy for this composition and the composition weighted average of those for the equilibrium compositions, $f(C)$, as shown in Figure 4.15 by the relation

$$(dC/dx) = [\Delta f(C)/K]^{1/2}$$

where K is a positive definite quantity representing a material parameter defined by

$$K = -dK_1/dC + K_2$$

with $K_1 = \partial f(C)/\partial(\nabla^2 C)$ and $K_2 = \partial^2 f(C)/\partial(\nabla C)^2$. Thus, the thickness can be written as

$$l = (C''_e - C'_e)/(f(C)/K)^{1/2}.$$

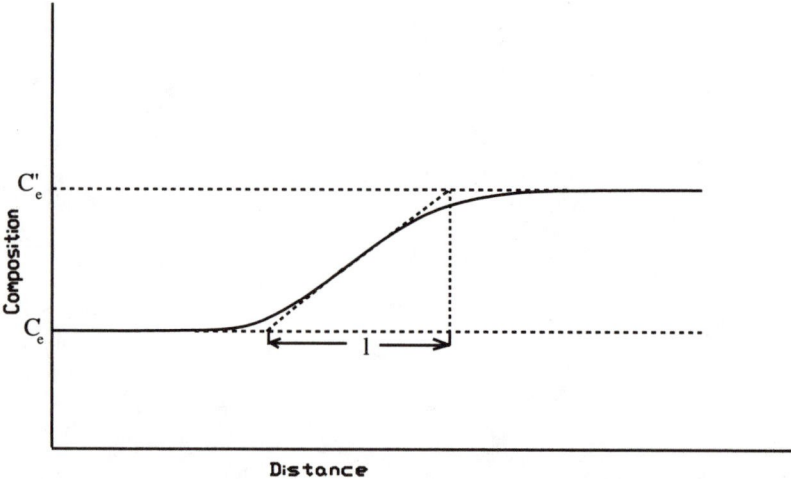

Figure 4.16. Illustrating the change in composition with distance at an interface between two phases having the same crystal structure.

For a miscibility gap system corresponding to the free energy curve shown in Figure 4.15, Cahn and Hilliard have evaluated the difference quantities in numerator and denominator of this relation for interface thickness. In the vicinity of the critical temperature, if $f(C)$ can be expanded as a Taylor's series about T_c and C_c then

$$f(C) = -b(T_c - T)[(C - C_c)^2 - (C'' - C_c)^2] + g[(C - C_c)^4 - (C'' - C_c)^4] + ...$$

and $[(C'' - C')/2]^2 = b(T_c - T)/2g$, while

$$f(C) = g\{[(C'' - C')/2]^2 - (C - C_c)^2\}^2$$

Substitution of these expressions into the relation for the interface thickness, l, yields in the vicinity of the critical temperature

$$l = 2[2K/b(T_c - T)]^{1/2}.$$

Thus, as the temperature approaches the critical temperature from below, the interface thickness increases and approaches infinity.

The interface energy of the interface defined in Figure 4.16 can be evaluated. It equals the difference between the total Helmholz free energy of the system less the Helmholz free energy the system would have if the properties of each homogeneous contiguous phase were continuous. This statement implies that the interface region is divided up into these two homogeneous regions. The principle of division is that in each differential volume of the transition region the fraction of each phase present is determined by the conservation of atoms and the compositions: C of the differential volume, and C' and C'', the equilibrium compositions of the contiguous phases. The specific Helmholz free energy of this mixture of contiguous phases is just that give by the intersection of the lowest common tangent to the free energy-composition curve (see Figure 4.15) with a vertical line at the composition C. Analytically, this specific free energy is given by

$$C^e\mu_2 + (1 - C)^e\mu_1$$

The actual specific Helmholz free energy of this differential volume is given by[13]

$$\Delta f(C) + K(dC/dx)^2$$

where K is defined above. Thus, the specific interface energy is obtained by subtracting the specific free energy of the mixture of coexisting phases from the actual free energy, integrating over the volume and dividing by the area of the interface to yield

$$\sigma = N_v \int_{-\infty}^{\infty} [\, \Delta f(C) + K(dC/dx)^2]dx$$

In the above, the compositions C are in atomic fraction, the free energies are in units of energy per atom and N_v is the number of atoms per unit volume.

In all of this chapter it has been assumed that the radii of curvature of the interfaces are large with respect to the thickness of the respective interfaces.

11. METHODS OF MEASURING INTERFACE ENERGIES.

The methods for measuring the liquid-vapor interface energy have been described in detail[14] and need not be repeated here. Reviews of the measurements of liquid-solid, solid-vapor and solid-solid interfaces are rarer and, hence, some space will be devoted to brief descriptions of the methods used. The main technique for measuring solid-vapor interface energy for the case of metals and alloys is the Zero Creep Method.[14] This technique involves the equilibration of a fine filament, both with respect to its environment and with respect to a load that tends to extend the filament at an elevated temperature. By systematically testing filaments with successive loads and measuring the extension rates of the filaments it is possible to deduce the load corresponding to zero creep rate of the filament. Since the surface energy of the filament itself acts to contract the filament, zero creep rate corresponds to a balance of the rate of change of $2 \pi r\sigma$ with respect to length and the load, at constant volume, from which it is possible to calculate the surface energy in the event that the filament is a single crystal. Usually, the filament is a polycrystal in the form of a bamboo structure, in which the grain boundaries are normal to the filament's axis. Correction must be made for the contribution of the energies of the grain boundaries to the change in shape of the filament. In order to be able to deduce a value for the surface energy it is necessary to have a knowledge of the ratio of the energy of such grain boundaries to that of the surface. This ratio can be deduced from measurements of the contact angle of the thermal grooves developed where these grain boundaries intersect the surface.

Measurement of solid surface energies for non-metallic, brittle materials is accomplished using a controlled cleavage technique where the force required to form new surface by extension of a crack is balanced against the surface energy of the new surfaces produced by the crack extension.[15] The controlled cleavage technique has also been use to measure grain boundary energies in brittle materials.[16] Modern evaluation of this technique is in terms of fracture mechanics considerations in which the crack propagation resistance G_C is measured, where G_C is proportional to the specific energy of the surface created in an ideally brittle material.Another method of measuring solid surface energies is heat of solution calorimetry. Grain boundary energies are generally measured by equilibrating the grain boundaries and the surface and evaluating the groove angles. Finally, solid-liquid energies are measured by equilibrating a grain boundary in contact with the liquid phase in a temperature gradient. The theory of this technique may be obtained in reference 17. Thus, it is apparent that most of the values depend in an absolute sense on the values of the solid surface energies.

12. PSEUDOMORPHIC STABILIZATION OF METASTABLE PHASES IN THIN FILMS.

The fact that the geometric component of the interface energy in coherent type interfaces is negligible is put to use in the pseudomorphic stabilization of metastable phases. Since, the chemical component of the interface energy can be adjusted to be negative in coherent interphase interfaces we have another parameter that can be controlled in the attempt to achieve this objective. For certain combinations of substrate and thin film material it is possible to demonstrate that this system will have a lower free energy if the thin film has a metastable crystal structure, suitably related to the crystal structure and parameters of the substrate, than if it had its stable crystal structure. For example, alpha-Sn is metastable with respect to the metallic beta Sn above about 13.2 °C. Alpha Sn has the diamond cubic structure with lattice parameter equal to 6.489 Angstroms at 25 °C. A candidate substrate material with the same crystal structure and lattice parameter(6.4798) is zinc-blend type InSb. The beta Sn has a tetragonal crystal structure with lattice parameters of a=5.8311 and c=3.1817. Thus, an appreciable elastic distortion is required for the beta-Sn lattice to match coherently with that of InSb. This elastic distortion has an associated energy which adds to the free energy of the substrate/beta-Sn film system an amount proportional to the thickness of the film. When this total distortional energy per unit area of the film exceeds the interface energy associated with the non-coherency between the beta-Sn film and the substrate then there will be a driving force for the film/substrate interface to generate misfit dislocations and non-coherency. On the other hand, the InSb substrate/alpha-Sn film system has no strain energy and the interface energy is nil or negative. The only positive contribution to the free energy **relative to the InSb/beta-Sn system** is the lattice stability energy (i.e. the difference in free energy between the alpha and beta polymorphs of Sn), which is also proportional to the thickness of the film.

These considerations lead to the possibility of stabilizing the metastable structure as described schematically in Figure 4.17. In this figure we have assumed that the free energy comparison is between a stable β phase

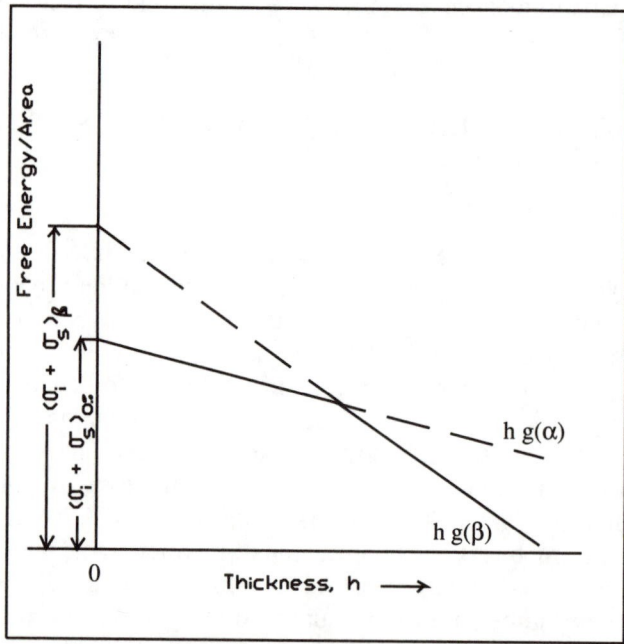

Figure 4-17. Illustrating the relative free energies of coherent metastable α and non-coherent stable β phases as a function of film thickness. Subscripts i and S represent interface and surface , respectively.

having a non-coherent interface with the substrate and a metastable α phase having a coherent interface with the substrate. Thus, the sum of the interface and surface specific energies is larger for the stable phase than for the metastable phase as indicated in Figure 4.17. Hence, there is a range of film thickness over which the coherent α phase is stable with respect to the non-coherent β phase. Let us also consider the free energy comparison for the case that the interface with the substrate is coherent for both phases. In this case, if the strain energy per unit volume induced in the stable phase is less than the lattice stability energy, the strained, but coherent stable phase will be stable with respect to the coherent metastable phase. On the other hand, when the strain energy per unit volume exceeds the corresponding lattice stability energy then again the coherent metastable phase will have a range of stability up to the same thickness defined in Figure 4.17. These are simplified concepts. A more sophisticated approach has been undertaken by Bruinsma et al[18]

Summarizing, solid surface energies reveal minima, in the form of cusps, in the Wulff plot of surface energy versus orientation. Orientations in the vicinity of these cusp orientations tend to have surfaces comprising cusp orientation surfaces separated by ledges of low energy orientation. The arrangement of atoms at surfaces or interfaces need not mimic the arrangement in bulk. Indeed, reconstruction of the surface or interface region to produce surface or interface phases can and does occur for many materials. Capillarity effects due to the small size of particles lead to driving forces for atom transfer from small particles to large particles. Solute tend to partition in an equilibrium manner between the homogeneous bulk and interface regions according to the Gibbs adsorption equation. The relative values of interface energies can and do have important consequences. Local equilibrium between interfaces lead to the shapes of interfaces found in microstructure. The gradient implicitly associated with an interface can be treated by gradient thermodynamics to yield the thickness and energy of the interface. The excess energy of interfaces provides the basis for pseudomorphic stabilization of metastable phases in thin films.

REFERENCES

1.J.Friedel, Acta Met $\underline{1}$,79(1953).
2. J.J. Gilman, J.Appl.Phys.$\underline{31}$,2208(1960).
3.C.Herring, Phys.Rev. $\underline{82}$,87(1949).
4.J.K.Mackenzie, J.Phys.Chem.Solids $\underline{23}$, 185(1962).
5.R. Trivedi in LECTURES ON THE THEORY OF PHASE TRANSFORMATIONS , ed. H.I.

Aaronson, AIME,1975.

6.E.A.Guggenheim, THERMODYNAMICS, North-Holland Publ., Amsterdam, 1967.

7. R. Defay, I. Prigogine, A. Bellemans and D. H. Everett in SURFACE TENSION AND ADSORPTION, Longman's, Green & Co., London 1966.

8. E.D. Hondros and M.P. Seah in PHYSICAL METALLURGY, eds.R.W. Cahn and P. Haasen, 3rd edition, North-Holland Physics Publ., NY, 1983.

9. D.P. Woodruff, THE SOLID-LIQUID INTERFACE, Cambridge University Press, London, 1973.

10. J.W. Cahn in INTERFACIAL SEGREGATION,ASM, Metals Park, Ohio, 1979.

11. J. Wetzel and E.S.Machlin, Surface Science $\underline{144}$,124(1984).

12.J.H. van der Merwe, Proc.Phys.Soc. $\underline{63A}$,613(1950);in TREATISE ON MATERIALS SCIENCE AND TECHNOLOGY,vol.2, Academic, New York, 1973.

13. J.W. Cahn and J.E. Hilliard, J. Chem Phys.$\underline{28}$,258(1958).

14. L.E. Murr, INTERFACIAL PHENOMENA IN METALS AND ALLOYS, Addison-Wesley, Reading, Mass.,1975.

15. J.J. Gilman, J.Appl. Phys.$\underline{31}$,2208(1960).

16. W.H.Class and E.S. Machlin, Am.Cer.Soc.Jl.$\underline{49}$,306(1966).

17. M. Gunduz and J.D. Hunt, Acta Met.$\underline{33}$,1651(1985).

18. R. Bruinsma and A. Zangwill, J. de Physique $\underline{47}$, 2055(1986).

BIBLIOGRAPHY.

1. J.W. Gibbs, THE SCIENTIFIC PAPERS OF J. WILLARD GIBBS,vol.1, Dover, New York, 1961.

2. C. Herring in STRUCTURE AND PROPERTIES OF SOLID SURFACES, R.Gomer and C.S. Smith, eds., University of Chicago Press, Chicago,1953.

3. W.W. Mullins in METAL SURFACES,ASM, Metals Park, Ohio, 1963.

4. GRAIN BOUNDARY STRUCTURE AND PROPERTIES, G.A Chadwick and D.A. Smith, eds.,Academic Press, New York, 1976.

5.GRAIN-BOUNDARY STRUCTURE AND KINETICS,ASM, Metals Park, Ohio, 1980.

6. C.Herring in PHYSICS OF POWDER METALLURGY,ed W.E.Kingston, McGraw-Hill, New York,1951.

7. G.A. Somorjai, CHEMISTRY IN TWO DIMENSIONS:SURFACES, Cornell University Press, Ithaca, 1981.

PROBLEMS

1. On the assumption of unreconstructed surfaces, evaluate the relative surface energy for silicon (diamond cubic) of the 100, 110, and 111 surfaces. Take into account that the number of dangling bonds can vary depending between which of the possible parallel planes the surface is formed.

2. Why, at equilibrium, is it possible to form faceted surfaces on solids even though this corresponds to an increase in surface area per unit volume relative to a smoothly curved surface?

3. Why do surface torques exist near low index surface orientations that tend to rotate the surface to another orientation?

4. Describe the dependence of equilibrium vapor pressure on the radius of the particles in equilibrium with the vapor?

5.What is the driving force for Ostwald Ripening?

6.In what way does the concentration at a two phase boundary in the phase diagram depend upon the size of one of the phases?

7. If increase in solute concentration acts to decrease the surface tension will the solute segregate to the surface or desegregate from the surface over the range of concentration within the spinodal?

8. If increase in the solute concentration acts to decrease the average of the solidus and liquidus temperatures relative to the concentration weighted average of the melting points of the pure components will the solute segregate or desegregate from interfaces in this dilute alloy?

9. If the surface of an elemental solid is considered to be a separate phase will it melt at a lower or higher temperature than the bulk phase? Justify your answer. Is it valid to consider the surface to be a separate phase in this context?

10. Will surface stress and surface tension differ at a temperature more than about 0.5*the absolute melting temperature?

11. What is the relation between the surface tensions of the interfaces at local equilibrium for a drop of liquid on a flat solid surface, both in contact with a vapor phase.

12. In a two dimensional thin polycrystalline film, what shape will the grains assume to produce a state of metastable equilibrium? (i.e. any small deviation of the shape of the grains will increase the free energy.) Hint. Consider the grain boundary energy to be independent of orientation. In this case, what inclusive angles do three grain boundaries that intersect along a line define ?

13. What group of elements in the periodic table will tend to segregate at the surfaces of metallic solids? Hint. Consider the physical basis for the excess energy associated with the surface to provide a clue to one group of such elements.

14. A solder must wet the parts it joins. In order to achieve such wetting (as small a contact angle as possible) the surface energy of the solder is usually much smaller than that of the parts it joins. Why?

15. Suppose it is desired to prevent phase separation in a thin film at a composition of the film where a two phase mixture is stable at the deposition temperature. Not only is it desired to prevent phase stabilization, but it is also desired to stabilize the solid solution at a particular composition in the two phase region. How would you use pseudomorphic stabilization to accomplish this objective?

16. Will particles of bismuth, smaller than 10^{-5} cm and at equilibrium with respect to change of phase, have the same crystal structure as bulk bismuth? Hint. See Figure 1.3 and Table 4.2 for the data required to solve the problem.

17. Estimate the grain boundary solute segregation ratio for the case of Ga as a solute in polysilicon. Describe the steps you took to obtain this estimate.

18. Derive an expression for the thickness of the liquid phase that may exist along a high angle grain boundary at a temperature difference ΔT below the bulk melting point. Estimate this thickness for copper using data listed in tables in this chapter, for $\Delta T = 10$ °C. Comment on the possibility that this effect may account for the frequent observation of an amorphous phase along grain boundaries in polycrystalline ceramics.

V-HETEROPHASE AND HOMOPHASE FLUCTUATIONS

INTRODUCTION

The bases for consideration of fluctuations of various kinds in otherwise homogeneous phases have been developed in the previous chapters. In particular, heterophase fluctuations involve interfaces between different phases, as well as changes in the free energy of a system. Homophase or, more accurately, homostructural fluctuations are fluctuations in composition and/or density or strain. At equilibrium, there are distributions of both such fluctuations, called embryos. Nucleation of product phases may occur by spontaneous growth of critically sized embryos belonging to an equilibrium distribution of embryos.

In the homophase system we can define a locus of spinodal compositions. Inside the locus of spinodal compositions, any composition fluctuation leads to the spontaneous decomposition of the metastable host phase into product phases having the same crystal structure, but different compositions. In this regime equilibrium distributions of embryos cannot exist in the metastable host. Indeed, in this regime the host is not metastable, but is thermodynamically unstable. A favored type of composition fluctuation is periodic or "wavelike" in nature, occurs throughout the parent volume and is called "spinodal decomposition". The locus of spinodal compositions lies within the miscibility gap. Analogous to the spontaneous reaction along a composition path yielding "spinodal decomposition" there is one along a strain path that can decompose a host phase into periodic strain waves at constant composition. We consider these concepts in the present chapter.

1. HETEROPHASE FLUCTUATIONS.

In Chapter I we defined a phase, at equilibrium, to be a homogeneous material, whether a solid, liquid or vapor. However, this statement needs to be modified in that we need to state the dimension of the phase under discussion. If the dimension is such that the volume of the phase contains more than about several thousand atoms or molecules then the concept of homogeneous needs no modification. If the dimension is smaller than this scale then deviations from homogeneity due to the temperature induced motions of the atoms can occur in density, composition or local arrangement of the atoms such that these intensive parameters can vary from one such volume of the phase to another such volume, or in the course of time within a given sub-microscopic volume. Further, these temperature induced deviations from homogeneity, called fluctuations, occur while the phase is in its thermodynamic equilibrium state. We will call fluctuations that yield atomic arrangements having a symmetry differing from the parent phase--heterophase fluctuations.

Consider a homogeneous host phase, alpha, containing N atoms. Suppose that the free energy of the system is increased by δg_n, when a particular small volume containing n atoms, where n<<N, fluctuates to a configuration characteristic of a new phase, beta. However, since this beta embryo could be formed at any of the N sites, the free energy of the system of embryos and host phase must include the contribution of the entropy of mixing of the embryos among the N possible sites. It will be shown that the contribution of this entropy of mixing heterophase embryos among the "boxes" centered about the host sites initially reduces the total free energy of the system.

Consider the free energy change in the host system due to these heterophase fluctuations.

$$\Delta G_{system} = N_n \delta g_n - Tk\ln W \qquad (5.1)$$

where N_n is the number of embryos of the phase, beta, which are distributed at random in the host alpha phase, each embryo containing n atoms; T is the absolute temperature; k is Boltzmann's constant; and W is the number of independent configurations of embryos and host atoms, each configuration having the same energy (see equation 1.2b).

Now, if the embryo has a spherical shape

$$W = N!/[N_n!(N-N_n!)!]$$ (5.2)

To obtain the stable state of the system consisting of these embryos of beta phase distributed in the host alpha phase we minimize the free energy of the system with respect to the number of such embryos as follows.

$$\frac{\partial G^{system}}{\partial N_n} = \delta g_n - kT\frac{\partial \ln W}{\partial N_n} = 0$$ (5.3)

But, by Stirling's approximation $N! \cong N\ln N - N$ for large N. Hence, substituting and taking the derivative we obtain

$$\frac{\partial\ G^{system}}{\partial N_n} = \delta g_n + kT\ln[N_n/(N-N_n)] = 0$$ (5.4)

We now perform the second derivative to determine whether (5.4) represents a minimum or a maximum and obtain

$$\frac{\partial^2 G^{system}}{\partial N_n^2} = kT[(1/N_n)-1/(N-N_n)]$$ (5.5)

Since $N_n << N$, the right hand side of (5.5) is positive definite and hence (5.4) represents a minimum in the free energy of the system of embryos and host. Solving (5.4) for N_n we obtain then that the number of embryos of phase beta containing n atoms at equilibrium is

$$N_n = N \exp[-\delta g_n/kT]$$ (5.6)

The dependence of the free energy of the system of embryos and host as a function of the number of embryos is shown in Figure 5.1. Thus, it is apparent that fluctuations initially act to decrease the free energy. To determine how many embryos exist at equilibrium it will be necessary to obtain an estimate for the free energy of formation of a particular embryo, δg_n. Consider a heterophase embryo. It will have an interface between it and the surrounding host phase, alpha, which has an associated specific interface free energy, $\sigma_{\alpha\beta}$. The beta phase will have a specific free energy (per unit volume), g_β, different from that of the host phase, g_α. Suppose, for the sake of simplicity, the embryo has a spherical shape of radius r. In this case, the increase in free energy in the formation of one embryo in a specific volume of the host phase is

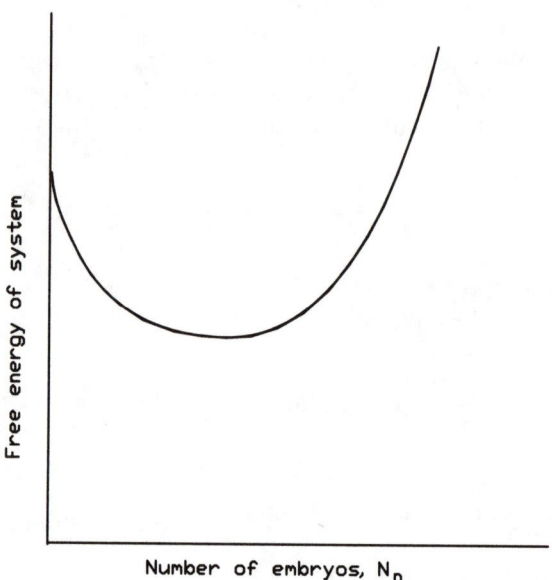

Figure 5.1. Dependence of free energy of system on number of embryos, each containing n atoms.

$$\delta g_n = (4/3)\pi\, r^3(g_\beta - g_\alpha) + 4\pi r^2\sigma_{\alpha\beta} \tag{5.7}$$

Here, we assume that there is no change in volume in the transformation.

If the host alpha is the stable phase at this temperature then δg_n has a positive value for all values of r (of n). If, on the other hand, alpha is a supersaturated or supercooled metastable phase relative to beta then δg_n will depend upon r as shown in Figure 5.2. It exhibits a maximum positive value at the radius r^*. This maximum value is called the free energy of nucleation and is denoted by the symbol, ΔG^*. For a spherically shaped embryo, differentiation of (5.7) with respect to r and appropriate rearrangement of terms yields that

$$\Delta G^* = 16\,\pi\sigma_{\alpha\beta}^3 /[3(g_\beta - g_\alpha)^2] \tag{5.8}$$

and
$$r^* = -2\,\sigma_{\alpha\beta}/(g_\beta - g_\alpha) \tag{5.9}$$

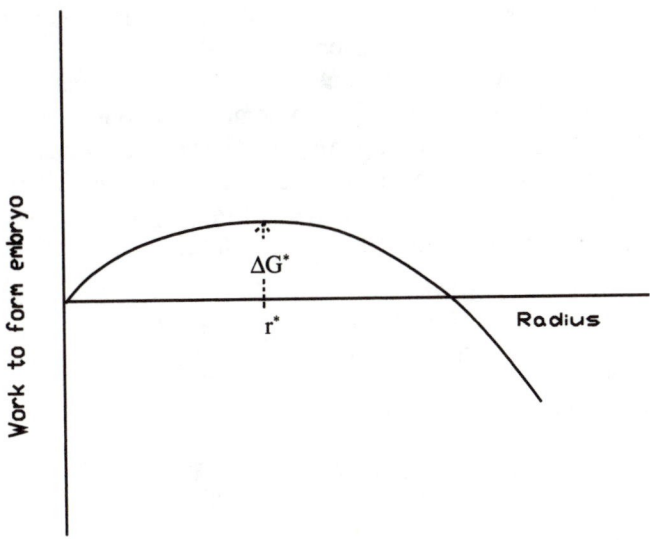

Figure 5.2. Dependence of the work to form an embryo on the radius of the embryo. At r* this work reaches a maximum and beyond this radius the embryo can grow spontaneously.

For any other shape of the embryo defined by the shape factor Q, where the surface area of the embryo is equal to $Q\Omega^{2/3}$ and Ω is the volume of the embryo, the free energy of nucleation is given by

$$\Delta G^* = 4Q^3 \, \sigma_{\alpha\beta}^3 \, /[27(g_\beta - g_\alpha)^2] \qquad (5.10)$$

with

$$\Omega^* = [-2Q\sigma_{\alpha\beta} \, /\{3(g_\beta - g_\alpha)\}]^3 \qquad (5.11)$$

The reason why the critically sized embryo is called a nucleus follows from the fact that as the nucleus grows the free energy of the system decreases. Since the critically sized embryo can spontaneously grow into a particle of the stable product phase it is the nucleus of the growing particle. However, for any embryo smaller than the nucleus the free energy of the system must initially increase if it is to grow.

The classical concept of nucleation is based on the assumption that fluctuations develop and maintain an equilibrium distribution of sub-critically sized heterophase embryos. As nucleation drains off critically sized embryos, new ones are produced at a rate sufficiently fast as to maintain the equilibrium distribution. We shall consider the rate of nucleation and embryo formation in a later section.

It is instructive to insert values for the quantities that enter into the free energy of nucleation (or to form an embryo) to obtain a sense of the liklihood that such nuclei (or embryos) can exist in a host phase. If the case being considered is the formation of a solid from a liquid host phase, then we may use Richard's approximation to estimate $g_\beta - g_\alpha$ in the vicinity of the equilibrium melting temperature, T_m, as follows.

$$g_\beta - g_\alpha = (h_\beta - h_\alpha) - T(s_\beta - s_\alpha) \tag{5.12}$$

However, at T_m, the difference in free energy corresponding to the left hand side of this equation must equal zero. Hence,

$$h_\beta - h_\alpha = T_m(s_\beta - s_\alpha) \tag{5.13}$$

Substituting (5.13) into (5.12) yields

$$g_\beta - g_\alpha = (T_m - T)(s_\beta - s_\alpha) \tag{5.14}$$

By Richard's rule $(s_\beta - s_\alpha) = 2$ cal/mol/°C. Suppose we take $|T_m - T|$ equal to 10 °C. Also, from the table in the previous chapter, we obtain an estimate of a typical solid/liquid interface energy to be about 100 ergs/cm². Substituting these values, using a typical molar volume of about 10 cm³/mol, $N \approx 10^{23}$ and the appropriate conversion factors to achieve the same units we then obtain the following values

$$r^* = 2.39*10^{-6} \text{ cm} = 23.9 \text{ nm}$$

$$\text{and } N^* = 10^{-7509} /\text{cm}^3$$

This number of embryos is much smaller than one in a host volume of about 1 cm³ and hence quite unlikely to be observed. Obviously, for the circumstance being considered-that of a homogeneous distribution of embryos in the host phase-the only chance of observing embryos is when the barrier energy provided by the interface energy is very small or when the driving energy (the supercooling or supersaturation) is very large and $g_\beta - g_\alpha < 0$. Is the existence of heterophase embryos thus merely an academic question? We shall show in the next section that the existence of embryos at interfaces that act to catalyze their nucleation is highly probable.

1.1. Heterogeneous distributions of heterophase fluctuations.

We have considered homogeneous distributions of heterophase embryos and nuclei in the previous section. In this one we shall show that the presence of interfaces in a host phase, which the embryo phase tends to "wet", can act to catalyze the presence of heterophase embryos and nuclei. We shall investigate two cases. In the first, the interface is between a material that does not interact chemically either with the host phase or with the embryo phase. In the second, the interface is a grain boundary in the host phase. Consider the situation shown in Figure 5.3. Local interface equilibrium in the absence of "torque" contributions and chemical interactions between the particle phase, p, and either the host or embryo phases yields the following relation between the interface energies.

$$\sigma_{ph} = \sigma_{pe} + \sigma_{eh}\cos\theta \qquad (5.15)$$

Using this relation between the specific interface energies the change in

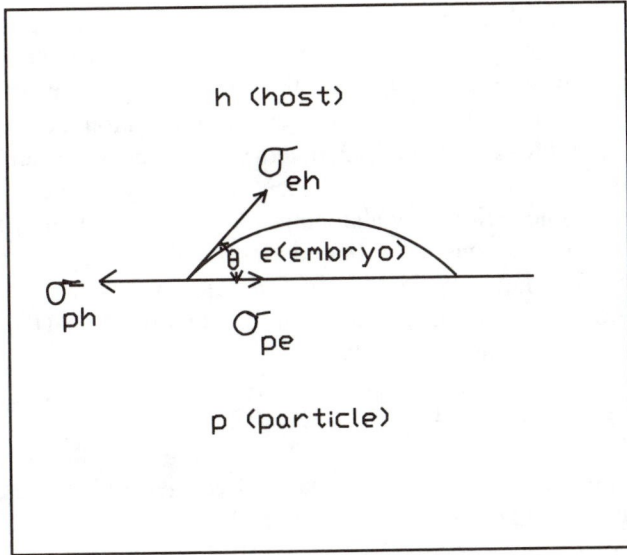

Figure 5.3. Illustrating an embryo formed at a particle (p)/host (h) interface and the local equilibrium at the host-particle-embryo intersection.

total interface energy due to the formation of one embryo at the original host-particle interface can be written as

$$\Delta G_s = 4\pi r^2 \sigma_{eh} f(\theta) \tag{5.16}$$

where $f(\theta) = (2 - 3\cos\theta + \cos^3\theta)/4$. It can be shown that the change in free energy in the volume occupied by the embryo is given by

$$\Delta G_b = (4\pi r^3/3)(g_e - g_h)f(\theta) \tag{5.17}$$

Now, the total change in free energy due to the formation of the embryo is the sum of ΔG_s and ΔG_b. To obtain the free energy of nucleation, we set the derivative of this sum with respect to the radius equal to zero with the result that

$$\Delta G^* = (16\pi \sigma_{eh}^3 /[3(g_e - g_h)^2])f(\theta) \tag{5.18}$$

and
$$r^* = -2 \sigma_{eh}/(g_e - g_h) \tag{5.19}$$

Thus, the radius of the spherical cap of the nucleus is unaffected by nucleation on the substrate. However, the free energy of nucleation can be markedly affected depending upon the value of the function $f(\theta)$. As the embryo phase more effectively wets the substrate, the contact angle θ approaches zero, $f(\theta) \to 0$, and the free energy of nucleation $\Delta G^* \to 0$. Thus, depending upon the contact angle θ, heterophase fluctuations and nucleation at inert interfaces in the host phase may or may not occur readily. Heterogeneous nucleation may also be produced at defects in metastable solids, such as dislocations, grain boundaries or point defects. Consider the case of grain boundaries as sites for the formation of heterophase fluctuations. Figure 5.4 defines the geometry of the situation. Local equilibrium at interface intersections requires that

$$\sigma_{hh} = \sigma_{gb} = 2 \sigma_{eh}\cos(\theta/2) \tag{5.20}$$

Under this constraint it can be shown that the free energy change associated with the interfaces upon formation of the embryo is

$$\Delta G_s = 4\pi\, r^2 \sigma_{eh} f'(\theta) \tag{5.21}$$

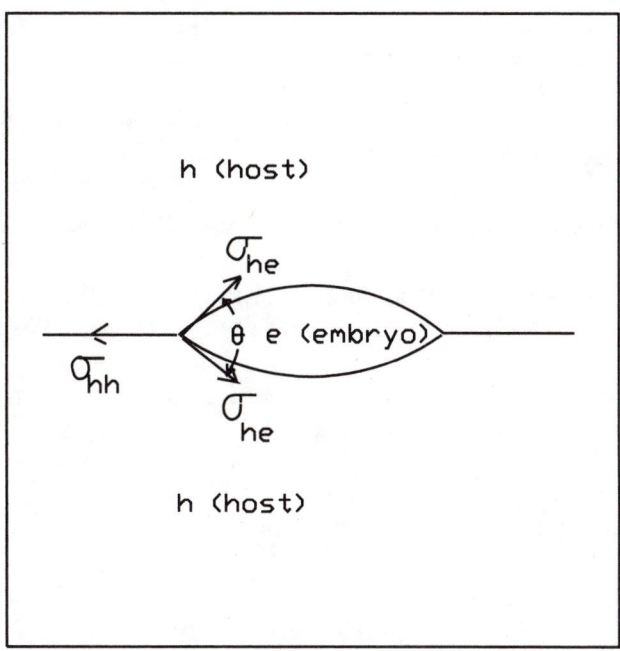

Figure 5.4. Embryo shown formed along a grain boundary with local equilibrium of the interface tensions.

where $f'(\theta)=f(\theta)/2$. Also, the free energy change in the volume occupied by the embryo is

$$\Delta G_b = [4\pi\, r^3(g_e - g_h)/3]f'(\theta) \tag{5.22}$$

Again, the total free energy change due to the formation of one embryo is the sum of ΔG_s and ΔG_b. The free energy of nucleation is obtained as the maximum of this sum with respect to variation in the radius of the embryo's spherical cap, r, and is given by (5.18) with $f(\theta)$ replaced by $f'(\theta)$. Thus, for embryo phases that wet grain boundaries in metastable host phases (i.e. for which $f'(\theta)\rightarrow 0$), the free energy to form such embryos can be sufficiently small as to make the appearance of heterophase fluctuations and nucleation of the stable phase at these grain boundaries highly probable.

Technologically, the result we have just obtained has great significance. For example: inert particles that are wetted by the product stable phase are used to seed supersaturated clouds in order to produce rain, to catalyze the production of fine-grained castings in supercooled metallic melts, etc.; wetted inert interfaces are the location at which nucleation of boiling occurs in superheated liquids; grain boundaries are one of the sites of nucleation of solid state transformations. Indeed, in practise, heterogeneous nucleation is the rule.

1.2. Effect of stress on fluctuation probability and embryo shape.

If the parent and product phases are solids then an additional contribution to the increase in free energy due to the formation of one embryo arises either because the specific volumes of the two phases differ or because in the presence of a coherent interface between the host and embryo phase the lattice parameters of the two phases parallel to the interface differ. In the case of an incoherent interface between embryo and parent phase, Nabarro[1] has evaluated the increase in free energy due to the strain energy induced in both embryo and host. He has also evaluated the effect of shape of the embryo on this increase in free energy on the assumption that all the strain energy is borne by the host phase with the result that the strain energy per atom of embryo is

$$\Delta g_e = [2\mu_h(\Omega_e - \Omega_h)^2/3\Omega_e]E(y/r) \tag{5.23}$$

where $E(y/r)$ is a function of the ratio of the half thickness y of a spheroid shaped embryo to its radius r and hence of the shape of the embryo, μ_h is the shear modulus of the host and Ω is the atomic volume. This function is illustrated in Figure 5.5. From the dependence of $E(y/R)$ on y/R shown in this figure it is apparent that the minimum strain energy occurs when the embryo has the shape of a disc. However, the disc shape has a higher surface energy for the same volume as compared to a sphere, which, as shown, is associated with a higher strain energy. Thus, the shape adjusts to some compromise between a disc and a sphere to minimize the total overall increase in free energy on formation of the embryo.

If the parent and product phases are such as to accommodate a coherent interface between the two, then the possibility exists that for small embryos the total free energy of formation of an embryo having coherent interfaces with the parent phase will be less than that for the case where these interfaces are incoherent. Although there is an enhanced strain energy for the embryo with coherent interface relative to the one with non-coherent interfaces, the decrease in surface free energy due to the lower energy of the

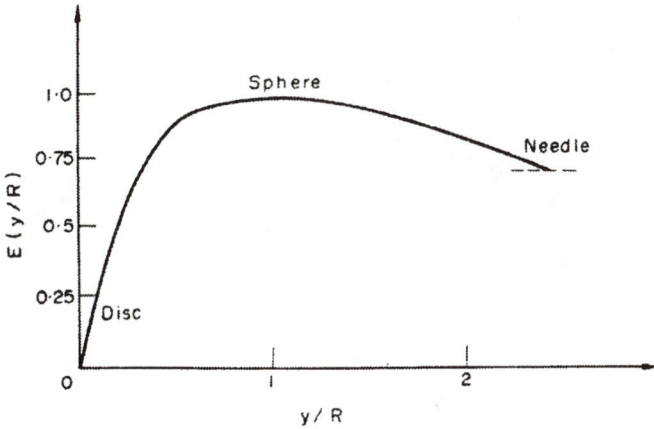

Figure 5.5. The function E(y/R) as calculated by Nabarro[1].

coherent interface (25 ergs/cm^2) as compared to a non-coherent interface (>200 ergs/cm^2) is more significant than the increase in strain energy, at least for sufficiently small embryos. This concept is illustrated in Figure 5.6.

For the simple case of equal elastic moduli of host and embryo, similar atomic arrangements in both, and disregistry along only one atomic direction, the strain energy per unit volume of embryo is independent of the shape and is given approximately by

$$\Delta g_e = \mu \delta^2/(1-\nu) \qquad (5.24)$$

where the relative disregistry in atomic spacing parallel to the coherent interface is

$$\delta = (a_h - a_e)/a_e \qquad (5.25)$$

. The concepts just described hold as well for heterogeneous type nucleation. For example, at grain boundaries, it is possible for the embryo to have a coherent interface with at least one of the contiguous grains. Also, considering the plenitude of low energy boundaries at special orientations, it is conceivable that the interface with the other grain, if not a coherent boundary is, at least, a low energy boundary.

The existence of coherent interfaces between host and product

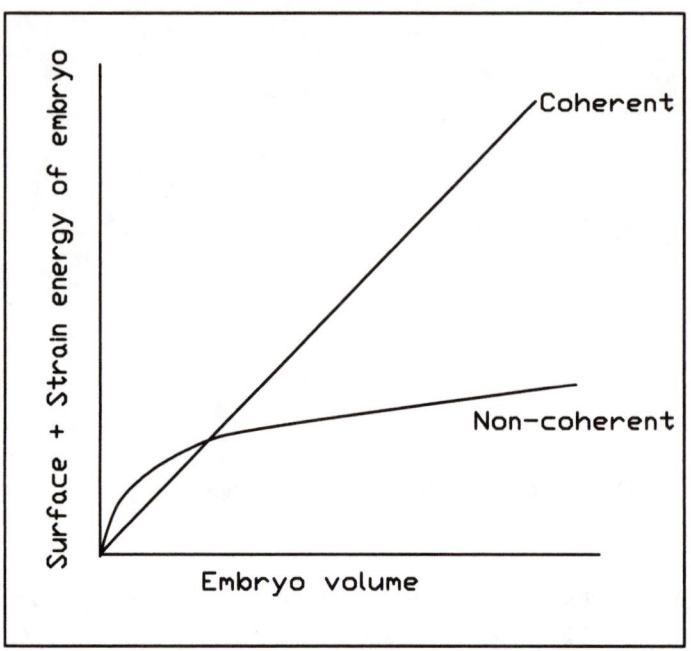

Figure 5.6. Showing that for small volumes coherent embryos have lower free energy per embryo of given volumethan non-coherent ones.

phases implies that there is a unique crystallographic relation between host and product phases. However, there is another possible reason for the observance of crystallographic relations between host and product phases which is based on the kinetics of growth of the product phases. Thus, observation of unique crystallographic relations between host and product phases may not be used as a proof of the mode of nucleation.

2. HOMOPHASE FLUCTUATIONS OF COMPOSITION IN A METASTABLE HOMOGENEOUS PHASE.

Consider a metastable phase of total volume V corresponding to Figure 5.7 at the composition C(atom fraction) and temperature T'. We wish to know the change in free energy associated with a fluctuation in composition in some small volume v from C to C+ ΔC. Let the free energy per unit volume of the phase be g. Thus, the work to form this composition fluctuation is

$$\Delta G = g(C+\Delta C)v + g[C- \Delta C\{v/(V-v)\}](V-v)-g(C)V \qquad (5.26)$$

But, a Taylor's expansion yields

$$g(C- \Delta Cv/(V-v)) = g(C) +(\partial g/ \partial C)[- \Delta Cv/(V-v)] +...$$

and by substitution then

$$\Delta G=[g(C+ \Delta C)-g(C)-(\partial g/ \partial C)\Delta C]v \qquad (5.27)$$

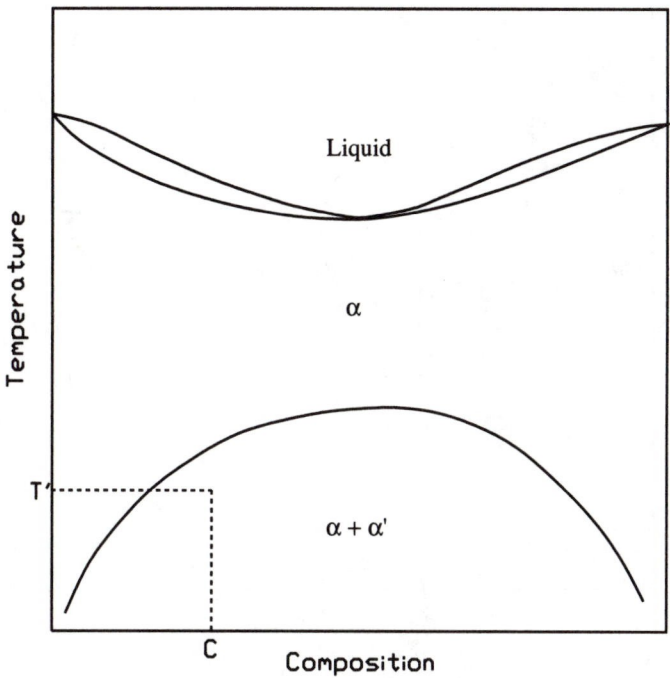

Figure 5.7. Phase diagram of miscibility gap system defining the metastable phase discussed in the text.

The graphical interpretation of (5.27) is shown in Figure 5.8. When the composition of the host phase is such as to fall in the range where the free energy-composition curve has a positive curvature then as the magnitude of the composition fluctuation increases the magnitude of ΔG increases until it has a maximum value, ΔG^*, at some composition. When the composition of the host phase falls in the range of compositions where the curvature of the free energy-composition curve has a negative value then ΔG has no maximum and is always negative in value. The turning point between the two cases is defined by the inflection point in the free energy-composition curve at which the second derivative

$$\partial^2 g / \partial C^2 = 0 \qquad (5.28)$$

Equation (5.28) defines the conditions associated with what is called the spinodal. (The origin of the term "spinodal" is due to van der Waal. As

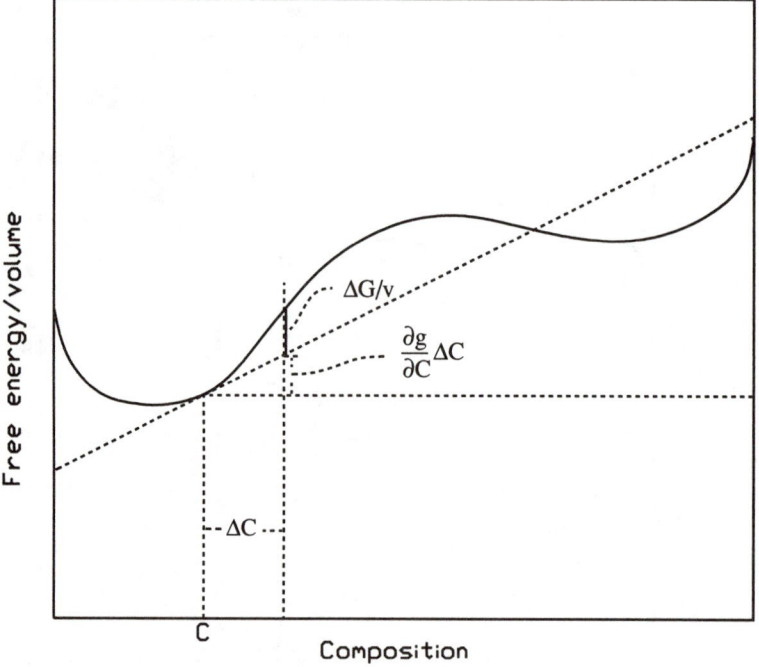

Figure 5.8. Graphical interpretation of the parameters in equation (5.27).

shown in Figure 5.9, in a plot of the grand potential (−PV) against the chemical potential the spinodal occurs at the tip of a spine while in a plot of free energy against composition it satisfies (5.28)!)

ΔG in (5.27) does not represent the total change in free energy due to the composition fluctuation because the contribution of surface energy to this change has not been included. In the previous chapter, we noted that for the situation under consideration there must be an interface associated with the gradation of composition from that of the host to that in the volume v. The evaluation of this interface energy has been accomplished by Cahn and Hilliard and was described in the previous chapter.

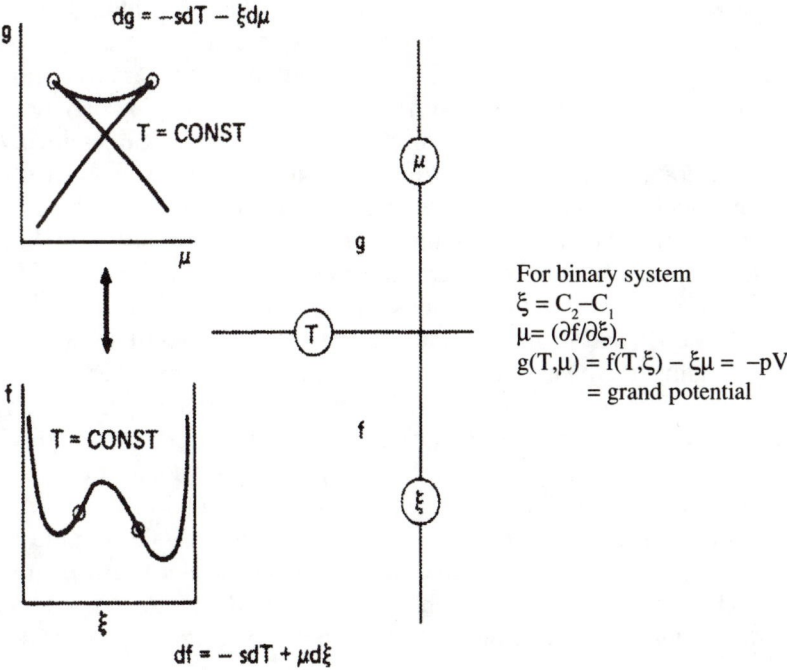

For binary system
$\xi = C_2 - C_1$
$\mu = (\partial f/\partial \xi)_T$
$g(T,\mu) = f(T,\xi) - \xi\mu = -pV$
$\quad$ = grand potential

Figure 5.9. Schematic illustration of spine in plot of grand potential versus the chemical potential at position corresponding to zero value of 2nd derivative in plat of free energy versus composition. Note that the g in this figure does not correspond to the g in the text.

3. HOMOPHASE, PERIODIC "WAVELIKE" MODE OF DECOMPOSITION OF UNSTABLE PARENT SOLID SOLUTION—SPINODAL DECOMPOSITION.

In the above we have considered one type of concentration fluctuation which consists of a change in composition from that of the host C to C+dC in some small volume, v, and an unspecified continuous "interface" region in which the composition degrades from C+dC to C. We could follow the procedure used in the section on heterophase embryos to obtain an expression for the equilibrium distribution of the type of composition fluctuations we are considering presently. Of course, to accomplish such a calculation it would be necessary to specify both the magnitude of the fluctuation (to evaluate the volume related work to form the fluctuation) and the parameters defining the interface (to evaluate the interface energy). In this way it should be possible to evaluate an equivalent work of nucleation for such a composition fluctuation. An alternative mode of decomposing an unstable solution has been recognized in modern times, as a consequence of the work of Cahn and Hilliard. This mode does not involve the development of an equilibrium distribution of embryos of the stable product and in this sense represents a radical departure from the classical treatment of nucleation represented by the work of Volmer and Weber, and described in a simplified manner in the previous sections. Cahn and Hilliard consider composition fluctuations that can be described in terms of a Fourier series and its components. Ultimately, as we will show in a later chapter, this treatment leads to the prediction that one Fourier component, which has a periodic nature (wave length and amplitude), grows at a faster rate than other components and, consequently, is the component observed upon the decomposition of the unstable parent solid solution. Thus, basically, the Cahn-Hilliard treatment involves fluctuations that consist of periodic "waves" of composition throughout the whole volume of the parent phase and thus, by their definition of an interface, of a continuous periodic interface, as well. This periodic mode of decomposition, as contrasted to the nucleation and growth mode, is called "spinodal decomposition".

The basic assumption of the Cahn and Hilliard treatment is that the free energy per molecule is dependent not only on the magnitude of the intensive parameters but also on the spatial gradients of these parameters. Thus, for the case under consideration, the specific free energy is assumed to be a function not only of the composition but also of the spatial gradients of the composition. The result obtained for the Helmholz free energy of a non-homogeneous system but one in which the molar volume does not vary with composition is

$$F = \int_V [f'(C) + K(\nabla C)^2]dV \qquad (5.29)$$

where $f'(C)$ is the specific Helmholz free energy (per unit volume) of a homogeneous phase of composition C and K is a positive definite material parameter already defined in the previous chapter.

We are interested in the increase in free energy due to a composition fluctuation in a metastable host system of original composition C_o. Cahn and Hilliard suggested that this composition fluctuation be described in terms of its Fourier components. Such components are orthogonal and consequently the total change in free energy due to these components is the sum or integral over the free energy changes accompanying each Fourier component. Therefore, if any Fourier component leads to instability in the free energy change due to that component, the solution is itself unstable. Thus if (5.29) is evaluated for some Fourier component and if the difference

$$\Delta F = \int_V [f'(C)-f'(C_o)+K(\nabla C)^2]dV < 0 \qquad (5.30)$$

for all C then the solution is unstable. It is convenient to use the composition defined by

$$C-C_o = A\cos\beta x \qquad (5.31)$$

where $\beta=2\pi/\lambda$ and λ is the wave length of the composition wave. Expanding $f'(C)$ via a Taylor's expansion, substituting (5.31) and the appropriate derivatives into (5.30) it can be shown that the change in free energy per unit volume, in the volume over which the integration is accomplished is given by

$$\Delta F/V = (A^2/4)[\partial^2f'/\partial C^2|_C + 2K\beta^2] \qquad (5.32)$$

The instability of the metastable solid solution with respect to a composition fluctuation is governed by the sign of the term in the square brackets of 5.32. Examination of this term shows that the effect of composition gradients is to stabilize the solution (i.e. the term $2K\beta^2$ is positive definite.) Thus, the solution no longer is unstable at the spinodal composition, but at some composition at which the second derivative term is sufficiently negative so as to overcome the barrier term. A comparison of the lines defined by (5.28) and (5.32) relative to the solvus lines in a miscibility gap phase diagram is shown in Figure 5.10.

At composition-temperature points encompassed by the line defined by (5.32) the metastable solid solution is unstable with respect to any arbitrary composition fluctuation that leaves the molar volume unchanged, i.e. composition fluctuations in the metastable host solution spontaneously generate and grow. Setting the term in the square brackets in (5.32) equal to zero defines a critical value of the wave length, λ_c, such that for $\lambda > \lambda_c$, the metastable solid solution is unstable.

The effect of molar volume variation with composition leads to the incorporation of another positive definite term inside the square brackets of (5.32). In the situation where the host is elastically isotropic this term is

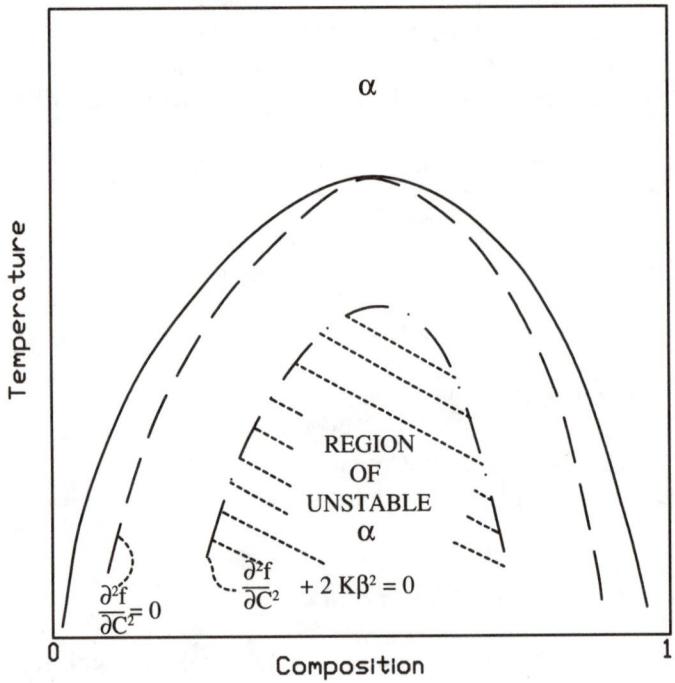

Figure 5.10. The various spinodal loci are shown relative to the solvus of a miscibility gap phase diagram.

given by

$$2\eta^2 Y/(1-\nu) \hspace{3cm} (5.33)$$

where η is the linear expansion of the host per unit change in composition, Y is Young's modulus and ν is Poisson's ratio. This term, normally called the "coherency strain energy" term, enhances the stability of the metastable solid solution with respect to a composition variation. For example, calculation shows that the critical temperature T_c should be depressed by $40°C$ when $\eta=0.0257$ in the Al-Zn system and by $2000 °C$ when $\eta=0.15$ in the Au-Ni system. Inclusion of the effect of elastic anisotropy in cubic crystals yields that when the elastic modulus has its lowest value parallel to <100> then the coherency strain energy term will be least for {100} plane waves. This result applies to most cubic materials. One exception is Mo for which the elastic modulus is smallest parallel to <111> and in this case {111} plane waves are observed. Spinodal decomposition is not limited to metals. It has been observed in glasses[3] and polymers[4], as well.

4. COHERENT EQUILIBRIUM.

Concomitant with spinodal decomposition as inherent characteristics are the existence of coherent interfaces between the developing phases and the existence of stresses set up by the difference in lattice parameters between these phases. Spinodal decomposition may lead to a metastable equilibrium, such as that between the coherent and stressed phases, where each phase has a constant composition. Such metastable coherent equilibrium does not obey the same rules as do the "normal" incoherent phase equilibrium we have studied up to this point. In particular, away from critical points, where the compositions of the coexisting phases differ, it is no longer true that the solvus compositions are defined by the equality of the chemical potentials of each species! This statement is a consequence of the coherency stresses, the fact that transfers of atoms from one phase to the other change the stress fields, and the fact that the total free energy depends upon both the composition and stress fields. For incoherent equilibrium, these transfers of atoms occur at constant intensive parameters, such as pressure, leading to the equality of chemical potentials corresponding to these constant intensive parameters. Transfer of atoms at constant stress between coherent coexisting phases is not a realistic condition.

Coherent phase equilibrium above a temperature called the Williams point exhibits no two phase equilibrium. Below this temperature there is a two phase equilibrium, but the compositions of the coexisting phases ("solvus") can be within and outside those for the stable incoherent equilibrium. However, not all average compositions between the "solvus" compositions yield a two phase equilibrium. These are strange relations to anyone who views the rules of stable equilibrium as normal. The effect of stress on coherent equilibrium is currently being actively researched. These concepts are likely to find significant application in the area of strained layer superlattices (quantum well devices) and the like. The interested reader is referred to the bibliography listing for coherent equilibrium at the end of this chapter.

5. HOMOPHASE STRAIN "WAVES" IN A METASTABLE HOST PHASE.

In a situation completely analogous to the case of spinodal decomposition treated in the previous section, it is possible to have a metastable host phase produce a strain "wave" which decreases the free energy of the system. In this case, instead of the specific free energy being a function of the composition it is a function of the strain and strain gradient. The equation governing the change in free energy of the system is the same as (5.32) with strain replacing the composition in the equation. The dependence of the specific free energy on the strain obviously is a function of the tensor character of the strain. We have mentioned "soft phonon modes" in a previous chapter. Such modes are likely to be involved in the embryos being considered. The strain leads to a decrease in the free energy because it produces a configuration of the atoms that has a lower free energy than that of the metastable host phase.

Evaluation of the liklihood that homophase fluctuations of the latter type can exist requires some way of evaluating the gradient factor K. Recent first-principle calculations suggest that K may be small enough to allow for a reasonable probability of detecting such fluctuations in strain.[5] Indeed, there is a well known phenomenon, called "tweed", that may well be a manifestation of these fluctuations, although it has not yet been proved that any of the "tweed" phenomena represent homophase fluctuations of the type under consideration.

6. SUMMARY

In principle, fluctuations of phase in homogeneous stable or metastable hosts initially lead to a lowering of the free energy of the system before increase in their population increases the latter. However, in practise, the equilibrium concentration of heterophase fluctuations, called embryos, is so small as to be less than unity in a macroscopic specimen, except for the special cases of either a very small embryo-host interface energy (<20 ergs/cm^2) or a very high supercooling or supersaturation of the host phase. Heterophase embryos are more likely to exist at interfaces, dislocations or other defects that lower the barrier to their formation.

It may or may not be possible to develop observable homophase concentration fluctuations in metastable solid solutions. The liklihood of producing such fluctations increases the closer is the composition of the solid solution to the composition at which the solid solution becomes unstable. Contrary to the case of heterophase fluctuations, there exist conditions for the spontaneous generation of composition fluctuations that may grow. One mode of growth involves periodic "waves" of composition and leads to the spinodal decomposition of unstable solutions. This mode does not involve growth of "critically" sized embryos that exist in thermodynamic equilibrium. Rather, the decomposition occurs uniformly throughout the unstable host solution and proceeds by the increase in wave amplitude (the composition deviation from that of the host solution) with time. Analogously, spinodal type strain fluctuations leading to more stable atomic configurations than the host may occur in principle with diffuse interfaces involving gradients in the strain. In practise, some of the phenomena known as "tweed" may be a manifestation of such fluctuations.

REFERENCES.

1. F.R.N. Nabarro, Proc.Roy.Soc.A175,519(1940);Proc. Phys. Soc.52,90(1940).
2. L.E.Murr,INTERFACIAL PHENOMENA IN METALS AND ALLOYS, Addison-Wesley,

Reading, Mass,1975, Table 3.8,p.142.

3. M. Tomozawa, R.K. McCrone and H. Herman, Phys. Chem. Glasses 11,572(1970).

4. H.L. Snyder and P. Meakin, J. Chem. Phys. 79,5588(1983).

5. G.B.Olson and M. Cohen,J. Metals 37,36(Abstracts)(1985).

BIBLIOGRAPHY.

1.M.Volmer and A.Weber, Z.Phy. Chem.119,277(1925).

2.R.Becker and W.Doring, Ann.Phys.24,719(1935).

3.J. Frenkel, KINETIC THEORY OF LIQUIDS, Oxford University Press, 1946; Dover, New York,1955.

4. J. Cahn and J.E. Hilliard, J.Chem.Phys 28, 258(1958); 31, 688(1959); Acta Met.9, 795(1961); 10, 179(1962).

COHERENT EQUILIBRIUM

5. R.O. Williams, Metall.Trans.11A, 247(1980).

6. R.O. Williams, Calphad 8, 1(1984).

7. J.W. Cahn and F. Larche, Acta Met.32, 1915(1984).

8.F.C. Larche, Ann. Rev. Mat. Sci.20, 83 (1990).

PROBLEMS

1. In the heterogeneous nucleation of a stable phase from a supersaturated or supercooled parent phase what energies act as barriers to the nucleation process?

2. Why does the radius of the critical spherical nucleus depend upon temperature?

3. How does the number of embryos of a given radius depend upon temperature?

4. On the assumption of isotropic elastic properties and an incoherent interface between nucleus and matrix explain why certain nuclei are plate shaped and others are spherically shaped.

5. Do microscopic precipitates necessarily have the same shape as the nuclei from which they formed? Explain your answer.

6. What is the chemical potential of a vacancy at equilibrium? Is there a real chemical potential or is it an erroneous concept? Explain your answer.

7. Why are grain boundaries preferred sites of nucleation in solids?

8. In the catalysis of nucleation in supersaturated or supercooled matrices why are certain catalyst particles more effective than others?

9. If at the temperature of precipitation, the curvature of the free energy-composition function of the supersaturated matrix phase at its corresponding composition is positive will there be a barrier to the development of composition fluctuations? Explain your

answer.

10. Give the analytic definition of the spinodal.

11. In a miscibility gap binary system, why does the minimum possible thickness of the interface between the two terminal phases depend upon the maximum difference between the free energy of the supersaturated solution and that given by the common tangent to this free energy-composition curve representing the free energy of the mixture of equilibium compositions?

12. What is the basic assumption concerning the free energy of a non-homogeneous phase that allows an analytic description of the energy of a diffuse interface?

13. If we define the generalized spinodal to represent the temperature-composition function in a phase diagram that separates the region where a composition fluctuation in the supersaturated solid solution increases the free energy of the system from that in which such a fluctuation decreases this free energy how will the temperature corresponding to a given composition at the generalized spinodal boundary behave a) as the wave length of the composition fluctuation decreases and b) as the linear expansion per unit increase in composition decreases?

14. How will elastic anisotropy of a supersaturated solid solution affect its spinodal decomposition?

15. Consider the possible "spinodal transformation" where the dependence of free energy on composition is replaced by one on lattice strain. For the host to be able to spontaneously transform at constant concentration to a more stable structure, what condition must be satisfied? If none of the elastic moduli of the parent phase equals or is less than zero, can the parent phase spontaneously transform to the product phase spinodally? Justify your answers.

16. Using the material parameters in the example illustrated on p.142, at what degree of supercooling would more than one solid nucleus exist at equilibrium in a metastable liquid sample of 1 cu.cm.?

VI-THERMODYNAMICS OF DEFECTS

INTRODUCTION.

Defects in solids exert significant effects in kinetic phenomena, as well as on the thermodynamic properties of semiconductors and ionic materials. We consider the thermodynamics of defects in this chapter. Apllications involving defects are considered in later chapters.

1. MONATOMIC SOLIDS.

1.1. Point defects.

The defects that we consider in this section are vacancies and interstitials. Figure 6.1 defines these defects and also makes the point that a vacancy is sometimes called a Schottky defect while the combination of the two defects (Figure 6.1 c) is called a Frenkel defect. Let us determine the concentration of vacancies that may be expected in a monatomic solid containing N atoms at equilibrium. By the definition of equilibrium, the Gibbs free energy must be at a minimum with respect to a variation in the concentration of vacancies. Hence, we formulate the Gibbs free energy of the monatomic solid containing vacancies. If the increase in free energy of the solid on the introduction of one vacancy into some particular site in the lattice by transfer of the atom at this site to a surface site[*] is g_v, then the free energy of the system containing N_v vacancies is

---

[*] The surface site must be at an inexhaustable source or sink for vacancies and yields the lowest value of g_v for such sites.

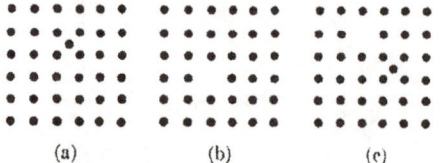

Figure 6.1. Examples of point defects: a) interstitialcy; b) vacancy(Schottky defect); c) displacement (Frenkel defect).

$$G = N_v g_v + Ng - TklnW \qquad (6.1)$$

where g is the free energy per atom in the absence of vacancies and W is the number of different complexions of N_v vacancies and N atoms (see equation 1.2b and note the change in symbol to prevent misinterpretation of the symbols).

Now, for a similar situation we have shown that

$$W = (N+N_v)!/[N_v!N!] \qquad (6.1a)$$

Since we are concerned with the equilibrium situation we must minimize the free energy of the system of atoms and vacancies, noting that vacancies are formed at constant N. Hence,

$$\partial G/ \partial N_v|_N = 0 = g_v + kTln[N_v/(N+N_v)] \qquad (6.2)$$

In obtaining this result we have made use of Stirling's approximation for the factorial of a large number, substituted the result into (6.1) and then carried out the differentiation. The reader's attention is called to the fact that the left hand side of (6.2) is just the definition of the chemical potential of a vacancy and hence it has just been shown that the latter is equal to zero at equilibrium.* Since $N_v \ll N$ then (6.2) yields the following result for the number of vacancies at equilibrium in a monatomic solid

$$N_v = N \exp(-g_v/kT) \qquad (6.3)$$

The work to form a single vacancy has been measured for many monatomic solids. Table 6.1 lists many of these values. Substitution of these values and the corresponding melting temperatures yields the result that the equilibrium concentration of vacancies in monatomic solids at the melting point is

---

* This statement is not strictly true as will be shown later is this section.

Table 6.1.

Vacancy Formation Enthalpies in Metals
(eV)

Metal	Enthalpy (±0.05, except where indicated)
Al(fcc)	0.71
Cu	1.22
Au	0.95
Ni	1.67 ±0.12
Pb	0.57 ±0.07
Pt	1.32 ±0.17
Ag	1.14
Fe(bcc)	1.5 ±0.1
Mo	3.1 ±0.1
Nb	2.65
Ta	2.95 ±0.15
W	3.5 ±0.5
Cd(hcp)	0.44
Co	1.34
In	0.49 ±0.1
Mg	0.58
Zn	0.52 ±0.02

between 0.0001–0.001.

A relation equivalent to (6.3) holds also for the number of interstitialcies in monatomic solids. However, it is changed slightly from (6.3) because the number of sites that interstitialcies can occupy is not equal to the number of atoms. In general there will be greater than one interstitialcy site per atom. Let this number be p. Thus, (6.1a) must be replaced by

$$W = (pN)!/[n!(pN-n)!]$$

where n is the number of interstitialcies. This leads to the following result for the number of interstitialcies at equilibrium

$$n = pN \exp[-g_i/kT] \qquad (6.4)$$

There are fewer of these defects than vacancies for the simple reason that the work to form interstitialcies, g_i, by exchange with an atom at a surface site is larger than that to form vacancies in these materials.

The presence of vacancies affects not only the free energy but also all the other thermodynamic quantities. It would require a much more detailed discussion than is warranted by the scope of this book to describe all these relationships. A thorough discussion of them can be found, however, in reference 1. Divacancy "molecules" may also be an equilibrium entity in monatomic solids. Indeed, it is believed that divacancies contribute measureably to diffusion in the vicinity of the melting point. We may derive their concentration at equilibrium using (6.1). However, the expression given above for W must be altered in the present case because a given divacancy has several possible distinguishable orientations of one vacancy about the other fixed vacancy member of the divacancy pair. Thus, for the divacancy case

$$W = (zN/2+N_{2v})!/[(N_{2v})!(zN/2)!]$$

where z is the coordination number. This result then yields for the number of divacancy pairs, to a good approximation,

$$N_{2v} = zN/2 \exp(-g_d/kT) \qquad (6.5)$$

at equilibrium in a monatomic solid (i.e. $N_{2v} \ll zN/2$). It should be noted that the atoms involved in the formation of a divacancy exchange with vacancies at sites on the surface.

In the above derivation we have not allowed the divacancies to occupy any of the lattice sites already occupied by single vacancies. Thus, the total crystal contains $N+N_v+2N_{2v}$ total number of sites and (6.3)-(6.5) are in error to the extent that the extra number of sites in the crystal differs with respect to the total number of atoms. We shall neglect this error because it is small. Nevertheless, this physical situation suggests that we must reexamine our definition of chemical potential of a vacancy. It should now be apparent that the definition of a chemical potential of a species requires the variation of the free energy with respect to the number of that species at constant number of other species and other independent variables, such as T and P. However, the species in question must be an independent variable. Since, vacancies and clusters of vacancies can all exist in a crystal specimen of N atoms, they are no longer independent variables (i.e. the presence of one diminishes the number of sites available to the other by the number of sites that are nearest-neighbors to the former defects.) Nevertheless, it is useful to consider these partial derivatives with respect to a defect species as virtual chemical potentials because they obey certain relations characteristic of chemical potentials. In particular, the virtual chemical potentials of the vacancy and vacancy cluster defects are equal to zero at equilibrium.

It is possible for two vacancies to become nearest-neighbors and hence form a divacancy pair. At equilibrium, this reaction obeys the mass action law, according to which for dilute concentrations of defects,

$$(N_{2v}/N)/(N_v/N)^2 = K,$$

where K is an equilibrium constant for the reaction. Substitution from (6.3) and (6.5) yields

$$K = (z/2)\exp[-(g_d - 2g_v)/kT]$$

where the expression in parentheses in the exponent represents the binding free energy of the divacancy.

It should be noted that in the above the vacancies and interstitials were formed by exchange with sites on surfaces. For such sites the free energy of formation of the defects represents the minimum possible <u>for an inexhaustable type of surface site</u> and hence these sites are likely to control the actual number of such defects present in crystals. These defects could have been formed by exchange with sites(atoms) in the vapor state. It is apparent that the work to form the defects for the latter case is larger per atom in the defect by the free energy of sublimation (i.e. the free energy required to remove an atom at a kink site along a ledge on the surface and transfer it to the vapor.)

Many other defects and defect clusters can exist at equilibrium in monatomic solids. For example, rather than there being a well defined interstitialcy, relaxation can lead to the formation of a split interstitial, which is a defect that distributes the distortion due to an interstitialcy over two atoms centered about a lattice site. In solutions, there will be association of solute atoms and defects. In general, the numbers of such defects will be given by an equation of the form of (6.4) or (6.5), where N is multiplied by some geometric factor.

1.2. Electronic defects.

1.2.1. Intrinsic semiconductors.

Semiconductors and covalent bonded monatomic solids may also, in addition to the defects noted above, contain charged entities as defects, at equilibrium. In this case, there exists the constraint of overall electroneutrality for the solid. In a pure semiconductor, the intrinsic number of conduction electrons and holes must therefore balance. The number of conduction electrons at equilibrium is affected by the Pauli exclusion principle, in that the energy levels are quantized and no more than two electrons (of opposite spin) can have the same energy. Thus, it is no longer possible to use equation 1.2b to evaluate the entropy contribution due to the conduction electrons, but instead it is necessary to use equation 1.2a. However, it is also necessary to make use of band theory to aid in solving (1.2a). In particular, it is necessary to know how many energy states $(N(E)dE)$ there are between an energy E and E+dE. This number corresponds to the number of boxes available to be filled by electrons having this energy. However, each energy level may be filled by no more than two electrons of opposite spin.

It is convenient to consider the problem of determining the configurational entropy of the conduction electrons using the electrons of one spin quantum number only. The total number of ways of mixing such electrons so that there is no more than one electron per level is given by

$$W = \prod_i N_i!/[n_i!(N_i-n_i)!]$$

where $N_i=N(E_i)dE$ and n_i is the number of electrons of one type spin in the ith increment of dE at E_i.

To obtain the equilibrium distribution of electrons we now maximize the entropy at constant volume, V, energy of the system , U, and total number of electrons, n, i.e.

$$dS = kdlnW = 0 = kln(N_i/n_i-1)dn_i$$

where we have made use of Stirling's approximation. But, from thermodynamics

$$0 = dS = \partial S/\partial U|_{V,n}dU + \partial S/\partial n|_{U,V}dn = (1/T)dU - (\mu/T)dn$$

where μ is the chemical potential of the electrons in this case. Hence, we may write

$$0 = dS = k\ln(N_i/n_i-1)dn_i = (1/T)dU - (\mu/T)dn$$

or $$k\Sigma\ln(N_i/n_i-1)dn_i-(1/T)dU+(\mu/T)dn = 0$$

But, since $U = \Sigma n_i E_i$ and $n = \Sigma n_i$ we have $dU = \Sigma E_i dn_i$ and $dn = \Sigma dn_i$. Substituting, we then obtain the result that

$$k\Sigma[\ln(N_i/n_i-1)-E_i/T+\mu/T]dn_i = 0$$

But, this can only be satisfied for all variations of n_i by

$$n(E)dE = N(E)dE/[1+\exp(E-\mu/kT)] \tag{6.6}$$

where $n(E)dE$ has been put in place of n_i and $N(E)dE$ has been substituted for N_i (i.e. the subscript i has been removed.) Here $n(E)dE$ is the number of electrons of one type spin that have an energy in the range between E and E+dE. For $(E_c-\mu)>>kT$ (see Figure 6.2 for definition of E_c) then the total number of electrons (of both types of spin) in the conduction band can be shown to equal

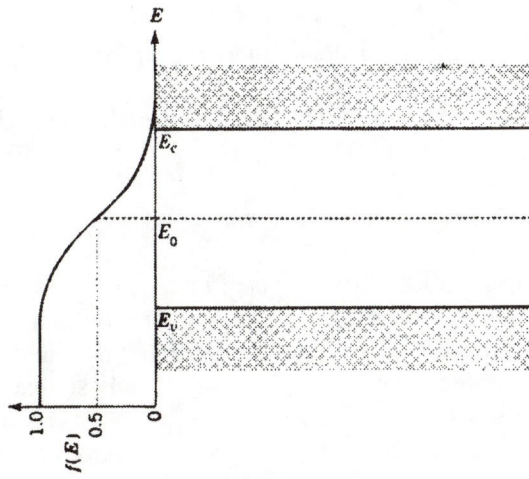

Figure 6.2. Illustration of energy levels in intrinsic semiconductor. After L.V. Azaroff and J.J. Brophy, ELECTRONIC PROCESSES IN MATERIALS, McGraw-Hill, N.Y. 1963.

$$n = N_c\exp[-(E_c-\mu)/kT] \qquad\qquad (6.7)$$

where $N_c=2(2\,\pi m_c kT/h^2)^{3/2}$, m_c is the effective mass of the electron in the conduction band, and h is Planck's constant. If $\exp(E_c-\mu)/kT)$ is on the order of unity, then the 1 in the denominator of (6.6) cannot be neglected and (6.7) must be modified. In the latter case, the electron distribution is said to be degenerate.

An equivalent relation can be derived for the number of holes, n_+, by recalling that a hole corresponds to an empty state and hence that the probability that a state is empty is equal to $1-F(E)$, where $F(E)$ is given by

$$F(E) = 1/[1+\exp(E-\mu/kT)]$$

(Reference to (6.6) indicates that $n(E)dE=F(E)N(E)dE$.) Thus, replacing $F(E)$ by $1-F(E)$ in the latter relation yields a relation similar to (6.7) except that the term in the parenthesis in the exponent is replaced by $(\mu-E_v)$ and the effective mass of the electron in the conduction band is replaced by that in the valence band. We may now relate the chemical potential of the electrons to the Fermi level by noting that the latter is defined to equal the energy E when $F(E)=0.5$. Substitution in $F(E)$ then yields that the Fermi energy $E_f=\mu$. (See Figure 6.2.)

To determine the actual position of the Fermi level we equate the total number of electrons to the total number of holes. This yields

$$E_f = (E_v+E_c)/2 + (3kT/4)\ln(m_v/m_c)$$

If the two effective masses are equal then the Fermi level is in the middle of the gap.

1.2.2. Extrinsic semiconductors.

In extrinsic semiconductors it is of interest to know how the Fermi level depends upon the donor and/or acceptor concentrations. We will find that the limit of solubility of a solute is dependent upon the position of the Fermi level. We assume that classical statistics apply so that we can make use of the law of mass action. In effect, this assumption corresponds to the case where the difference in energy between the donor or acceptor levels and the Fermi energy is much greater than zero. In this case, the relations are much simplified. In any case we have the requirement of electrical neutrality, according to which the number of electrons plus ionized acceptors equals the sum of the number of holes and ionized donors. Now, it can be shown that the number of ionized species is given by

$$n_{s*} = n_s / [2\exp\{(E_s - E_f)/kT\} + 1] \qquad (6.8)$$

where E_s is the energy level of the donor or acceptor, as s may represent.

We can consider three limiting cases, as follows.

1. Both acceptor and donor impurities present, but weakly ionized.—In this case, setting the electoneutrality condition yields

$$E_f = (E_d + E_a)/2 + (kT/2)\ln(m_v/m_c) + [(E_c - E_d) - (E_a - E_v)]/2 \qquad (6.9)$$

For the assumed case in which classical statistics applies the third term on the right hand side is small compared to the first term and if the effective masses are equal then the Fermi level is near the center of the gap as in an intrinsic semiconductor.

2. Only Donor Solute Present.— In this case, for the assumed conditions, the number of electrons is given almost wholly by the number of ionized donors, i.e. the contribution of intrinsic electrons from valence band excitation will be negligible. Thus, setting (6.7) equal to (6.8) and solving for the Fermi level yields

$$E_f = E_c + kT\ln(n_d / \{2[2 \pi m_c kT/h^2]^{3/2}\}) \qquad (6.10)$$

where n_d is the concentration of donors. Thus, the Fermi level is close to the conduction level provided that the temperature is not too high.

3. Only Acceptor Type Solute Present.——Proceeding in a manner analogous to the case just described, the following result for the Fermi level is obtained

$$E_f = E_v - kT\ln(n_a / \{2[2 \pi m_v kT/h^2]^{3/2}\}) \qquad (6.11)$$

where n_a is the concentration of acceptors. Hence, in this case the Fermi level is close to the valence level, if the temperature is not too high. Figure 6.3 shows how the Fermi level varies with temperature in n- and p-type semiconductors.

Suppose that we have a semiconductor that is equilibrated with respect to a reservoir of solute that maintains the chemical potential of the latter constant. At equilibrium, the change in free energy of the system must be equal to zero. Hence, for transfer of dn_s solute to the semiconductor,

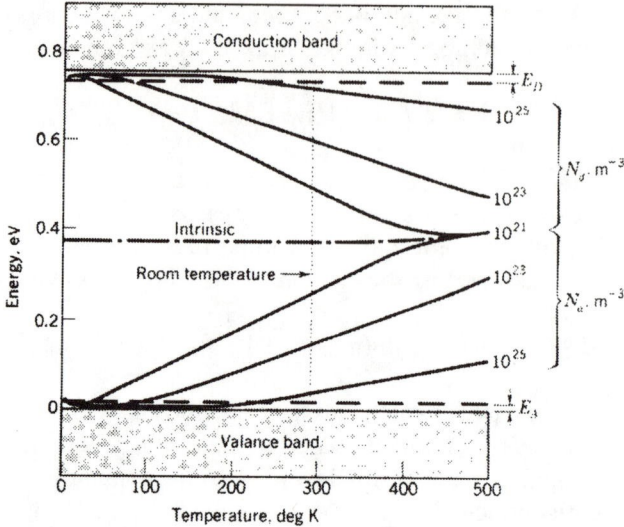

Figure 6.3. Illustrating the temperature dependence of the Fermi energy in an extrinsic semiconductor. After L.Azaroff and J.J. Brophy, ELECTRONIC PROCESSES IN MATERIALS, McGraw-Hill, NY, 1963.

which produces for the case of a donor solute the same number of ionized solute atoms and electrons, we have

$$\mu_s^o = \mu_s^+ + \mu_{el}$$

Now, μ_s^o is maintained constant. Consequently, any change in μ_{el}, which is equal to the Fermi level must bring about the opposite sign change in μ_s^+. Since the latter is monotonically related to the concentration of the solute in solution, the latter must be similarly changed by the change in Fermi level, i.e. an increase in the Fermi level will decrease the concentration of the solute that is in equilibrium with an external reservoir and vice versa. In particular, adding a donor solute to a solid solution that already contains acceptors at the solid solubility level will raise the Fermi energy and decrease the solid solubility of the acceptor type solute. On the other hand, adding an acceptor type solute to a solution that already contains donor type solute at its solid solubility level will decrease the Fermi energy and thereby raise the solid solubility of the donor type solute. It can be shown that the ratio of the solubility of a donor solute in the presence of ternary solutes, X_d, to that in the absence of ternary solute, X_d^o, is given by

$$X_d/X_d^\circ = \exp[(E_F^\circ - E_F)/kT]$$

Similarly, for an acceptor solute, this relation becomes

$$X_a/X_a^\circ = \exp[-(E_F^\circ - E_F)/kT]$$

Only a slight change in the Fermi level can exert orders of magnitude change on the solubility.

We do not present here a complete description of the thermodynamics of the electrical defects in semiconductors for it involves matters outside the scope of this book. The interested reader is referred to reference 1, the book by Swalin and the articles by Reiss cited in the bibliography.

2. COMPOUNDS.

2.1. Defects in stoichiometric compounds.

In ionic compounds there are two constraints governing the defects that may form in such materials. One is that the ratio of the number of cation to anion lattice sites is a constant and equal to the stoichiometric ratio of cations to anions, even for non-stoichiometric compositions. Thus, for the compound MX, the ratio of cation to anion lattice sites is unity even for off stoichiometric compositions. The second constraint is that the solid must be electrically neutral. To achieve electroneutrality the total number of positive charges must equal the total number of negative charges.

In ionic compounds the equivalent of the vacancy defect is the Schottky defect, which consists of a combination of cation and anion vacancies obeying the electroneutrality condition. This configuration obeys the electroneutrality requirement because the effective positive charge on the anion vacancies is balanced by the effective negative charge on the cation vacancies. The effective charge on a vacancy is due to the absence of the charge on the lattice site that would normally be present if an ion occupied that site. Thus, the sign of the effective charge of an anion vacancy is positive. Schottky defects can form, as in metals, by exchange of a lattice sited cation and anion with equivalent vacant surface lattice sites.

The equivalent of the interstitialcy is the Frenkel defect. This defect consists of a vacancy-interstitial pair, which can be formed by the jump of an ion from a lattice site into a neighboring interstitial site. This defect is obviously electrically neutral.

Let us consider the number of Schottky pair defects and Frenkel pair defects that may exist at equilibrium in a stoichiometric crystal of formula MX. For the case of the Schottky defect consider that there are a total of 2N atoms and that the Schottky defect is produced by exchange of interior ions with surface sites. Thus, mixing of vacancies having a given sign with their equivalent ions on the sites belonging to these ions contributes

$$\Delta S = k \ln[(N+n)!/\{n!N!)]$$ (6.12)

to the entropy of the system. Equation (6.12) also represents the contribution of mixing the vacancies having the opposite sign on their sites. Hence, the total change in entropy for the Schottky defects is twice that given by (6.12). If g_p is the work to form one Schottky pair of vacancies then the change in free energy of the system due to the production of n Schottky pairs randomly distributed in the lattice is

$$\Delta G = ng_p - T2 \, \Delta S$$

where ΔS is given by (6.12). Minimizing the free energy with respect to n and using Stirling's approximation yields

$$n = (N+n) \exp(-g_p/2kT)$$ (6.13)

for the equilibrium number of Schottky pairs.

For the case of Frenkel defects, the mixing entropy involves a somewhat different relation to that for the Schottky defect. Again assume that there are N ions, some of which will move to interstitial sites to form n Frenkel defect vacancy-interstitial pairs. In this case, the contribution to the mixing entropy due to the vacancies is

$$S_v = k \ln\{N!/[(N-n)!n!]\}$$ (6.14)

If there are p interstitial sites per Frenkel ion site then the contribution to the mixing entropy due to the interstitial ions is

$$S_i = k \ln\{(pN)!/[(pN-n)!n!]\}$$ (6.15)

Proceeding as for the case of the Schottky defect leads to the following result for the number of Frenkel defects at equilibrium

$$n = [(N-n)(pN-n)]^{1/2}\exp(-g_F/2kT) \tag{6.16}$$

where g_F is the work to form one particular Frenkel defect pair.

The Schottky and Frenkel defects are the ones usually found in stoichiometric ionic compounds. However, other types of defects can also occur in such materials. For example, there can be an exchange of ions such that cations sit on anion sites and vice versa, or if one type of ion sits on a wrong site it is compensated by a vacancy on its own site or by the other type of ion in an interstitial site, etc. Indeed, additional types of complex defects have been found and with time and continuing research it is likely that new defects will be discovered.

Some measure of the liklihood of the existence of a given type of defect in some stoichiometric compounds can be obtained from the data given for the enthalpies of formation of the various defects, which are listed in Table 6.2. If the enthalpies of a given defect have been measured for some compound it is reasonable to assume that they are the majority defect in that compound.

Electronic defects will be present in compound as well as monatomic insulators or semiconductors. The relations governing their concentration have been derived in a previous section. It is of interest to note for the case of small concentrations of electrons and holes, that from equation (6.7) and the equivalent one for holes, the product of the electron and hole concentrations will be exponentially dependent upon the energy gap. Table 6.3 lists the energy gap values for some pure stoichiometric compounds and the corresponding concentrations of electrons and holes.

2.2. Non-stoichiometric compounds.

Nonstoichiometry can be achieved by having an excess (or deficiency) of some component relative to the stoichiometric composition. Because the stoichiometric ratio of lattice sites is conserved, nonstoichiometry is equivalent to the presence of point defects. Electroneutrality in such a case is preserved by the formation of complementary electronic defects. For example, an anion deficient compound can have either anion vacancies or cation interstitials maintain the ratio of lattice sites for the compound. If anion vacancies are formed then their effective positive charge is balanced by extra electrons associated with the cations. However, if cation intersti-

Table 6.2
Enthalpies of Defect Formation in Ionic Crystals

	Enthalpies of Defect Formation[a]		
Crystal	ΔH (kJ/g-atom)	Crystal	ΔH (kJ/g-atom)
Schottky Defects			
LiF	225–258	KI	154
LiCl	212	CsCl	179
LiBr	173	CsBr	193
LiI	129	CsI	183
NaCl	210–229	TlCl	125
NaBr	166	$PbCl_2$	150
KCl	218–222	$PbBr_2$	135
KBr	221–244		
Frenkel Defects (Cation Sublattice)			
AgCl	119, 139	AgBr	102
Frenkel Defects (Anion Sublattice)			
CaF_2	220–270	BaF_2	180
SrF_2	220		

[a] From tabulation by L. W. Barr and A. B. Lidiard, in *Physical Chemistry, An Advanced Treatise*, Vol. X, Academic Press, New York, 1970.

tials are formed, then the extra electrons, necessary to balance the effective positive charge of the cation interstitials, are associated with the lattice sited cations. In such a case, the number of point defects is not determined solely by thermal equilibrium of the defects, but by the nonstoichiometry as well. Let us consider one example to illustrate the thermodynamics of one such situation.

We consider the case of an anion deficient compound for which the anion vacancy is the most probable associated defect. If the nonstoichiometry is accomplished by equilibium of the compound with some partial pressure of the anion vapor and if the compound has the formula MX and the molecules of X in the vapor phase are normally diatomic (i.e. X_2), then the transfer of an anion from a lattice site to the vapor produces an anion vacancy. The equation for this reaction can be written as follows

Table 6.3

Band Gap° and Approximate Concentrations of Electrons and Holes in Pure, Stoichiometric Solids

Crystal	E_g (eV)	$n \approx 10^{19} \exp\left[-\dfrac{E_g}{2kT}\right]$ electrons/cm^3			Temp (°K)
		Room Temp	1000°K	Melting Point	
KCl	7	10^{-40}	20	150	1049
NaCl	7.3	10^{-43}	4	70	1074
CaF$_2$	10	10^{-66}	10^{-6}	10^3	1633
UO$_2$	5.2	10^{-25}	10^6	10^{15}	3150
NiO	4.2	10^{-16}	10^8	10^{13}	1980
Al$_2$O$_3$	7.4	10^{-44}	2.0	10^{11}	2302
MgO	8	10^{-49}	0.01	10^{12}	3173
SiO$_2$	8	10^{-49}	0.01	10^8	1943
AgBr	2.8	10^{-5}	10^{12}	10^9	705
CdS*	2.8	10^{-5}	10^{12}	10^{15}	1773
CdO*	2.1	20	10^{13}	10^{16}	1750
ZnO*	3.2	10^{-8}	10^{11}	10^{14}	1750
Ga$_2$O$_3$	4.6	10^{-20}	10^7	10^{13}	2000
LiF	12		10^{-11}	10^{-8}	1143
Fe$_2$O$_3$*	3.1	10^{-7}	10^{11}	10^{14}	1733
Si	1.1	10^{10}	10^{16}	10^{17}	1693

°Most of the data are based on the optical band gap, which may be larger than the electronic band gap.

*Sublimes or decomposes.

$$X_X <=> V_X + 1/2 X_2 \qquad (6.17)$$

In this equation the anion vacancy is neutral. Depending on the temperature, the trapped electrons associated with the vacancy may be excited and freed from the vacancy, with the result that the following equilibrium reactions are obeyed

$$V_X <=> V_{X\cdot} + e' \qquad (6.18)$$

$$V_{X\cdot} <=> V_{X\cdot\cdot} + e' \qquad (6.19)$$

(The absence or presence of dots (..) associated with the defect indicates the absence or presence of effective positive charge. Thus, with the absence of effective positive charge the defect is electrically neutral. Also, the presence of a prime (') indicates the presence of an effective negative charge.) The free electrons indicated in these equations are associated with the cations on their normal sites and hence these equations may be rewritten as

$$M_M + V_X \iff V_X{}^{\cdot} + M_{M'} \tag{6.20}$$

$$M_M + V_X{}^{\cdot} \iff V_X{}^{\cdot\cdot} + M_{M'} \tag{6.21}$$

In the latter equations M type cations are changed to M' cations with one effective negative charge. The valence of the M cations is changed from +2 to +1 in these reactions.

In addition to the above equations, another is involved in the intrinsic formation of electronic defects, namely, the formation of electrons and holes by excitation of electrons from the valence band to the conduction band of the solid, which we have considered in a previous section.

All of the above reactions are governed by equilibrium constants and the mass action law. The latter corresponds to equations of the type illustrated by (6.13) and (6.16), whereas the equilibrium constants are given by the reciprocal of the exponential terms in these equations. Thus, the concentration of the various defects denoted by the terms in brackets [] are

$$[V_X](p(X_2))^{1/2} = K_1[X_X] \tag{6.22}$$

$$[V_X{}^{\cdot}]n = K_2[V_X] \tag{6.23}$$

$$[V_X{}^{\cdot\cdot}]n = K_3[V_X{}^{\cdot}] \tag{6.24}$$

where n equals the concentration of electrons. If, as assumed, anion vacancies and the complementary electrons are the most numerous defects then electroneutrality requires

$$n = [V_X{}^{\cdot}] + 2[V_X{}^{\cdot\cdot}] \tag{6.25}$$

Substitution into the above relations and solution for separate variables yields for the electron concentration the equation

$$n^3 = K_1 K_2 (p(X_2))^{-1/2} (2K_3 + n) \tag{6.26}$$

There are two limiting solutions for n as follows

for $n \gg 2K_3$, $\qquad n = (K_1 K_2)^{1/2} [p(X_2)]^{-1/4}$ (6.27)

and for $n \ll 2K_3$, $\qquad n = (2K_1 K_2 K_3)^{1/3} [p(X_2)]^{-1/6}$ (6.28)

Similarly, there are several limiting solutions for the total number of anion vacancies depending upon which of the vacancies is the predominant species, i.e. doubly charged, singly charged or neutral ones. For the case that neutral vacancies predominate then the total anion vacancy concentration is given by $K_1 [p(X_2)]^{-1/2}$. If singly charged vacancies predominate then the total vacancy concentration is equal to $(K_1 K_2)^{1/2} [p(X_2)]^{-1/4}$. If, on the other hand, doubly charged anion vacancies predominate then the total vacancy concentration is given by $[(1/4)K_1 K_2 K_3]^{1/3} [p(X_2)]^{-1/6}$. Examination of these results shows that both the electron concentration and the anion vacancy concentration are dependent on the partial pressure of the anion vapor in equilibrium with the compound.

It is possible to consider other limiting conditions, such as the case of an anion deficient compound in which the predominant defect is the Frenkel defect, or the case of a cation deficient compound in which one or the other defect species predominates, etc.. It is obvious that there are many possible sets of conditions that can exist in any particular material. Table 6.4 lists the types of defects that can be found in oxides. Table 6.5 gives the equilibria describing the formation of typical point defects in oxides based on the criteria of electroneutrality, conservation of mass and maintenance of the proportion between cation and anion sites.

For the case where the mass action law can be applied to the defect reactions, the partial pressure of oxygen in the non-stoichiometric range of composition becomes proportional to $p(O_2)^{-1/n}$. Table 6.6 summarizes this dependence for the various defects. In order to simplify the situation diagrams, called Brouwer or Kroger-Vink plots have been prepared showing the behavior to be expected in a compound that adheres to the conditions assumed. Figure 6.4 illustrates several such diagrams. It is to be emphasized, however, that these diagrams apply only for the assumed conditions. If there is association between defects or there are impurities present or there are multiply charged defects, etc., the diagrams will be affected.[2]

The tendency for the existence of a range of solubility about the stoichiometric composition is affected by the ionization potential of the cation. For cations with low ionization potential there will be a tendency for an extensive region of solubility. For cations with high ionization

Table 6.4

Defects in Nonstoichiometric Oxides

	Defects	Defect type	Compensating Species	Oxide system
Point defects	Oxygen vacancies	V_O^X, $V_O^{\cdot}$, $V_O^{\cdot\cdot}$	M_M'	MO_{2-x}
	Metal vacancies	V_M^X, V_M', V_M''	$M_M^{\cdot}$	$M_{1-y}O$
	Interstitial oxygen	O_i^X, O_i', O_i''	$M_M^{\cdot}$	MO_{2+x}
	Interstitial metal	M_i^X, $M_i^{\cdot}$, $M_i^{\cdot\cdot}$	M_M'	$M_{1+y}O$
	Substitutional disorder	M_M''	$V_O^{\cdot\cdot}$	
Extended defects	Clusters	Combination of point defects		$Fe_{1-x}O$ UO_{2+x}
	Shear planes - see Chapter (16)	Elimination of point defects		TiO_{2-x} WO_{3-x}

Table 6.5

Defect Equilibria

	Defect reaction	Oxide
$V_O^{\cdot\cdot}$	$2M_M + O_O \rightarrow V_O^{\cdot\cdot} + 2M_M' + \tfrac{1}{2}O_2$	any
V_M''	$2M_M + \tfrac{1}{2}O_2 \rightarrow V_M'' + 2M_M^{\cdot} + O_O$	MO
O_i''	$2M_M + \tfrac{1}{2}O_2 \rightarrow O_i'' + 2M_M^{\cdot}$	any
$M_i^{\cdot}$	$M_M + O_O \rightarrow M_i^x + \tfrac{1}{2}O_2$ $M_i^x + M_M \rightarrow M_i^{\cdot} + M_M'$ <hr> $2M_M + O_O \rightarrow M_i^{\cdot} + M_M' + \tfrac{1}{2}O_2$	MO
$M_i^{\cdot\cdot\cdot}$	$M_M + 2O_O \rightarrow M_i^x + O_2$ $3M_M + M_i^x \rightarrow M_i^{\cdot\cdot\cdot} + 3M_M'$ <hr> $4M_M + 2O_O \rightarrow M_i^{\cdot\cdot\cdot} + 3M_M' + O_2$	MO_2

Table 6.6

Defect Formation: p_{O_2} Dependence

$$M_i^x, \ V_O^x \ \alpha \ p_{O_2}^{-1/2}$$

$$M_i^{\cdot}, \ V_O^{\cdot} \ \alpha \ p_{O_2}^{-1/4}$$

$$M_i^{\cdot\cdot}, \ V_O^{\cdot\cdot} \ \alpha \ p_{O_2}^{-1/6}$$

$$O_i^x, \ V_M^x \ \alpha \ p_{O_2}^{1/2}$$

$$O_i', \ V_M' \ \alpha \ p_{O_2}^{1/4}$$

$$O_i'', \ V_M'' \ \alpha \ p_{O_2}^{1/6}$$

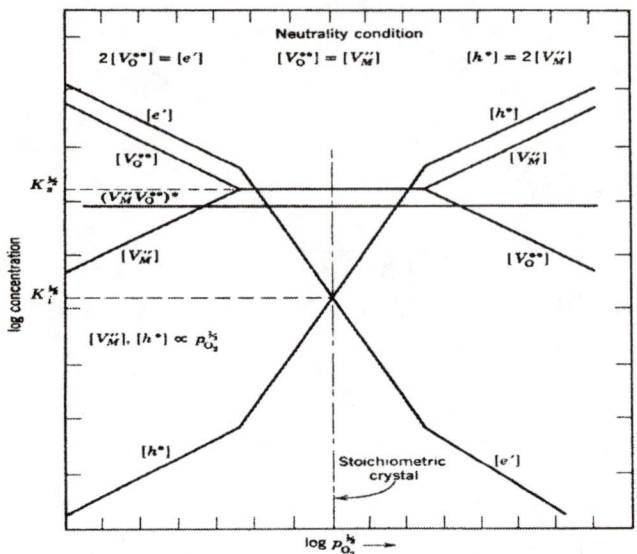

Schematic representation of defect concentrations as a function of oxygen pressure for (a) a pure oxide which forms predominantly Schottky defects at the stoichiometric composition.

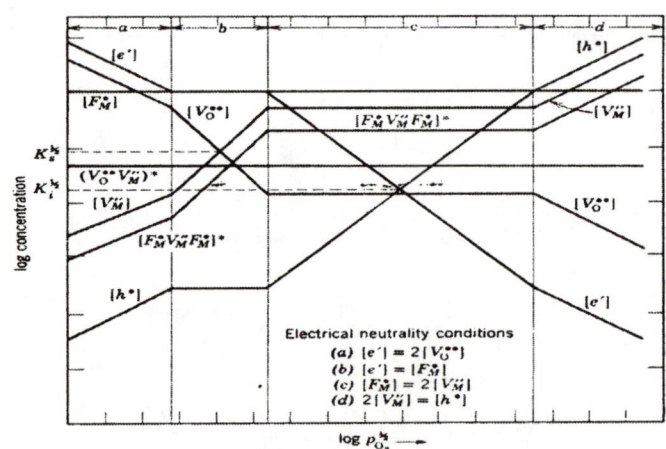

(contd.) (b) An oxide which forms Schottky defects but contains cation impurities $[F''_M] > K^{1/2}$

Figure 6.4 After W.D. Kingery, H.K. Bowen and D.R. Uhlmann, INTRODUCTION TO CERAMICS, J. Wiley & Sons, NY,1976.

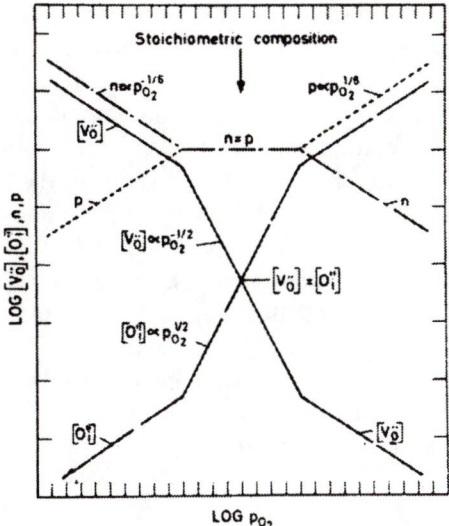

Brouwer plot for $V_O^{\cdot\cdot}/O_i''$ system with intrinsic ionization dominating near the stoichiometric composition.

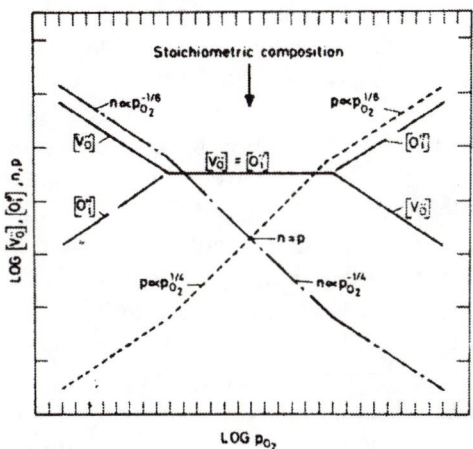

Figure 6.4 cont'd   Brouwer plot for $V_O^{\cdot\cdot}$ system with atomistic disorder dominating near the stoichiometric composition.

After P. Kofstad,NON-STOICHIOMETRY, DIFFUSION AND ELECTRICAL CONDUCTIVITY IN BINARY METAL OXIDES, Wiley-Interscience, NY,1972.

potential the range of solubility about the stoichiometric composition will be limited and non-stoichiometry in these materials is often related to impurity content.

The range of defect possibilities in compounds is enormous. We have alluded to only a few of them. Defects can be ordered, clustered, complexed, sheared, etc..[3] It is beyond the scope of the present simplified discussion to consider them in detail. We have not considered molecular solids in this section, in which another type of defect involving rotational disorder of molecular units can occur, as, for example, is found in polymers.[4] Nor have we considered those solids in which orientational disorder can occur, as, for example, is found to take place in KCN and other salts that contain polyatomic ions.[5] The interested reader must be prepared to devote sufficient time to develop expertese in the subject of defects in solids, which is in the stage of active development.

3. COMPARISON OF DEFECTS IN METALS, SEMICONDUCTORS AND IONIC CRYSTALS

Metals are characterized by having one predominant equilibrium point defect-the vacancy. This situation exists because the work to form a single vacancy is much smaller than that for any other defect or defect complex. In some metals, the work to form a divacancy pair is not much larger than that for the single vacancy and divacancies then complement the single vacancies at high temperature. In semiconductors, the situation is more complicated in that the possible equilibrium point defects can exist in both an electrically neutral and charged condition and in that the lattice is more open than for the other solids. Thus, not only can charged and neutral vacancies exist but because of the open nature of the lattice the work to form an interstitial type defect may not be significantly higher than that for a vacancy type defect. Finally, in ionic solids the complexity of the equilibrium point defect spectrum increases significantly.

The previous discussion suggests that some quantitative knowledge of the work to form point defects can be helpful in ascertaining the equilibrium population of the competing defects. Indeed, although much effort has been expended to distinguish between the defects experimentally, most of the observations cannot be unambiguously interpreted. Thus, a systematic attack on the problem of modelling the point defects energetically has been begun. To date, the most success has been achieved in modelling the defects in ionic solids.[6] Unfortunately, for the case of metals, calculations based on rigorous interatomic potentials have been limited to a few metals to which pseudopotential theory could be applied. Nevertheless, the use of empirical interatomic potentials has been helpful in the discovery of new modes of impurity-point defect complexing and the like.[7]

Summarizing, we have developed the statistical thermodynamic bases for the equilibria governing some of the principal point defects in monatomic and compound solids.

REFERENCES

1. L.A. Girifalco, STATISTICAL PHYSICS OF MATERIALS, J.Wiley, New York, 1973.
2. F.A. Kroger and H.J. Vink in SOLID STATE PHYSICS 3,317(1956); F.A. Kroger,THE CHEMISTRY OF IMPERFECT CRYSTALS, North-Holland, N.Y., 1956.
3. O.T. Sorensen in MASS TRANSPORT IN SOLIDS, eds. F.Beniere and C.R.A. Catlow, Plenum Press, N.Y., 1981;ibid, C.R.A. Catlow.
4. P. Flory, STATISTICAL MECHANICS OF CHAIN MOLECULES. Hanser, N.Y., 1989.
5. N.G. Parsonage and L.A.K. Staveley, DISORDER IN CRYSTALS, Clarendon Press, Oxford, 1978.
6. C.R.A. Catlow and W.C. Mackrodt, eds.,COMPUTER SIMULATION OF SOLIDS, Lecture Notes in Physics 166. Springer-Verlag, Berlin, 1982.
7. P.C. Gehlen, J.R. Beeler,Jr. and R.I.Jaffee, INTERATOMIC POTENTIALS AND SIMULATION OF LATTICE DEFECTS Battelle Inst. Materials Science Colloquia, Plenum, N.Y., 1971.

BIBLIOGRAPHY.

1. R.A. Swalin, THERMODYNAMICS OF SOLIDS, J.Wiley, New York,1962.
2. F.A.Kroger, F.H.Stieltjes and H.J.Vink, Philips Res.Repts 14, 557(1959).
3. P. Kofstad, NONSTOICHIOMETRY, DIFFUSION AND ELECTRICAL CONDUCTIVITY IN BINARY METAL OXIDES, Wiley-Interscience, New York, 1972.
4. H. Reiss, C.S. Fuller and F.J. Morin, Bell System Tech. J.35, 535(1956) and later publications.

PROBLEMS

1. Suppose that the work to form a vacancy by transfer of an atom from a particular lattice site to a kink site along an edge dislocation line is more (less) than that corresponding to a surface site. What would you expect to occur to the dislocations if the specimen is equilibrated at elevated temperature?

2. Evaluate the binding free energy to form a divacancy if the concentration of divacancies is to be the same as that for single vacancies at the melting point of aluminum (660°C).

3. Why is the concentration of vacancies at equilibrium at 300°C in aluminum likely to be different from that of divacancies even if the assumption of question 2 regarding the equality of their concentrations at the melting point is valid?

3. Derive the expression for the equilibrium concentration of solute-vacancy pairs in terms of the work to form one such pair.

4. Suppose that the substitutional solute B present is not mobile at temperature T while the interstitial solute C is mobile at this temperature. Also suppose that the binding energy of a B-C pair is E_b. Calculate the number of such pairs that will exist at this temperature if the atomic fractions of B and C per substitutional atoms present are X_B and X_C, respectively, in a lattice containing N total substitutional atoms.

5. Suppose that the addition of a ternary solute increases the Fermi energy by 0.1eV. What is the ratio of the new solubility to the solubility at 300°K before the addition of the ternary solute, of an acceptor solute equilibrated with respect to a constant chemical potential external reservoir?

6. For the case of an oxide in which oxygen Frenkel defects occur, the oxygen concentration varies over a range of stoichiometry, and the electron and hole concentrations are appreciable, determine the oxygen partial pressure dependence of the various defects under the assumption that a) the concentration of electronic defects is substantially greater than that of Frenkel defects and b) vice versa.

7. Relate the mass action law to the "quasi-chemical" approximation considered in Chapter 2. Indeed, derive this law for the $2V \rightarrow V_2$ reaction using the statistics involved in the Q-C approximation. If quantum statistics are applicable, can the mass action law be used?

8. Suppose the addition of a ternary acceptor type solute changes the Fermi energy by 0.03 eV. Evaluate the change in solubility of a donor solute at room temperature.(1 eV/molecule=23,050 cal/mol)

VII-CONCEPTS IN KINETICS

INTRODUCTION.

In this chapter we begin our study of kinetic phenomena in materials. We should become knowledgeable about concepts common to all kinetic phenomena in condensed matter. In gases, molecules come into contact by collision processes. In condensed matter, the atoms or molecules oscillate or rotate about a stable position or orientation, at any finite temperature. Even at 0 K they perform a zero point motion. These oscillations or rotations make possible the existence of kinetic phenomena in condensed matter. Each such phenomenon involves the motions of atoms or molecules in some particular way, which we will describe as a motion along a reaction path. If the energy of the system passes through a maximum as these particles move along the reaction path, then we describe this maximum energy as an "activation energy", which the system must exceed for it to proceed from its initial state to a different state. We shall explore the deeper significance of the activation energy in this chapter.

Gas phase kinetics has also taught us about different orders of reaction. We review this subject and various empirical relations for the time and temperature dependence of the fraction transformed from the initial to final state in the remainder of the chapter.

1. ACTIVATION ENERGY

Let us consider a process taking place in a condensed phase in which reactants existing in some metastable state proceed to a more stable product state. Atom or molecule arrangements define the reactant and product states. The transition between the initial and final states involves rearrangements of atoms or molecules and is called a reaction path. Since the initial and final states represent either metastable or stable states, the energy of the system must increase along any reaction path between them, as illustrated schematically in Figure 7.1. The assumption has been made in the foregoing that any atomic or molecular arrangement characterisitic of a position along a reaction path, the equivalent of an excited state, has a unique energy associated with it. However, the energy is not the only measure of the state of the system that is important in the problem of describing the transition from the reactant state to the product state. It is also necessary that the set of momentum vectors associated with the atoms or molecules obey some criteria, to be defined, if the transition from the reactant state to the product state is to be possible. As will be shown, the free energy is the quantity that describes both these factors.

The maximum energy along the reaction path relative to the energy of the reactant state is called the activation energy. See Figure 7.1. Thus, if the system of reactant atoms is to procede from the reactant to the product state it must somehow be excited at least to the energy corresponding to the activation energy. The problem of obtaining an analytic relation for the rate at which this system can overcome the activation energy can be approached via several levels of sophistication. At the lowest level, the assumption can

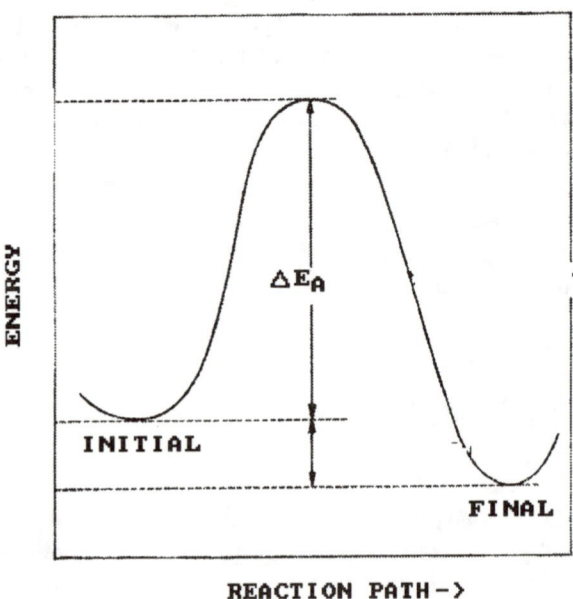

Figure 7.1. Energy of system as a function of reaction path from initial to final state.

be made that the energy states of the system obey classical statistical mechanics. We may then calculate the probability that the system can have an energy that exceeds the activation energy. To obtain the rate at which the system procedes from the reactant to the product state, we multiply this probability by the frequency with which the system makes attempts to move along the reaction path from the reactant to the product state. It may be a surprising fact, but this crude concept is usually adequate. Higher levels of sophistication in treating this problem act to confirm the result given by the simple concept that the rate is proportional to the exponential term, $\exp(-E_A/kT)$ and to yield a quantitative expression for the frequency factor, which in most cases cannot be experimentally validated due to the lack of knowledge of some of the factors involved.

Let us therefore make use of the simplest method of obtaining an analytic formulation for the rate at which a system can overcome an activation energy barrier. We assume the case where the energy levels of the system are equally spaced such as in an harmonic oscillator, for which the ith level has the energy $(i + 1/2)h\nu$, where ν is the frequency of the oscillator. The partition function for the oscillator, in analogy with that developed for a set of classical states in Appendix 1 of Chapter 1 is

$$Z = \sum_i \exp[-(i+1/2)h\nu/kT] = \exp(-h\nu/2kT)_i\exp[-ih\nu/kT] \qquad (7.1)$$

Let the activation energy, E_A, correspond to say the nth level, i.e. $E_A=nh\nu$. Then the fraction of the number of systems having energy higher than or equal to $nh\nu$ is thus

$$p_i = Z^{-1}\sum_{i=n}\exp[-(i+1/2)h\nu/kT] = \exp(-nh\nu/kT) = \exp(-E_A/kT) \quad (7.2a)$$

The fractional rate of a reaction having an activation energy E_A is then the product of the frequency with which the reactants tend to overcome the activation energy barrier, ν, by the fraction of these attempts that have the correct momentum vectors, $P\exp(-E_A/kT)$, where P represents the probability that these vectors are directed within some solid angle favorable for the reaction to take place or

$$R/N = \nu P\exp(-E_A/kT) \qquad (7.2b)$$

Wiener[1] has given a rigorous treatment of the development of a rate equation for the case of impurity diffusion in a single degree of freedom situation. It is somewhat more physical than the previous treatment and will be repeated below. Consider a collection of impurity atoms. The phase space applicable involves specification of the positions and momenta of these atoms. The positions need only be specified relative to the equilibrium positions of the atoms. The positions $x=\pm b$ represent the points where the energy of the oscillating atoms relative to that in the equilibrium position equals the activation energy. The coordinates of the phase space are thus as shown in Figure 7.2.. Now, atoms that have momentum between p and dp and will cross the line at $x=+b$ in a time dt are represented by points that occupy the shaded region in this diagram. We wish now to determine the fraction of impurity atoms that have coordinates within this region. To accomplish this we first define a distribution function f such that $\int_R f(x,p)dxdp$ equals the fraction of the impurity atoms that have coordinates in an arbitrary region R of phase space. The function f must satisfy the normalization condition that

$$\int_{-b}^{b} dx \int_{-\infty}^{\infty} dff = 1 \tag{7.3}$$

It is assumed that the distribution function is not dependent upon time. Indeed, if the impurity atoms do not interact with each other and are in thermal equilibrium with a thermal bath at temperature T, then f is given by the canonical distribution function

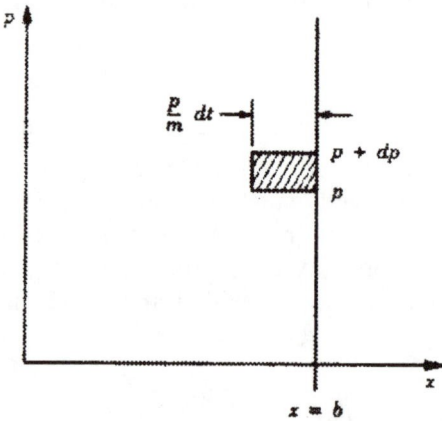

Figure 7.2. Phase space of momenta versus positions. After Wiener[1].

$$f = Cexp(-E/kT) = Cexp(-p^2/2m + Kx^2/2) \qquad (7.4)$$

where the latter quadratic term is an approximation to the potential energy of an harmonic oscillator, m is the mass of the impurity atom oscillator and C satisfies the normalization condition of (7.3).

Now the fraction of impurity atoms having coordinates within the shaded region of Figure 7.2 is given by

$$f(b,p)pdtdp/m$$

Division by the time interval dt and integration over all positive velocities, yields the desired fractional rate, R/N, with which impurity atoms cross the actvation energy barrier at x=b;

$$R/N = (1/m) \int_0^\infty pf(b,p)dp \qquad (7.5)$$

Substituting and solving yields $C = v_o/kT$, where $v_o = (K/m)^{1/2}/2\pi$ is the natural frequency of oscillation, and

$$R/N = v_o exp(-E_A/kT) \qquad (7.6)$$

Because the above relation was derived for a single degree of freedom the factor P in (7.2) equals 1 in this case.

In the theory of absolute reaction rates[2] the fractional rate is given by

$$R/N = (kT/h)exp[\Delta S_A]exp[-\Delta H_A]$$

This theory has been generalized to include the case of many degrees of freedom[3], with the result that the coefficient is no longer proportional to temperature, but has the form of equation 7.2b, with the coefficient having the significance of a ratio of the product of N normal frequencies of the entire system at the starting point of the transition to the product of the N–1 normal frequencies of the system constrained in the saddle point configuration. The fractional rate then becomes this ratio of products, v^*, multiplied by $exp[-E_A/kT]$. It seems reasonable to forego a discussion of this matter at this point, except to make the point that the activation energy in the exponent may be replaced by the free energy of activation with a consequent alteration of the coefficient of the exponential term.

2. EMPIRICAL RATE RELATIONS.

2.1. Homogeneous Reactions.

Reactions may take place in solutions between atomic (or molecular) components of the solution. The rate of a reaction will usually depend upon the concentrations of the reactants in the solution. Thus, if the decrease in the concentration, dc, of a species in time dt depends upon the first power of the concentration of that species, i.e. $-dc/dt=k^*c$, the reaction is called a first-order reaction. If it depends upon the second power ($-dc/dt=k^*c^2$) then it is called a second order reaction, etc..

Although collision processes can physically lead to the simple kinetics just described, the existence of such simple kinetics does not assure that the physical process that gave rise to them involves collisions between reactants.

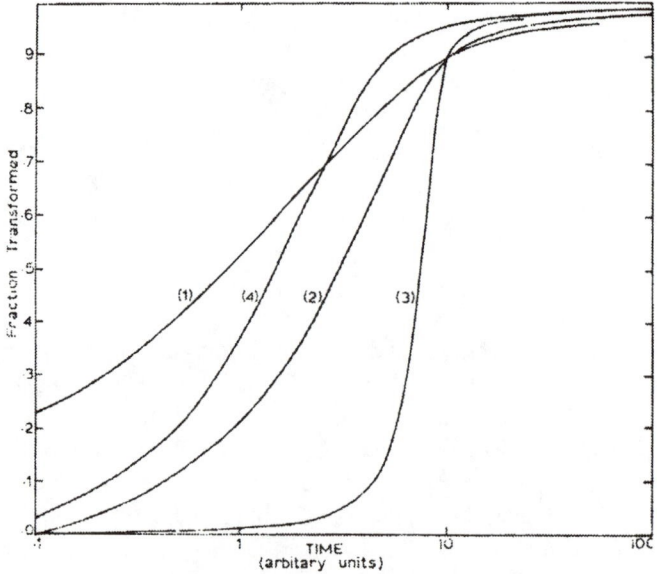

Figure 7.3. Reaction curves corresponding to Johnson-Mehl-Avrami equation. Curves (1), (2) and (3) have the same value of k and n=0.5, 1 and 4, respectively. Curve (4) has n=1 and k half the value of the other curves. After J. Burke,THE KINETICS OF PHASE TRANSFORMATIONS IN METALS,Pergamon Press, Oxford, 1965.

2.2. Heterogeneous Reactions.

An equation that describes an heterogeneous reaction often found is of the form

$$dy/dt = k^{*n}t^{n-1}(1-y) \tag{7.7}$$

where y represents the fraction of the metastable phase that is transformed to the product phase. If k^* and n are independent of y and t, then integration yields

$$\ln[1/(1-y)] = (k^*t)^n \tag{7.8}$$

This relation yields a sigmoidal curve for y as a function of t as shown in Figure 7.3. Assuming k^* and n to be constants (7.8) can be integrated to yield

$$y = 1 - \exp(-[k^*t]^n) \tag{7.9}$$

This equation is known as the Johnson-Mehl-Avrami equation and is most applicable for y<0.5. It is not necessarily true that k^* is a rate constant in this relation, which brings up the matter of how to derive an activation energy from empirical data. As a general conclusion the validity of an activation energy obtained from a plot of log rate versus reciprocal temperature or a similarly derived plot using a relation such as (7.9) depends upon the validity of the atomic model used to describe the process.

There are, of course, other empirical relations describing the fraction transformed as a function of time, some of which we will encounter in later chapters. The unique aspect of (7.9) is its range of applicability to diverse processes. Table 7.1 lists some of the processes for which it is a valid empirical description with the associated values of the exponent n. It is not wise to use an empirically determined value of n to conclude that the process is some particular one in the table.

3. COMPETING PROCESSES

Suppose that independent processes for the decomposition of a metastable phase may compete in the sense that they are physically possible

Table 7.1. Values of n in Equation 7.9

(a) Polymorphic changes, discontinuous precipitation, eutectoid reactions, interface controlled growth, etc.

Conditions	n
Increasing nucleation rate	>4
Constant nucleation rate	4
Decreasing nucleation rate	3–4
Zero nucleation rate (saturation of point sites)	3
Grain edge nucleation after saturation	2
Grain boundary nucleation after saturation	1

(b) Diffusion controlled growth

Conditions	n
All shapes growing from small dimensions, increasing nucleation rate	$>2\frac{1}{2}$
All shapes growing from small dimensions, constant nucleation rate	$2\frac{1}{2}$
All shapes growing from small dimensions, decreasing nucleation rate	$1\frac{1}{2}-2\frac{1}{2}$
All shapes growing from small dimensions, zero nucleation rate	$1\frac{1}{2}$
Growth of particles of appreciable initial volume	$1-1\frac{1}{2}$
Needles and plates of finite long dimensions, small in comparison with their separation	1
Thickening of long cylinders (needles) (e.g. after complete end impingement)	1
Thickening of very large plates (e.g. after complete edge impingement)	$\frac{1}{2}$
Precipitation on dislocations (very early stages)	$\sim\frac{2}{3}$

processes. If these processes can occur in parallel and not require one to occur before another then it is reasonable to expect that the fastest process will be the one to produce the product phase and hence will be responsible for the observed transformation kinetics. This statement should not be taken to mean that for a given process, if their is some ambiguity about the boundary conditions and different rates of growth are presumeably possible, that the fastest mode will be the one observed. The meaning of the latter qualification will be clarified in the chapter concerned with the dendritic mode of growth.

If the approach to equilibrium involves a series sequence of processes, then it is the slowest process that controls the overall rate. Usually, in this circumstance, the boundary conditions are not fixed but are dependent upon the processes themselves. Hence, the processes involved in such a series sequence are not strictly independent of each other.

4. THERMODYNAMIC THEORY OF IRREVERSIBLE PROCESSES

The thermodynamic based consideration of irreversible processes dates back to 1854 when Thomson(Lord Kelvin) analyzed various thermo-electric phenomena.[5] It wasn't until 1931 when Onsager derived his "reciprocal relations" that the modern theory of irreversible processes had its rebirth and it has had an active development to date beginning with the systematic contributions of Prigogine[6], DeGroot[7] and others.

We must first understand that the theory applies only to a limited range of irreversible phenomena that can be described in terms of the existence of linear relations between fluxes and forces. Although efforts are being made to apply the theory to the non-linear regime it is too early at this time to consider these results. Thus, the theory cannot be used to mandate that the relation between force and flux must be linear. Instead, only where this relation is linear can the theory be applied. We describe the generalized linear relation by the following equation:

$$J_i = \sum L_{ik} X_k \qquad \text{(for i = 1 to n)} \qquad (7.10)$$

where the L_{ik} are coefficients not dependent on the forces X_k. The fluxes and forces in these linear relations have the same tensor rank, which is a statement of the Curie principle that in an isotropic system, forces and fluxes of different tensor rank do not couple.

The fluxes and forces in (7.10) have not been defined. Whatever their definition they must yield the result that the sum over the product of the conjugate forces and fluxes equals the rate of production of entropy. Here we have used a concept that we have not yet defined, i.e. the rate of production of entropy. Let us consider the meaning in "the change in entropy of a system". We need first to define the system. For an open system that can exchange mass and heat with its surroundings we split the change in entropy in a process in the system into two parts: the change in entropy due to flow of matter and energy from or into the system, ΔS_e, and change of entropy due to production inside the system, ΔS_i. Since the change in entropy for the system can be evaluated from a knowledge of its state variables before and after the process, and the entropy associated with the exchange of matter and energy with the surroundings can also be evaluated, then the entropy change due to entropy production inside the system can also be evaluated. In general, the latter will be positive definite. Let the entropy production in the system be designated by $\Delta_i S$. Then our previous statement leads to the relation that the proper fluxes and forces in (7.10) satisfy

$$\Delta_i S/dt = \sum_i J_i X_i \tag{7.11}$$

Fluxes and forces that satisfy (7.11) will then have coefficients in (7.10) that satisfy Onsager's reciprocal relations, i.e.

$$L_{ik} = L_{ki} \tag{7.12}$$

The matrix of phenomenological coefficients in (7.10) is thus symmetric. Onsager's reciprocal relations follow from time reversal invariance of the microscopic equations of motion of the system of particles, which is a statement of the principle of "microscopic reversibility", and responsible for the detailed balancing of chemical reactions. Let us consider the latter for the simple case of reactions in which the rates of change of concentrations of the reactants is proportional to the concentrations of all the reactants. The reactions considered are schematically illustrated below.

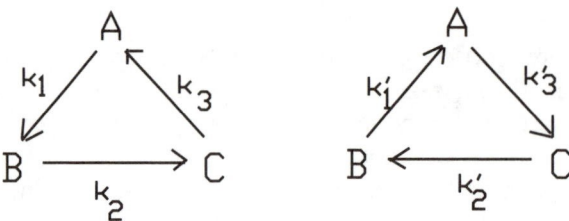

Then

$$dc_A/dt = -(k_1 + k'_3)c_A + k'_1 c_B + k_3 c_C$$

$$dc_B/dt = k_1 c_A - (k'_1 + k_2)c_B + k'_2 c_C \tag{7.13}$$

$$dc_C/dt = k'_3 c_A + k_2 c_B - (k_3 + k'_2)c_C$$

At equilibrium, the time derivatives equal zero and the equations can be rearranged to give

$$c_{Ae}/c_{Be} = [k'_1(k_3 + k'_2) + k_2 k_3]/[k_1(k_3 + k'_2) + k'_2 k'_3]$$

$$c_{Be}/c_{Ce} = [k'_2(k_1 + k'_3) + k_1k_3]/[k_2(k_1 + k'_3) + k'_1k'_3]$$

But, we can consider each separate reaction as well and for these separate reactions at equilibrium we would find

$$k_1c_{Ae} = k'_1c_{Be}$$

$$k_2c_{Be} = k'_2c_{Ce} \qquad (7.14)$$

$$k_3c_{Ce} = k'_3c_{Ae}$$

Now, set $y_A = c_A - c_{Ae}$ and, hence $dy_A/dt = dc_A/dt$. Substituting, the latter relation in the first of the equations (7.13) with the corresponding expressions for y_B and y_C we obtain

$$dy_A/dt = -(k_1 + k'_3)y_A + k'_1y_B + k_3y_C - (k_1 + k'_3)c_{Ae} + k'_1c_{Be} + k_3c_{Ce}$$

The sum of the last three terms in the latter equation is equal to dc_{Ae}/dt and hence equals zero. Thus,

$$dy_A/dt = -(k_1 + k'_3)y_A + k'_1y_B + k_3y_C. \qquad (7.15)$$

Now the chemical potential of A at concentration c_A can be related to that at c_{Ae} by $\mu_A - \mu_{Ae} = RT \log(c_a/c_{Ae}) = RT\log(1 + y_A/c_{Ae})$. For small deviations from equilibrium the logarithmic term is simply replaced by y_A/c_{Ae}. Thus, rearranging we find

$$y_A = -(\mu_A - \mu_{Ae})c_{Ae}/RT$$

Carrying out similar calculations to obtain expressions for y_B and y_C and substituting for these quantities in (7.15) and in the equivalent expressions for dy_B/dt and dy_C/dt then

$$dy_A/dt = \{[(k_1 + k'_3)c_{Ae}/RT][-(\mu_A-\mu_{Ae}) + [-k'_1c_{Be}/RT][-(\mu_B-\mu_{Be})] + [-k_3c_{Ce}/RT][-(\mu_C-\mu_{Ce})]\}$$

$$dy_B/dt = \{[-k_1c_{Ae}/RT][-(\mu_A-\mu_{Ae})] + [(k'_1 + k_2)c_{Be}/RT][-(\mu_B-\mu_{Be})] + [-k'_2c_{Ce}/RT][-(\mu_C-\mu_{Ce})]\}$$

$$dy_C/dt = \{[-k'_3 c_{Ae}/RT][-(\mu_a-\mu_{Ae})] + [-k_2 c_{Be}/RT][-(\mu_B-\mu_{Be})] + [(k_3 + k'_2)c_{Ce}/RT][-(\mu_C-\mu_{Ce})]\}$$

These relations can be compared to the linear relations consistent with the thermodynamic theory of irreversible processes

$$dy_A/dt = L_{11}X_1 + L_{12}X_2 + L_{13}X_3$$

$$dy_B/dt = L_{21}X_1 + L_{22}X_2 + L_{23}X_3$$

$$dy_C/dt = L_{31}X_1 + L_{32}X_2 + L_{33}X_3$$

Thus, $L_{12} = -k'_1 c_{Be}/RT$ and $L_{21} = -k_1 c_{Ae}/RT$, etc. But, at equilibrium, by detailed balancing of the individual reactions, as indicated by (7.14): $k_1 c_{Ae} = k'_1 c_{Be}$, etc. Hence, we find that $L_{12} = L_{21}$, etc. The Onsager relations hold for the linearized chemical reactions.

When the boundary conditions on a system are maintained constant with time the system will develop a steady state condition in which the state variables at a point do not vary with time. For example, when a bar is maintained at temperature T_1 at one end and temperature T_2 at the other end, a steady state distribution of temperature will develop along the bar. A consequence of the thermodynamic theory of irreversible processes (linear relations, Onsager relations) with constant values for the phenomenological coefficients L_{ik} is that the steady state corresponds to the state of minimum entropy production rate. There has been a tendency in the literature to claim that all irreversible processes occur such that the rate of production of entropy is a minimum. This statement cannot be true because there are many irreversible processes that do not involve linear relations between fluxes and forces, nor have constant L_{ik} even if they do.

Linear relations between fluxes and forces are applicable to some of the irreversible processes we will consider in the remainder of this book. The prime example of such a process is that of diffusion. Some growth situations obey linear laws, some do not. We will examine both cases and show where the thermodynamic theory can be of use and where not.

5. SUMMARY

We have considered the meaning of activation energy, order of reaction, various empirical relations describing kinetics of reactions in condensed phases and the Thermodynamic Theory of Irreversible Processes.

REFERENCES

1. J.H. Wiener, Proc. Sixth U.S. National Congress of Applied Mechanics, A.S.M.E., New York, 1970, pp 62-77.
2. S. Glasstone, K.J. Laidler and H. Eyring, THE THEORY OF RATE PROCESSES, McGraw-Hill, NY, (1941).
3. G.H. Vineyard, J. Phys. Chem. Solids 3, 121(1957).
4. N.B. Slater, THEORY OF UNIMOLECULAR REACTIONS,Cornell University Press, Ithaca, (1959), p. 105.
5.W. Thomson, Proc.Roy.Soc Edinburgh 3, 225(1854).
6. I. Prigogine, ETUDE THERMODYNAMIQUE DES PHENOMENES IRREVERSIBLES, Dunod, Paris and Desoer, Liege, 1947.
7.S.R. DeGroot, THERMODYNAMICS OF IRREVERSIBLE PROCESSES,North-Holland, Amsterdam, 1951.

BIBLIOGRAPHY

1. J.W. Christian, THEORY OF TRANSFORMATIONS IN METALS AND ALLOYS, Pergamon Press, New York, 1965, pp. 16-22, 80-94.
2. J.H. Wiener, STATISTICAL MECHANICS OF ELASTICITY, John Wiley, New York, 1983, Chapter 7.
3. S.R. DeGroot and P. Mazur, NON-EQUILIBRIUM THERMODYNAMICS, North-Holland, Amsterdam, 1962.
4. Y.L. Yao, IRREVERSIBLE THERMODYNAMICS, Science Press, Beijing, 1981.

DIFFUSION

INTRODUCTION.

Most kinetic phenomena in condensed matter involve diffusion, or a unit step very similar to that operating in diffusion. It is appropriate therefore to start our study of kinetic phenomena with a study of the fundamentals involved in the diffusion in various materials. We cannot provide a complete description of diffusion because, as with nearly every other chapter in this book, such a complete description requires a book in itself. This chapter begins with a phenomenological treatment of linear processes that is based on the thermodynamic theory of irreversible processes. This very general approach is made specific by application to diffusion in a binary alloy.

Diffusion in ionic crystals and semiconductors follows on the basis of the relations developed for metals and the thermodynamics of defects in these materials.

Finally, diffusion along high diffusivity regions, such as grain boundaries, is discussed.

1. PHENOMENOLOGICAL BASIS

1.1. Intrinsic diffusivities

It is often possible to relate fluxes of matter, heat and charge to corresponding forces linearly, as shown in equation (8.1)

$$J_i = \sum_k L_{ik} X_k \qquad (i=1...n) \qquad (8.1)$$

where J_i is the number or quantity of i passing unit area per unit time, L_{ik} is a conductance relating the kth generalized force X_k to the flux J_i. In the case of matter flow, the generalized force is the negative gradient of the chemical potential (i.e. $X_k = -\nabla\mu_k$). (In a network solid, $-J_i = \sum_j B_{ij} X_{jk}$ (i=1...n) with X=M, the diffusion potential.) Onsager has shown that the matrix of conductance coefficients in equation 8.1 is symmetric.

$$L_{ik} = L_{ki} \tag{8.2}$$

We will use a binary alloy as an example to illustrate the use of these relations to describe matter transport in the remainder of this section. In these equations, we assume that vacancies are the species denoted by the subscript $_3$, and the other two components are the atom species in a binary system.

$$^*J_1 = {}^*L_{11}X_1 + {}^*L_{12}X_2 + {}^*L_{13}X_3$$

$$^*J_2 = {}^*L_{21}X_1 + {}^*L_{22}X_2 + {}^*L_{23}X_3 \tag{8.3}$$

$$^*J_3 = {}^*L_{31}X_1 + {}^*L_{32}X_2 + {}^*L_{33}X_3$$

We now act, by setting $\sum J_i = 0$, to define a coordinate system relative to which the J's are measured. Using this definition, then

$$^*L_{11} + {}^*L_{21} + {}^*L_{31} = 0$$

$$^*L_{12} + {}^*L_{22} + {}^*L_{32} = 0$$

$$^*L_{13} + {}^*L_{23} + {}^*L_{33} = 0$$

by the independence of the X_i's. Also, from (8.2) and substitution into (8.1) yields

$$^*J_1 = {}^*L_{11}(X_1 - X_3) + {}^*L_{12}(X_2 - X_3)$$

$$^*J_2 = {}^*L_{21}(X_1 - X_3) + {}^*L_{22}(X_2 - X_3)$$

$$^*J_3 = {}^*L_{31}(X_1 - X_3) + {}^*L_{32}(X_2 - X_3)$$

But, $X_1 = -\nabla\mu_1$, $X_2 = -\nabla\mu_2$ and we make the assumption that there are sufficient sources and sinks for vacancies to maintain $X_3 = 0$, i.e. $X_3 = -\nabla\mu_{vacancy}$. Further, we assume constant molar volume, i.e. $C_1 + C_2 = C$ is independent of position. Then, if we consider a variation of composition only in one direction x, we obtain

$$^*J_1 = [^*L_{11}d\mu_1/dC_1 - ^*L_{12}d\mu_2/dC_2](-dC_1/dx) = D_1(-dC_1/dx)$$

$$^*J_2 = [^*L_{22}d\mu_2/dC_2 - ^*L_{12}d\mu_1/dC_1](-dC_2/dx) = D_2(-dC_2/dx) \qquad (8.4)$$

$$^*J_3 = [^*L_{31}d\mu_1/dC_1 - ^*L_{32}d\mu_2/dC_2](-dC_1/dx) = -(^*J_1 + ^*J_2) = Cv$$

This relation in which the flux is proportional to the negative gradient of the concentration is known as Fick's first law of diffusion and was an empirical discovery long before the thermodynamic theory of irreversible processes was developed. We shall now interpret these results via independent considerations.

Let there be a diffusing system in which there are inert markers (i.e. markers that do not interact with the matrix in which they are embedded.) Let us now measure the flux of components 1 and 2 relative to these markers and denote these by $^*J_1$ and $^*J_2$. Now, these markers may move with respect to a fixed laboratory coordinate system with a velocity v. We conserve matter, assuming $C_1 + C_2 = C$, that is a constant, independent of the distance x. Suppose that there is a concentration gradient in the diffusing system, that a marker at the distance x is moving with a velocity v to the right, and the instantaneous concentration of species at this point is C_1, as shown in Figure 8.1. If we now consider a differential volume of unit area normal to x and thickness dx, that is fixed in the laboratory coordinate system at x, and consider the change in the concentration of 1 that takes place in the time dt, then we obtain

$$(\partial C_1/\partial t)dxdt = [^*J_1(x) - (^*J_1(x) + (\partial ^*J_1/\partial x)dx)]dt - (\partial (vC_1)/\partial x)dxdt$$

Substitution of $^*J_1 = -D_1(\partial C_1/\partial x)$ yields

$$(\partial C_1/\partial t) = \partial[D_1(\partial C_1/\partial x) - vC_1]/\partial x \qquad (8.5a)$$

Proceeding in a similar manner for component 2, we can obtain

$$(\partial C_2/\partial t) = \partial[D_2(\partial C_2/\partial x) - vC_2]/\partial x \qquad (8.5b)$$

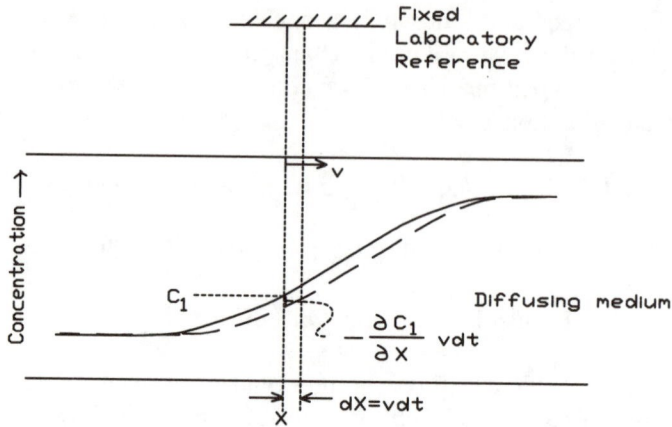

Figure 8.1. Illustration of change in composition, as observed in fixed laboratory frame of reference, due to motion of a diffusing medium.

Summing (8.5a) and (8.5b) yields

$$\partial(C_1+C_2)/\partial t = 0 = \partial[-(^*J_1+^*J_2)-vC]/\partial x$$

This result holds for all x and hence the integration constant must equal zero (i.e. as x-> ± infinity, $^*J_1$, $^*J_2$, v->0.) Hence,

$$vC = -(^*J_1+^*J_2)$$

Comparison with (8.4) yields the interpretation that the coordinate system relative to which the $^*J$'s are measured at a given position, in equations (8.1)-(8.4), is fixed relative to the inert marker at that position, which itself moves with respect to the laboratory coordinate system! In this case, note that $D_1 \neq D_2$. These diffusivities, D_1 and D_2, are called intrinsic diffusivities. Further, note that when $D_1 \neq D_2$, there is a net flux of vacancies in the same direction as the marker motion, i.e. $^*J_3 = vC$.

The effect that has just been described may be given a physically based interpretation as follows. Vacancies are generated on one side of the marker and they annihilate on the other side, to which they move. To conserve molar volume, the marker moves in the direction of the vacancy annihilation, i.e. atoms move in the opposite direction.* The vacancy sinks and sources may be considered to be edge dislocation jogs, so that the

---

* The diffusion couple is fixed relative to the laboratory system, far from the interface, where the concentration gradients equal zero.

annihilation of vacancies causes extra half planes of dislocations to disappear, while their generation causes such half planes to grow, corresponding to the process of dislocation climb. The disappearance of half planes on one side of the diffusion couple and growth on the other side results in a bulk movement of the matter in between in the direction of the side in which the half planes disappear. This phenomenon, which was discovered by E. Kirkendall has been called the Kirkendall Effect.

It is worth calling attention to the fact that in the above, although the concentration of vacancies was assumed to be the equilibrium concentration everywheres (i.e.X_3=0), nevertheless there exists a non-zero value of the vacancy flux $^*J_3$. The existence of vacancy sources and sinks, distributed throughout the diffusion couple, is implicitly assumed in the above analysis. Without such distributed sources and sinks, it would not be possible to maintain the vacancy concentration at the equilibrium value, and the above equations would have to be modified to take this fact into account. Usually, any development of a supersaturation in the vacancy concentration will lead to a tendency to nucleate voids, except in very clean and dislocation free single crystals. Such voids can be found in actual diffusion couples when the difference in the intrinsic diffusivities leads to the pumping of vacancies into a region at a rate faster than they can be annihilated at dislocation sinks. The detailed equations for vacancy migration must therefore include divergence terms to account for the generation and annihilation of the vacancies. In the above, we have avoided this problem by use of the assumption of constant molar volume, and consideration of the diffusion of the matter species only.

1.2. Chemical Diffusivity

If we had measured the flux of the species 1 and 2 relative to the fixed laboratory coordinate system, instead of with respect to the markers, then we would succeed in defining new diffusivities. To obtain these new diffusivities let us proceed as follows.

$$^0D_1 = -J_1/(dC_1/dx); \quad ^0D_2 = -J_2/(dC_2/dx)$$

where the J's are measured relative to the fixed laboratory coordinate system, and the ends of the diffusion couple are fixed relative to the laboratory coordinate system (as they were in the previous case as well, in the regions where the gradients in the composition were equal to zero.) Note, that to distinguish these fluxes from the previous ones they do not have asterisks. In the present case, $J_1 + J_2 = 0$, the net number of 1 type atoms moving to one side past a plane fixed relative to the laboratory coordinate system equals the net number of 2 type atoms moving past this plane in the opposite direction. We continue the assumption of constant molar volume. Thus, since $C_1 + C_2 = C = \text{constant}$ then $(\partial C_1/\partial x) = -(\partial C_2/\partial x)$. Since the sum of the fluxes of 1 and 2 equals zero, then by the above we obtain $^0D_1 = {}^0D_2 = {}^0D$. Consider now the change in C_1 in a region of volume of unit area normal to the diffusing direction x and distance dx in the time dt

$$(\partial C_1/\partial t)dxdt = [J_1(x) - (J_1(x) + (\partial J_1/\partial x)dx]dt \quad \text{or}$$

$$(\partial C_1/\partial t) = \partial(^0D(\partial C_1/\partial x))/\partial x \qquad (8.6)$$

We shall make use of this relation in a little while. First, let us substitute in (8.5a) the following relation for the marker velocity v

$$v = -(^*J_1 + {}^*J_2)/C = -[D_1(- \partial C_1/\partial x) + D_2(- \partial C_2/\partial x)]/C$$
$$= [D_1 - D_2](\partial C_1/\partial x)/C$$

to obtain

$$(\partial C_1/\partial t) = \partial[D_1(\partial C_1/\partial x) - C_1 v]/\partial x$$
$$(\partial C_1/\partial t) = \partial[D_1(1 - C_1/C)(\partial C_1/\partial x) - (C_1/C)D_2(\partial C_2/\partial x)]/\partial x$$

But, $C_1/C = N_1$, the mole fraction of component 1, and $\partial C_1/\partial x = -\partial C_2/\partial x$. Hence,

$$\partial C_1/\partial t = \partial[(N_2 D_1 + N_1 D_2)(\partial C_1/\partial x)] \, \partial x \qquad (8.7)$$

Comparison of (8.6) and (8.7) yields the result

$$^0D = N_2 D_1 + N_1 D_2 \qquad (8.8)$$

The diffusivity ^0D is called the chemical diffusivity and (8.8) relates it to the intrinsic diffusivities.

Let us consider how these diffusivities are measured in an alloy system. First, let us consider the measurement of ^0D. Fick's second law of diffusion, as given in (8.6), which is a second order partial differential equation, can be solved by using the Boltzmann integration factor $\lambda(C) = x/t^{1/2}$. Note that

$$(\partial C_1 / \partial t) = [dC_1/d\lambda](\partial\lambda/\partial t) = [dC_1/d\lambda](-\lambda/2t)$$

$$(\partial / \partial x) = [d/d\lambda](\partial\lambda/\partial x) = t^{-1/2}[d/d\lambda]$$

Substitution yields,

$$[(-\lambda/2)[dC_1/d\lambda] = [d/d\lambda](^0D[dC_1/d\lambda]),$$

a total differential equation in λ and C_1. We now integrate to obtain

$$-(1/2)\int_{C_{1o}}^{C_{1f}} \lambda dC_1 = {}^0D\{[dC_1/d\lambda]_{C_{1f}} - [dC_1/d\lambda]_{C_{10}}\}$$

For t constant, we replace λ to obtain

$$-(1/2t)\int_{C_{1o}}^{C_{1f}} x dC_1 = {}^0D\{[\partial C_1/\partial x]_{C_{1f}} - [\partial C_1/\partial x]_{C_{10}}\} \tag{8.9}$$

The significance of this equation can be appreciated on referring to Figure 8.2. To define the zero of x, we take the integration limits to equal zero, i.e. $[\partial C_1/\partial x]_{C_{1f}} = [\partial C_1/\partial x]_{C_{10}} = 0$. This serves to set the cross-hatched area to the left of x=0 equal to the cross-hatched area to the right of this interface, which is called the Matano interface. ^0D can now be evaluated using (8.9), except now the upper integration limit is allowed to float between C_{10} and C_{1f}, so that, as shown in Figure 8.3 and (8.10), all that is needed to define $^0D(C_1)$ is a measurement of the cross-hatched area in Figure 8.3, the slope $(\partial C_1/\partial x)$ at the composition C_1, and the diffusion time required to obtain the concentration-penetration profile shown in Figure 8.3.

$$-(1/2t)\int_{C_{1o}}^{C_1} x dC_1 = {}^0D\{[\partial C_1/\partial x]_{C_1} - [\partial C_1/\partial x]_{C_{10}}\} = {}^0D[\partial C_1/\partial x]_{C_1} \tag{8.10}$$

With ^0D measured, all that is necessary to obtain the intrinsic diffusivities is a measurement of the marker velocity corresponding to the concentration C_1. This can be obtained by measuring the distances between the various markers in a diffusion couple out to the ends, where the concentration gradients remain equal to zero throughout the experiment, as a function of time. In the above we have provided the theoretical bases for measurements of the intrinsic and chemical diffusivities in a binary alloy system that obeys the conditions explicitly stated in the assumptions. Next we shall relate the self-diffusivity of a pure metal to that in an alloy and later we shall provide

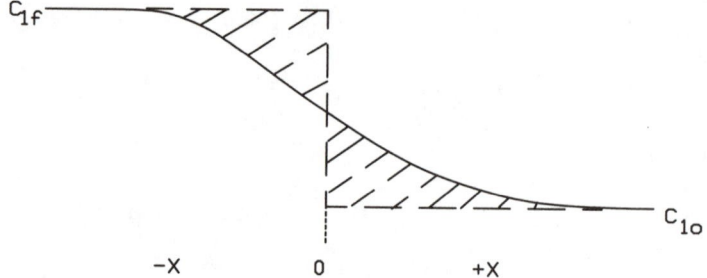

Figure 8.2. Illustrating the Matano interface at x=0. The cross-hatched areas are equal in area in the plot of concentration versus distance.

the bases for measurement of the diffusivities of the ionic species in an ionic crystal.

1.3. Darken's relations.

Let us extract from equations 8.4 the following expressions for the intrinsic diffusivities

$$D_1 = [^*L_{11}d\mu_1/dC_1 - ^*L_{12}d\mu_2/dC_2]$$

$$D_2 = [^*L_{22}d\mu_2/dC_2 - ^*L_{12}d\mu_1/dC_1] \qquad (8.11)$$

Since $\mu_i = \mu_i^\circ + kT\ln\gamma_i N_i$ then $d\mu_i/dC_i = C^{-1}d\mu_i/dN_i =$
$C^{-1}kT\{1/N_i + (1/N_i)d\ln\gamma_i/d\ln N_i\} = (kT/C_i)[1 + d\ln\gamma_i/d\ln N_i].$

But, $d\ln\gamma_1/d\ln N_1 = d\ln\gamma_2/d\ln N_2$ and setting $\varphi = 1 + d\ln\gamma_1/d\ln N_1$ then

$$D_1 = kT\varphi[^*L_{11}/C_1 - ^*L_{12}/C_2]$$

$$D_2 = kT\varphi[^*L_{22}/C_2 - ^*L_{12}/C_1] \qquad (8.12)$$

According to Howard and Lidiard[4] Darken's relations can be derived from the assumption that
$$<v_{i*}>/X_{i*} = J_{1*}/(C_{i*}X_{i*}) = J_1/(C_iX_i) = <v_i>/X_i$$

where v_{i*} is the average velocity of the i* isotope in a homogeneous alloy and v_i is that for component i at the same composition in a non-homogeneous alloy having the gradient $X_i = -d\mu_i/dx$. Substitution into the applicable

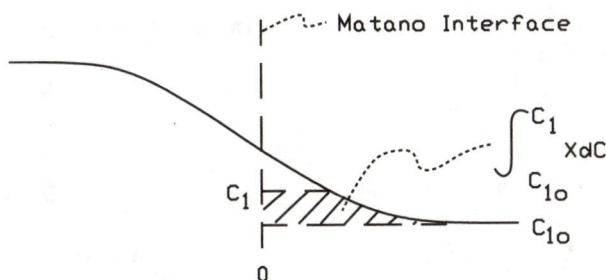

Figure 8.3. Illustrating equation (8.10).

phenomenological relations yield the Darken relations between the intrinsic diffusivities, D_i, and the tracer diffusivities, D_{i*},

$$D_i = D_{i*} \varphi \qquad (8.13)$$

It can also be shown that the second derivative of the molar free energy equals $(d\mu_A/dC_A)/C_B$. Hence, at inflection points in the molar free energy-composition dependence, the spinodal compositions, where the second derivative of the molar free energy equals zero, the Darken thermodynamic factor will equal zero. By equation 8.8 and the above expression for the intrinsic diffusivity in terms of the tracer diffusivity this result will lead to a zero value for the chemical diffusivity! Volkl and Alefeld[1] have found an experimental situation illustrating this effect.

If the cross-terms are not neglected Manning[2] has shown that in a random solid solution containing A and B atoms that an additional factor multiplies the right hand side of equation 8.8 given by

$$1 + \{2C_A C_B (D_{A*} - D_{B*})^2\}/[M_o (C_A D_{A*} + C_B D_{B*})(C_A D_{B*} + C_B D_{A*})] \qquad (8.14)$$

where M_o is defined by the self-diffusion correlation factor $f_o = M_o/(M_o + 2)$ for a vacancy mechanism. This multiplicative term is called a vacancy wind term since it reflects the coupling between the transport of components A and B through the vacancy flux.

1.4. Additional Driving Forces

In the above we have considered only the negative gradient of the chemical potential as a driving force for diffusion. However, other gradients can interact with the diffusing species. In particular, the negative gradients of the temperature and the electric potential provide driving forces for the flux of matter. The general phenomenological relations in these cases can be written as

$$J_1 = L_{11}[-\nabla(\mu_1/T)] + L_{1q}[-\nabla(1/T)] + L_{1\phi}[-\nabla(\Phi/T)]$$

$$J_q = L_{1q}[-\nabla(\mu_1/T)] + L_{qq}[-\nabla(1/T)] + L_{q\phi}[-\nabla(\Phi/T)] \qquad (8.15)$$

$$J_e = L_{1\phi}[-\nabla(\mu_1/T)] + L_{q\phi}[-\nabla(1/T)] + L_{\phi\phi}[-\nabla(\Phi/T)]$$

More detailed development of these relations can be found in the literature.[3]

A variety of phenomena are associated with different combinations of constraints on the fluxes and forces as illustrated in Table 8.1. We shall discuss the effect of electron drag and phonon drag in the production of failures in integrated circuits later. Thus, these cross effects can be significant in many practical situations.

1.5. Applications of phenomenological equations of irreversible thermodynamics to ionic crystals.

Howard and Lidiard[4] have given a description of diffusion in ionic crystals in terms of relations (8.1) and (8.2). Although the diffusivities applicable to a lattice depend on the nature of the associated defects, the

---

Table 8.1
Cross Effects

Constraints	Phenomena
$\nabla(\Phi/T)=\nabla(1/T)=0$	$J_q/J_1=L_{1q}/L_{11}=$Energy of Transfer
$\nabla(\Phi/T)=\nabla(\mu_1/T)=0$	$J_1/J_q=L_{1q}/L_{qq}=$Phonon Drag
$\nabla(\Phi/T)=\nabla(1/T)=0$	$J_1/J_e=L_{1\phi}/L_{\phi\phi}=$Electron Drag
$\nabla(\Phi/T)=\nabla(\mu_1/T)=0$	$J_e/J_q=L_{q\phi}/L_{qq}=$Thomson Coefficient
$\nabla(\mu_1/T)=\nabla(1/T)=0$	$J_q/J_e=L_{q\phi}/L_{\phi\phi}=$Thomson Heat
$J_e=0$	$|\nabla(\Phi/T)/\nabla(1/T)|=-L_{q\phi}/L_{\phi\phi}=$Seebeck Effect

(Seebeck effect is negative of Thomson heat(or Peltier heat) for a thermocouple).

Table 8.2
Relations between Diffusivities and Phenomenological Coefficients
for Ionic Crystals

For $q_A = q_B$, $D_B = [(L_{AA}L_{BB} - L_{AB}L_{BA})kT]/[NC_B(L_{AA} + L_{AB} + L_{BA} + L_{BB})]$ (1)

(When $L_{AA} \gg L_{AB}$ and L_{BB} then $D_B = kT L_{BB}/C_B$)

For $2 q_A = q_B$, $D_B = \dfrac{2kT(L_{AA}L_{BB} - L_{AB}L_{BA})(X'_V - X_k)}{N(L_{AA} + 2L_{AB} + 2L_{BA} + 4L_{BB})(2X_V C_B - X_k^2 - C_B^2)}$ (2)

(When $L_{AA} \gg L_{AB}$ and L_{BB} and $C_B \ll X_V$ then (2) simplifies to (1). When in other limit, $C_B \gg X_V$ then $D_B = 2kT L_{BB}/C_B(1+p)$, where p is the fraction of impurity ions associated with vacancies, i.e. $p = X_k/C_B$. Here q_A and q_B are the charges on the ions, X_V is the vacancy atom fraction, X_k is the concentration of impurity-vacancy pairs and C_B is the concentration of the impurity B.)

general result yielding Fick's law description is valid for ionic crystals. Table 8.2 collects some of Howard and Lidiard's results.

2. MECHANISMS OF DIFFUSION

In the above we have implicitly assumed that the mechanism of diffusion is via the exchange of atoms with thermal vacancies. Of course, this is not the only possible mechanism of diffusion. Among the others are direct exchange of neighboring atoms, cyclic exchange along a closed loop of nearest-neighbor atoms, interstitialcy movement in which an interstitial sited atom bumps a neighboring substitutional sited atom into an interstitial position with the stable arrangement of the interstitial defect being either a split interstitial (two atoms centered about a substitutional site) or a crowdion (n atoms along a close-packed row containing n-1 substitutional sites), vacancy aggregates which aid diffusion of atoms much as does the isolated vacancy by exchange of a component vacancy with a neighboring atom. In most metals we need only consider the vacancy and the divacancy as the major defects that control diffusion. However, in some metals that exhibit abnormally fast solute diffusion, some solute may diffuse interstitially or via some combination of interstitial and vacancy mechanisms. In ionic solids and semiconductors, we need to consider both vacancy and interstitial diffusion mechanisms.

2.1. Metals

We consider the mechanisms of diffusion in order to provide detailed expressions for the diffusivity that cannot be obtained from the phenomenological description of diffusion. The case of self diffusion in a pure metal is particularly instructive. Consider a crystal as shown in Figure 8.4 and diffusion in the x direction. The interatomic plane spacing in the x direction is d, the atomic concentration of radioactive solute A^* in the plane corresponding to position x is X_{A^*}. We make the assumption that the concentration of vacancies is everywheres that given by thermal equilibrium and that there is a gradient of A^* along the x direction. Figure 8.5 illustrates the dependence of the free energy in the course of exchange of the A^* atom in plane (1) with a nearest-neighbor vacancy in plane (2). From this figure the effective activation free energy for this exchange is

$$\Delta G_m + (1/2)(\partial\mu/\partial x)d \qquad\qquad (8.16)$$

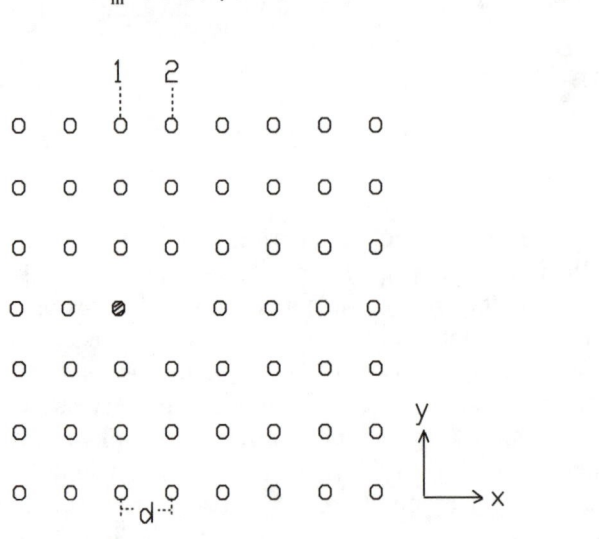

Figure 8.4. Schematic lattice defining atomic planes, interplanar spacing d and directions.

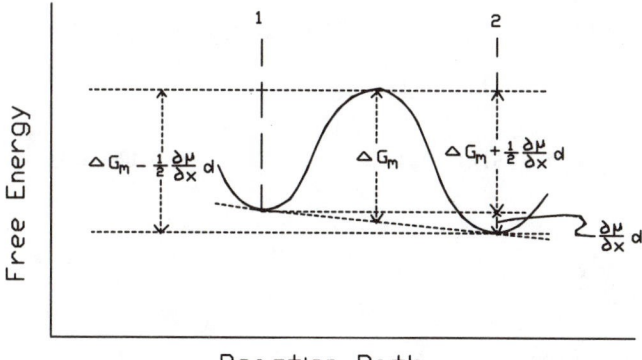

Figure 8.5. Defining free energy as a function of reaction path and activation energies for forward and backward jumps.

The flux of A^* atoms originating in plane (1) that traverse the dividing plane between the atomic planes (1) and (2) in moving to the right is given by

$$J_{A^*->} = C_{A^*}dmX_v vexp[-(\Delta G_m + (1/2)(\partial\mu/\partial x)d)/kT]$$

where C_{A^*} is the concentration of A^* atoms per unit volume at the position x, $C_{A^*}d$ is then the number of A^* atoms per unit area of plane (1), m is the number of nearest neighbors in plane (2) to an atom in plane (1), X_v is the equilibrium atom fraction of vacancies everywheres. In the above relation, $C_{A^*}dmX_v$ represents the number of A^*-V pairs, where the A^* atom is in plane (1) and the vacancy in plane (2), per unit area of plane, v is the frequency of attempts of such pairs to interchange sites and the exponential term is the fraction of such attempts which are successful. In a similar way, the flux of A^* atoms originating in plane (1) which exchange places with vacancies that lie in the plane to the left of plane (1) is given by

$$J_{A^*<-} = C_{A^*}dmX_v vexp[-(\Delta G_m - (1/2)(\partial\mu/\partial x)d)/kT]$$

The net flux of A^* atoms moving to the right out of plane (1) is then

$$J_{A^*,net} = J_{A^*->} - J_{A^*<-}$$

Collecting terms and neglecting second order effects then it can be shown that

$$J_{A*,net} = [d^2mvX_vC_{A*}/(kT)](\partial\mu_{A*}/\partial C_{A*})]exp[-\Delta G_m/kT](-\partial C_{A*}/\partial x)$$

$$= D_{A*}(-\partial C_{A*}/\partial x) \tag{8.17}$$

Thus, we have obtained an explicit relation for the intrinsic diffusivity, D_{A*}, namely

$$D_{A*} = (C_{A*}/kT)(\partial\mu_{A*}/\partial C_{A*})]d^2mvX_vexp[-\Delta G_m/kT]$$

For the case of metals we can make use of the result we have derived in Chapter VI for the temperature dependence of the equilibrium vacancy concentration $X_v = exp[-\Delta G_f/kT]$, where ΔG_f is the free energy of formation of a vacancy to obtain

$$D_{A*} = (C_{A*}/kT)(\partial\mu_{A*}/\partial C_{A*})]d^2mvexp[-\Delta G^*/kT] \tag{8.18}$$

where $\Delta G^* = \Delta G_f + \Delta G_m$. Also, since we are considering the diffusion of A^* tracer atoms in dilute solution in a matrix of pure A, then $\partial\mu_A^*/\partial C_A^* = kT/C_A^*$. Further, by using the relation $\Delta G^* = \Delta H^* - T\Delta S^*$, we obtain

$$D_A^* = D_o \, exp \, (-Q/RT) \tag{8.19}$$

where $D_o = md^2vexp(S^*/k)$ and $Q = N\Delta H^* = N(\Delta H_m + \Delta H_f)$.

   In the above derivation we have assumed that each atomic jump is independent of all jumps that precede it. Usually successive jumps are not independent and our result must be corrected for this neglect. The correction factor is known as the correlation factor, f, and has been extensively investigated.[5,6] The correlation factor multiplies the diffusivity to give a corrected diffusivity. Table 8.3 gives a collection of these factors for self-diffusion. Thus, to be strictly correct, D_0 should be multiplied by the correlation factor f. In the literature, the factor m is sometimes incorporated into the entropy factor S^*. Substituting reasonable values yields that the frequency factor D_0 should be on the order of 10^{-5} m²/s for the case of diffusion in metals. In fact, it is found to lie between 10^{-6} to 10^{-4} m²/s.
   Some metals exhibit a curvature in the Arrhenius plots of log diffusivity versus reciprocal temperature. If the activation energy is independent of temperature, as expected, then this observation may imply the existence of a contribution to atom transport of other than single vacancies.

Table 8.3

Correlation Factors for Self-Diffusion

Crystal Structure	Correlation Factor
Vacancy Mechanism, two-dimensional lattices	
Honeycomb lattice	0.5
Square lattice	0.46694
Hexagonal lattice	0.56006
Vacancy Mechanism, three-dimensional crystal structures	
Diamond	0.5
Simple cubic	0.65311
Body-centered cubic	0.72722
Face-centered cubic	0.78146
Hexagonal close-packed (with all	0.78121 normal to c-axis
jump frequencies equal)	0.78146 parallel to c-axis
Interstitialcy Mechanism (θ = angle between the displacement vectors	
of the atoms participating in the jump)	
NaCl, collinear jumps ($\theta = 0$)	0.666
NaCl, noncollinear jumps with	
$\cos \theta = 1/3$	0.9697
NaCl, noncollinear jumps with	
$\cos \theta = -1/3$	0.9643
Ca in CaF_2, collinear jumps	
($\theta = 0$)	0.80
Ca in CaF_2, noncollinear jumps	
with $\theta = 90°$	1.0

After J. R. Manning, *Diffusion Kinetics for Atoms in Crystals*, D. Van Nostrand, 1968, Chapter 3.

Indeed, this is currently the favorite explanation and the additional contribution to atom transport is believed to be provided by divacancy clusters. However, other explanations for the curvature in the Arrhenius plot have been offered. For example, it has been suggested that a strong thermal expansion coefficient for the vacancy leads to a temperature dependence of both D_0 and Q. Also, it has been suggested that dynamical correlation between successive vacancy jumps can contribute to this curvature. Despite these modifications there is no doubt that in normal metals the primary mechanism of atom transport is by vacancy-atom exchange jumps.

The approach we have taken in this section can be generalized to provide a relation for the one-dimensional diffusion of any species in terms of a jump frequency Γ, as follows:

$$D^* = (1/6)d^2\Gamma \tag{8.20}$$

where Γ is the jump frequency (number of jumps in all directions/time).

When correlation is taken into account then

$$D^* = (1/6)d^2\Gamma f \qquad (8.21)$$

Another way of arranging these quantities is useful when application is made to diffusion in ionic crystals and semiconductors. The relation of interest is

$$D^* = C_d D_d f \qquad (8.22)$$

where C_d is the concentration of the defect denoted by the letter d, D_d is the diffusivity of this defect, which for the above case of vacancy diffusion is given by

$$D_v = md^2v \exp(-\Delta G_m/kT) \qquad (8.23)$$

2.2. Ionic crystals

2.2.1. Stoichiometric-cation vacancy predominant defect.

For the case of a pure stoichiometric ionic solid (a rare species), if the major mechanism involved diffusion of vacancies on one type of site and the Schottky defect predominated, then the vacancy concentration corresponding to equation (6.13) and the vacancy diffusivity from (8.23) would be substituted into equation (8.22) to obtain the diffusion coefficient. The subscript m corresponds to the vacancy that has the least migration free energy, in the present case it is assumed to be the cation vacancy.

If soluble impurities were present in such a stoichiometric ionic solid to a larger concentration than the intrinsic concentration of the controlling vacancies. then it would be likely that the impurity concentration would determine the vacancy concentration. The resulting expression for the vacancy concentration would depend upon the valence of the impurity and its compensating defect. In this case, the vacancy concentration need not be temperature dependent and, in this event, the activation energy for diffusion will be that for migration of the vacancy.

2.2.2. General case.

Given a knowledge of the predominant defect and the identity of the most mobile defect, then (8.22) can be applied along with the relations developed in Chapter VI, for the case of a simple MO type oxide, to obtain the appropriate diffusivity. Normally, it is desired to obtain the dependence of the diffusivity on the partial pressure of O_2 for an oxide. In this case, the data in Table 6.6 can be used to obtain the partial pressure dependence of the defect concentration, which is the same as for the diffusivity. For example, for an hypothetical pure oxide of formula MO, in which the predominant mobile defect is the singly charged cation vacancy, the diffusivity will vary as the 1/4th power of the partial pressure of O_2. The order of the latter is usually reversed, i.e. the experimental partial pressure dependence of the diffusivity is used to obtain one limiting condition on the identity of the predominant defect. A real example illustrating this behavior is CoO, in which at 1200°C diffusion is by singly charged cation vacancies at high oxygen activity and doubly charged cation vacancies at low oxygen activity. At high oxygen pressure the cobalt tracer diffusivity is proportional to the 1/4th power of the oxygen activity.[7]

To obtain the relation between the partial pressure dependence of the predominant defect for the general case of an M_xO_y type oxide recourse should be made to the literature cited in the bibliography. It should come as no surprise, considering the number of alternative possibilities that exist in ionic crystals (e.g. defect type, charged impurities and their charge compensating defects, non-stoichiometry, contribution to diffusion from both cation and anion species, association of vacancies and solute, precipitation of solute, etc), that the mechanism of diffusion in ionic crystals is much more complicated than in metals. Indeed, one cannot speak of a mechanism for diffusion in ionic crystals. There are many possible mechanisms and some have been reliably identified only for a few ionic systems. The latter statement is especially applicable to oxides.

Another instructive example is that of diffusion in magnetite, Fe_3O_4, where more than one diffusion mechanism is operative. Magnetite has the inverse spinel structure, with Fe^{2+} on the octahedral sites and half of the Fe^{3+} on the octahedral and the other half on the tetrahedral sites at room temperature. At higher temperatures the distribution of these two cations is randomized over the two types of sites. There are four octahedral and eight tetrahedral sites/three cations. At low oxygen activities there is a cation excess with iron ions on interstitial sites as the dominant defect. At high oxygen activities the oxide is metal deficient with cation vacancies as the dominant defect. The relations governing the species in magnetite are

$$3Fe_{Fe} + 4O_O = 3Fe_i + 2O_2 \text{ (low oxygen activity)}$$

(electroneutrality is maintained by change in the valence of the appropriate cation)

$$8Fe_{Fe}^{2+} + 2O_2 = 8Fe_{Fe}^{3+} + 3V_{Fe} + 4O_O \text{ (high oxygen activity)}$$

(In both reactions the stoichiometric ratio of lattice sites of the two species is conserved.) Hence, the diffusivity should vary as the minus 2/3 rd power of the oxygen partial pressure at low oxygen activity and as the 2/3rd power of the partial pressure at high oxygen activity. The results shown in Figure 8.6 are consistent with these relations.

Diffusion in ionic crystals is particularly technologically significant in the process of oxidation and in superionic materials, as used in fuel cells and the like. Of course, it is also important in the control of the processing of ceramics and, hence, of their properties.

2.3. SEMICONDUCTORS.

The mechanisms of diffusion in semiconductors are more varied than found in metals. In particular, self-diffusion via interstitial motion occurs in silicon at high temperatures above about 1270 °C. Also, the ability to vary vacancy and self-interstitial concentrations by the addition of donors and acceptors can materially affect self diffusivity. The open lattice of co-valently bonded materials makes interstitial diffusion of solute much more common.

As for ionic crystals, the applicable diffusivity can be expressed by (8.22) and the defect concentration by the appropriate mass action relations. It has been found that the

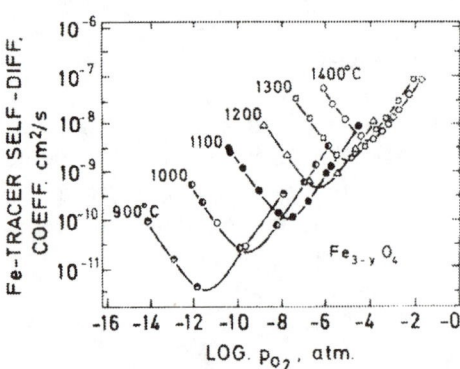

Figure 8.6. The iron tracer self-diffusion coefficient in magnetite as a function of oxygen activity. After Dieck-mann and Schmalzried,Ber/Bunsenges. Phys. Chem.81, 344(1977).

effect of solute on the tracer diffusivity of the host elemental semiconductor atoms can be expressed as the following linear relation

$$D(C_B) / C(0) = 1 + bC_B$$

where b is an enhancement factor that lies between −20 and +100.

Figure 8.7 shows solute diffusivities in silicon as a function of temperature. The more rapid diffusion of solutes compared to the self-diffusion of silicon is obvious. It appears that the Group III and group V solutes are incorporated solely on regular lattice sites and diffuse via the same mechanisms as self atoms. The much higher diffusivities of the metal solutes is likely to be due to the existence of these solute as interstitial atoms. A detailed explanation of these relative diffusivities is given in the article by Frank et al referenced in Figure 8.7.

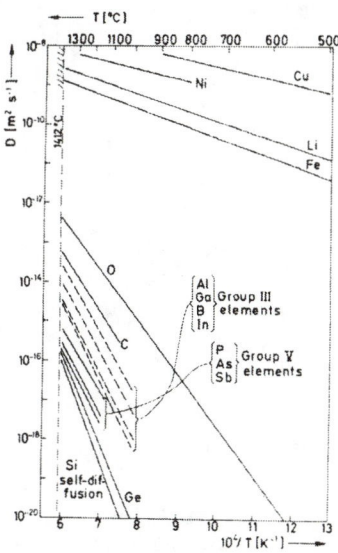

Figure 8.7. Diffusivities of foreign atoms in silicon. After Frank et al in DIFFUSION IN CRYSTALLINE SOLIDS, eds.,Murch and Nowick, Academic , 1984.

3.NERNST-EINSTEIN RELATION.

To derive the Nernst-Einstein relation we note that the equations for the flux of atoms given on p.211 can be written as

$$J_{i->} = \Gamma_o \exp(F_i d/2kT) \text{ and } J_{i<-} = \Gamma_o \exp(-F_i d/2kT)$$

yielding for the net flux

$$J_{net} = \Gamma_o F_i d/kT = <v_i>C_i$$

where $\Gamma_o = C_i dmv exp[- \Delta G^*/kT]$. But, the intrinsic diffusion coefficient is

$$D_i = \varphi d^2 mv exp[- \Delta G^*/kT]$$

Hence, $\Gamma_o = D_i C_i/\varphi d$ and, by substitution, $<v_i> = D_i F_i/\varphi kT = D_{i*} F_i/kT$. This equation is the Nernst-Einstein relation. Note that the diffusivity in this relation is the tracer diffusivity.

It is convenient to consider the various driving forces for diffusion which are

$F_i = d\mu_i/dx$ for a concentration gradient

$= d\Phi/dx = Z_i^* eE$ where E is the electric field for the latter as a driving force and Z_i^* is the effective valence

$= <Q_{ia}> d\ln T/dx$ for a temperature gradient, where $<Q_{ia}>$ is the effective heat of transport.[3]

4. EMPIRICAL RULES

Certain generalizations regarding diffusion have been discovered which are useful as follows.

Metals
1. The activation energy for self diffusion obeys the relation
   $Q = 34 T_m$ (cal/gatom/°C), where T_m is the melting temperature.
2. The activation entropy for self diffusion obeys the relation
   $\Delta S = Z\beta Q/T_m$, where Z is a constant that depends on the lattice and equals 0.55 for fcc and 1 for bcc, $\beta = -d(\mu/\mu_o)/d(T/T_m)$, μ is the shear modulus.
3. The activation volume for self diffusion obeys the relation
   $\Delta V = 4\chi Q$, where χ is the compressibility.

Alloys
1. The diffusion coefficient, D, at the solidus temperature is roughly a constant for a given structure and bond type, i.e. $D_{fcc}(T_{solidus}) = 5*10^{-9} cm^2/sec$; $D_{bcc}(T_{solidus}) = 3*10^{-8} cm^2/sec.$, except for the bcc alkali and actinide metals.
2. $Q/RT_{solidus} = 18(\pm 2)$.
3. Preexponential is roughly constant for a given structure and bond type. $D_o(fcc) = 0.3 cm^2/sec$; $D_o(bcc) = 1.6 cm^2/sec.$ with ex-

ceptions as noted in 1.

4.   In dilute alloys: addition of a faster(slower) diffusing species as an impurity causes an increase(decrease) in the rate of diffusion of both solute and solvent atoms; an increase in concentration of a solute that lowers(raises) the melting point increases(decreases) the diffusion rate of this solute at temperature T.

5.   In concentrated alloys:the tracer diffusion coefficients of the different components do not differ by more than an order of magnitude if the diffusion mechanisms of the species is the same.

6.   No generalizations exist concerning the diffusion of interstitial solute. Hence, some data are collected in Appendix 8.1.

Ionic Solids

The possibilities for not achieving reproducible results and the multitudes of possible mechanisms have worked against the discovery of similar reliable empirical rules in these materials.

5. SOLUTIONS TO DIFFUSION EQUATIONS.

Interdiffusion of component species is often involved in the processing of materials. For control of the processing, and consequently the properties of a material, it is desired to know the concentration of a species, C. as a function of position, time and temperature, i.e. know $C(x,t,T)$. For diffusion in three dimensions the applicable relation is Fick's 2nd law in the form

$$\partial C/\partial t = D[\partial^2 C/\partial x^2 + \partial^2 C/\partial y^2 + \partial^2 C/\partial z^2]$$

for D independent of composition. Solutions to this equation for a variety of boundary conditions are given in J. Crank, THE MATHEMATICS OF DIFFUSION, Oxford, 1956 and Ghez's book(see Bibliography).

For example, the solution for the one dimensional diffusion into a plate from a source maintaining a surface concentration C_s into a solid in which the concentration of the diffusing species is initially at the concentration C_o is

$$C = C_o + [(C_o - C_s)/2]\mathrm{erfc}(x/2[Dt]^{1/2})$$

where $erfc(u) = 2/\pi^{1/2} \int_0^u \exp(-w^2)dw$.

A very useful approximation stems from the integrating factor of Fick's 2nd law given by $\lambda=x/\sqrt{t}$. When in the form $\lambda' = <x>/\sqrt{Dt}$, then λ' is a dimensionless function of the concentration having a value on the order of unity for various boundary conditions in one dimensional diffusion. Thus, $<x>$ can be regarded as the penetration distance of some average concentration between that of the source and that of the host into which the species diffuses. The relation between the penetration distance, time and temperature has many applications in materials science and is probably the most widely used relation in the practical application of diffusion concepts, which may account for it being on the cover of Borg and Dienes's book.

6. HIGH DIFFUSIVITY REGIONS IN SOLIDS

It is a fact that polycrystals and even single crystals contain regions (dislocations and grain boundaries), where the diffusivity is higher than in the perfect crystal. We should like to answer two questions: 1) Can the intrinsic diffusivity of these regions be measured? 2) Can such measurements help define the nature of these imperfect regions?

The answer to the first question is that if the diffusivity of the imperfect region, at a constant temperature and pressure, is assumed to be constant over some volume associated with the imperfection, then it is a relatively simple matter to measure this "defined intrinsic diffusivity." The answer to the second question is still unknown, although it appears more likely that the reverse-that a knowledge of the imperfect regions will help define the mechanism of diffusion in them-will occur sooner.

At low temperatures the diffusivity in these imperfections is generally many orders of magnitude greater than that in the perfect lattice. When the product of the volume fraction by the diffusivity for each type region is of the same order of magnitude then it is important to consider the contribution of each to the measured transport of matter.

There are mainly two techniques for measuring the intrinsic self-diffusivities of defect regions. One involves plating a radioactive layer on the surface and then measuring the concentration of the radioactive species of sections parallel to the surface after a diffusion anneal. The radioactive material transported to each section by diffusion through perfect and imperfect material is then calculated on the basis of a model and measured values of the volume fraction of each region. The other technique requires the specimen to be a thin sheet in which the imperfect regions are mainly normal to the sheet surface so as to provide "short-circuit" paths from one surface of the sheet to the other. Then a radioactive layer is plated on a portion of one side of the sheet and the amount reaching the other side by "short-circuit" diffusion is measured after a diffusion anneal at a sufficiently low temperature. The former is called the sectioning technique and the latter the surface accumulation technique. In both of these cases, other sensitive techniques of measuring the concentrations of the diffusing species can be substituted for the techniques of using a radioactive species, i.e. in the surface accumulation technique AES may be used.

6.1. Models for Evaluating the Grain Boundary Diffusivity.

Fisher's approximate method.[8]

According to the model of Fisher for the sectioning technique, consider a grain boundary normal to the surface, of thickness δ, of a semi-infinite crystal. The boundary conditions for the diffusion are

$C=C_0$ for $y=0$ and $t \geq 0$; $C=0$ for $y>0$ and $t=0$.

Fisher assumes the concentration in the grain boundary to be independent of x, the distance in the direction normal to the grain boundary plane. Further, in going out of the grain boundary into the bulk in the x direction, in the case of self-diffusion in a pure element, there is no discontinuity in concentration. Conservation of matter in the grain boundary region of thickness δ and length dy then yields

$$\partial C/\partial t = [1.dy.\delta]^{-1}\{\delta(J_y - [J_y + (\partial J_y/\partial y)dy]) - 2dyJ_x\} = -(\partial J_y/\partial y) - (2/\delta)J_x$$

But, $J_y=-D_b (\partial C/\partial y); J_x=-D_l[\partial C/\partial x]_{\delta/2}$. Therefore

$$\partial C/\partial t = D_b (\partial^2 C/\partial y^2) + (2/\delta)D_l[\partial C/\partial x]_{\delta/2}.$$

Outside the grain boundary $(\partial C/\partial t)=D_l \nabla^2 C$.

Fisher assumes:

1) C_b is near to its final value during all of the time t of diffusion;

2) J_x is normal to the grain boundary, there is no J_y in the bulk lattice except that which initially diffused from the surface through the bulk.

These assumptions allow the problem to be divided into two parts: matter transported to a section at y either through the lattice or through the grain boundary and then along the x direction. For the latter we have the following solution.

$$C(x,y,t) = C_b(y)[1-\text{erfc}\{x/(2(D_l t)^{1/2})\}] \quad \text{where}$$

$$C_b(y) = C_o \exp[-(2y)^{1/2}/((\pi D_l t)^{1/4}(\delta D_b/D_l)^{1/2})]$$

Experimentally, the total amount of solute in a section dy thick is measured, i.e. $2\int_0^\infty C(x,y,t)dx$ plus that diffused in from the surface through the bulk. The latter contribution overwhelms the former until $D_b/D_l >> 5 \times 10^4$ when the reverse occurs.

The net effect of enhanced diffusion along grain boundaries in polycrystalline samples is to produce a deviation in the plot of log $D_{apparent}$ versus 1/T at a temperature below which $D_b/D_l >= 5*10^4$. The actual ratio at this point depends upon grain size.

The exact solution for the semi-infinite bicrystal and for the same boundary conditions has been given by Whipple[9] and by Suzuoka[10] for the case of an instantaneous source instead of a constant source.

6.2. Mechanism of diffusion along grain boundaries

Balluffi[11] has recently considered the possible atomic mechanisms for diffusion along grain boundaries and has come to the conclusion that vacancies in metals and alloys are most likely to be the defect responsible for the fast diffusion. Grain boundary dislocations are likely to be the sources and sinks for vacancies with the energy of formation of vacancies at grain boundary sites being somewhat larger than the migration energy of such vacancies along the boundaries, but with both being smaller than the corresponding values in the bulk lattice. These results are based on molecular dynamic simulation of grain boundary diffusion, which may or may not be sensitive to assumptions concerning the interatomic potential used in the calculations. Hence, it is desirable to have more evidence concerning the mechanism of grain boundary diffusion. Among such evidence that support the vacancy mechanism are measured values of activation volume that equal the atomic volume.

6.3. Empirical results

6.3.1. Metals.

1. For self-diffusion the apparent activation energy for diffusion along grain boundaries is from 0.4 to 0.6 that for bulk diffusion.

2. For solute diffusion the apparent activation energy includes the energy of segregation of the solute to the boundary.

3. In low angle boundaries the diffusion is anisotropic, being much faster along the direction of the tilt axis for tilt boundaries than perpendicular to it. For example, the ratio of activation energy for grain boundary diffusion to that for bulk lattice diffusion is 0.4 when the grain boundary diffusion is parallel to the <100> tilt axes, 0.6 when the grain boundary diffusion is parallel to the <211> tilt axis, and 0.7 when the grain boundary diffusion is parallel to the <111> twist axis, for both nickel and silver.

4. Diffusion is more rapid parallel to undissociated dislocation cores than to dissociated dislocation cores.

5. The activation energies in fcc metals for surface, grain boundary and dissociated dislocations are $13T_m$, $17T_m$ and $25T_m$, respectively, which are to be compared to that for lattice diffusion of $34T_m$.

6.3.2. Ionic crystals.

In principle, grain boundaries are also high diffusivity regions in ionic solids. However, because of the excess charge associated with grain boundaries there appear to be limitations on the species that can diffuse rapidly along them. This aspect of the high diffusivity problem has not yet been resolved. Nevertheless, there are many reported cases involving high diffusivity paths along grain boundaries in ionic crystals.

Grain boundary diffusion occurs in oxides and is believed to be responsible for the observed oxidation rates of many metals at temperatures less than about $0.5T_m$. This is an important result because many oxidation resistant metals are used in this temperature range. Among the oxides that reveal grain boundary diffusion at temperatures less than $0.5 T_m$ are NiO, Fe_3O_4, and ZrO_2.

6.4. Electromigration along grain boundaries

An important and sometimes limiting phenomenon found to occur in thin films used in integrated circuits is that of failure induced by electromigration. The failure is a consequence of the existence of divergences in the transport of matter along grain boundaries. For example, such divergences are present whenever three grain boundaries between differently oriented grains meet in a line because the grain boundary diffusivities of such boundaries are not likely to be equal! Further, neither will the forces be exactly equal because of the different angles made by the normals to such boundaries and the applied potential gradient. The latter can lead to very high current densities due to the small dimensions of the conducting lines in the integrated circuits, which can then lead to significant values of atom transport via the coupling between the two, i.e. the electron drag. The attempts to ameliorate this problem have focussed on means of decreasing the grain boundary diffusivities of the diffusing species and this was the driving force of the initial research carried out to solve the problem. Recently, improvements in the electromigration time to failure have been achieved empirically, but not by affecting the grain boundary diffusivity of the host species. With the continued drive for miniturization, electromigration failure remains a concern to integrated circuit manufacturers.

7. SUMMARY

Summarizing, we have derived the relations between the phenomenological relations and Fick's law of diffusion, relating the phenomenological coefficients to the diffusivity. These phenomenological relations are based on the "Thermodynamic Theory of Irreversible Processes" and obey Onsager's relation in that the matrix of phenomenological coefficients is symmetric. Many of the cross-effects have non-negligible practical consequences, e.g. electromigration. We found that definition of the diffusivity depends on the frame of reference relative to which the fluxes are measured and we explored the relationships between the various diffusivities. The phenomenological relations emphasize that the fundamental direct driving force for matter transport of a particular species is the gradient in chemical potential of that species and not the concentration gradient. We have also been introduced briefly to the concept of correlation, which describes the effect of prior jumps on successive jumps of atoms. Also, we considered briefly the mechanisms of diffusion in metals, ionic crystals and semiconductors. We found it possible to relate the self-diffusivity in all cases to the product of the concentration of the predominant defect active in diffusion and the diffusivity of the defect species. The Nernst-Einstein equation relates the average "velocity" of a species to the generalized force on that species and the tracer diffusivity for the species. We remarked on available solutions to Fick's second law for various boundary conditions and on the importance of the integrating factor $\lambda'(C) = x/(Dt)^{1/2} = $ constant (on order of unity). We considered transport along high diffusivity regions, such as grain boundaries and dislocations, and the significant contributions to be expected in polycrystalline solids at temperatures below about $T/T_m = 0.5$. Finally, some representative empirical results were given.

APPENDIX

Tables of data giving diffusivities and activation energies for interstitial diffusion of interstitially sited solute atoms in metals.

BCC Metals

metal	diff. elem.	$\frac{d}{a}$	D_{oi} $(10^{-6} m^2 s^{-1})$	Q_i (kJ/mole)	ref	$\frac{Q_i}{Q_{SD}}$	$\frac{S_M}{R}$	$\frac{Q_i}{RT_m}$
Ca	C	0,35	0,27	97,6	18		0	10,
β-Th	C	0,37	2,2	113	18	0,38	2,0	6,
	N	0,34	0,23	71	18	0,24	0,05	4,
	O	0,29	0,13	46	18	0,15	-0,2	2,
γ-U	C	0,44	21,8	123	22	1,07	4,4	10,
β-Ti	C	0,47	0,6	94,6	17	0,38	1,0	5,
	N	0,43	3,5	141,5	18	0,56	2,6	8,
	O	0,36	8,3	131	fig3	0,52	3,6	8,
β-Zr	C	0,43	24,5	150,3	21	0,55	4,4	8,
	N	0,39	1,5	128,5	18	0,47	0,5	7,
	O	0,33	98	172	18	0,63	5,9	9,
β-Hf	C	0,44	80	211	23		5,4	10,
	N	0,40	0,8	124	23		1,2	6,
	O	0,34	32	171	23		4,8	8,
V	C	0,51	0,88	97,6	18	0,32	1,5	5,
	N	0,47	5,02	151	28	0,49	3,1	8,
	O	0,40	2,66	125	28	0,41	2,6	6,
Nb	C	0,47	1,0	142	18	0,35	1,3	6,
	N	0,43	2,6	152,3	7	0,38	2,3	6,
	O	0,36	0,53	109,5	7	0,27	1,0	4,
Ta	C	0,47	0,67	162	18	0,39	0,9	5,
	N	0,43	0,52	158,5	15	0,38	0,7	5,
	O	0,40	1,05	110,4	15	0,27	1,6	4,
Cr	C	0,53	0,87	111	3	0,36	1,4	6,
	N	0,49	1,6	115	3	0,37	2,1	6,
	O	0,42	0,01	155	18	0,50	-3,0	8,
Mo	C	0,49	3,4	172	18	0,37	2,5	7,
	N	0,45	0,43	109	18	0,24	0,73	5,
	O	0,38	3,0	130	18	0,28	2,7	6,
W	C	0,51	1,2	188	18	0,32	1,4	6,
	N	0,47	2,4	119	18	0,20	2,4	4,
	O	0,40	0,013	100	18	0,17	-2,6	3,

FCC Metals

metal	diff. elem.	$\frac{d}{a}$	D_{o1} $(10^{-6} m^2 s^{-1})$	Q_i (kJ/mole)	$\frac{Q_i}{Q_{SD}}$	$\frac{S_M}{R}$	$\frac{Q_i}{RT_m}$
Ag	O	0,294	2,72	46	0,25	1,4	4,8
Co	C	0,434	50	159	0,56	3,8	10,8
Cu	O	0,332	1,7	67	0,33	0,8	5,9
γ-Fe	C	0,42	74	159	0,59	4,0	10,5
	N	0,39	91	169	0,62	4,3	11,2
	O	0,33	575	169	0,62	6,2	11,3
Ni	C	0,44	5	146	0,51	1,4	10,2
	O	0,34	$1,82 \cdot 10^6$	301	1,06	14,0	21
Pt	O	0,31	930	326	1,14	10,9	19,2
α-Th	C	0,3	2,7	159	0,53	0,4	9,5
	N	0,28	2,1	94	0,31	- 1,8	5,6
	O	0,24	1,3	205	0,68	- 0,3	12,2

HCP Metals

metal	diff. chem.	$\frac{d}{a}$	D_{o1} $(10^{-6} m^2 s^{-1})$	Q_i (kJ/mole)	$\frac{S_M}{R}$	$\frac{Q_i}{RT_m}$
Y	C	0,42	20	123	3,2	2,3
	N	0,39	0,1	251	-2,4	4,6
Re	C	0,56	10	222	2,5	7,7
	N	0,51	14	154	5,2	5,4
α-Ti	C	0,52	506	182	6,4	11,3
	N	0,48	1,2	189	0,5	11,7
	O	0,41	80	201	4,6	12,4
α-Zr	C	0,48	0,2	151,6	-1,4	8,6
	N	0,44	30	238,6	3,5	13,5
	O	0,37	235	213	5,8	12,0
α-Hf	C	0,48	$7,4 \cdot 10^3$	310	8,7	14,9
	O	0,38	66	213	4,4	10,3

REFERENCES

1. J. Volkl and G. Alefeld, in HYDROGEN IN METALS, eds. G. Alefeld and J. Volkl, Springer, Berlin, 1978.
2. J.R. Manning, DIFFUSION KINETICS FOR ATOMS IN CRYSTALS, Van Nostrand, Princeton, 1968.Ch. 1.
3. G. Bre'bec in DEFAUTS PONCTUELS DANS LES SOLIDES, Les Editions de Physique, Orsay, 1978, p.181.
4. R.E. Howard and A.B. Lidiard, Rep. Prog. Phys. 27, 161(1964).
5. J.R. Manning, DIFFUSION KINETICS FOR ATOMS IN CRYSTALS, Van Nostrand, Princeton, 1968.Ch.3.
6. A.D. LeClaire in PHYSICAL CHEMISTRY, AN ADVANCED TREATISE, eds. H.Eyring, D. Henderson and W. Jost, Academic, vol.10, 1970, Ch.5.
7. N.L Peterson and W.K. Chen, J. Physique, Colloq.C-6, 319(1980); R. Dieckmann, Z. Phys. Chem.Neue Folge 107, 189(1977).
8. J.C. Fisher, J.Appl.Phys. 22, 74(1951).
9. R.T.P. Whipple, Phil. Mag. 45, 1225(1954).
10. T. Suzuoka, Trans. Jap. Inst. Metals 2, 25(1961;J. Phys. Soc. Japan 19, 839(1964).
11. R.W. Balluffi in DIFFUSION IN CRYSTALLINE SOLIDS, eds G. Murch and A.S. Nowick, Academic Press, NY, 1984.

BIBLIOGRAPHY:

1. J.P.Stark, SOLID STATE DIFFUSION, Wiley, NY, 1976.
2. For diffusion data see Diffusion and Defect Data, eds. F.H. Wohlbier and D.J.Fisher, Trans. Tech. Publications, Aedermannsdorf, Switzerland.
3. G. Martin and B. Perrhaillon in GRAIN BOUNDARY STRUCTURE AND KINETICS, ASM, Metals Park, Ohio, 1979.
4.R.J. Borg and G.J. Dienes, AN INTRODUCTION TO SOLID STATE DIFFUSION, Academic, 1988.
5. I. Kaur and W. Gust FUNDAMENTALS OF GRAIN AND INTERPHASE BOUNDARY DIFFUSION. Zeigler Press, Stuttgart, 1989.
6. I. Kaur, W. Gust and L. Kozma, HANDBOOK OF GRAIN AND INTERPHASE BOUNDARY DIFFUSION DATA. Zeigler Press, Stuttgart, 1989.
7. R. Ghez, A PRIMER OF DIFFUSION PROBLEMS, J. Wiley & Sons, NY, 1988

Metals

J.L. Bocquet, G. Brebec, Y. Limoge in PHYSICAL METALLURGY, 3RD EDITION, eds. R.W. Cahn and P. Haasen, Pergamon, 1983.

Ionic crystals

P. Kofstad, NONSTOICHIOMETRY, DIFFUSION AND ELECTRICAL CONDUCTIVITY IN BINARY METAL OXIDES, Wiley, NY, 1972.
F.Beniere and C.R.A. Catlow, MASS TRANSPORT IN SOLIDS, Plenum Press, NY, 1981.
A. Atkinson, Rev. Mod. Phys.57, 437(1985).
P. Kofstad, HIGH TEMPERATURE CORROSION, Elsevier Applied Science, 1988.

PROBLEMS

1. Distinguish between how intrinsic diffusivities are measured and how chemical diffusivities are measured.

2. Can the chemical diffusivities of two species in a binary alloy differ? Explain your answer.

3. Will there be a Kirkendall effect if the intrinsic diffusivities in a binary diffusion couple are not equal?

4. In which direction will an inert marker move if the intrinsic diffusivity of component B is larger than that of component A? In this case, in which direction will there be a net drift of vacancies?

5. What is the solution for $C(r,t)$ for diffusion into a cylindrical bar from a thin plate on the surface of thickness ΔR, having a concentration C_i, when at time t_o the concentration everwhere in the bar is C_o?

6. What is the significance of the term phonon drag?

7. Why is it possible for voids to form in a diffusion couple in the absence of heat or current flow?

8. Give one reason for enhanced atom transport in a region under ion bombardment other than the formation of short circuit defects.

9. If one component of a solid solution is normally interstitially sited why would it be expected that its activation energy for diffusion will be less than that for the substitutionally sited component?

10. Why should you expect that the temperature required to change from a random solid solution to a short-range ordered solid solution to be less than that for the homogenization of a spinodally decomposed solid solution having a wavelike composition profile?

11. What would happen to the measured activation energy for diffusion if the vacancy concentration was maintained constant at some supersaturated level?

12. Within what range of values is the expected value for the pre-exponential coefficient of the lattice diffusivity? If the apparent value for the latter quantity falls outside this expected range how would you explain this result?

13. Suppose there is strong association between an impurity atom and a vacancy and that the frequency of vacancy-host atom interchange is greater than impurity-vacancy interchange, what value does the correlation factor have? If this inequality is reversed what does the correlation factor become and what frequency then controls the diffusion of the impurity?

14. The diffusion coefficient of the host atom at the temperature and composition of the solidus is roughly a constant value (e.g. for a fcc crystal it is about $5 \times 10^{-9} cm^2/sec.$) Provide an explanation of this empirical result.

15. Grain boundaries in metals vary in their associated values of grain boundary diffusivity from values even smaller than that for lattice diffusivity to those orders of magnitude larger than the latter. Explain this observation.

16. One consequence of the observation in the previous question is that at a grain boundary triple junction it is likely that $div(J_i) = /0$, where J_i represents a vector flux in the ith grain boundary, for a constant external driving force. What do you expect will occur at the triple junction in this event?

17. Describe several possible reasons why the gradient of concentration along a grain boundary will not equal a constant in the surface accumulation technique of measuring the grain boundary diffusivity.

18. How would a measurement of the activation volume for diffusion along grain boundaries help interpret whether the mechanism for grain boundary diffusion involves vacancy migration or not?

19.   a) Estimate the tracer diffusion coefficient for nickel at 900°C.

    b) What would this diffusion coefficient equal in a 90at%Ni-10at%Al alloy at 900 °C?

    c)Estimate the self-diffusion coefficient at a high angle grain boundary in nickel at 900° C.

20. Why should atom transport along grain boundaries contribute about the same fraction of the total flux as lattice diffusion at a temperature of about $T/T_m = 0.5$?

21. Given that the activity of component B increases with concentration of component C in a ternary alloy of A,B and C. Suppose a diffusion couple is constructed in which the concentration of component B is the same initially in both halves of the diffusion couple. Let the left half of the couple initially have C(C) = 5%, while the right half has C(C) = 10%. Further suppose that the diffusivity of C << than that of B. Show the concentrations of B and C along the diffusion couple assuming local equilibrium is reached at the original join, but that equilibrium is not attained away from this join upon heating for a finite time at some elevated temperature sufficient to achieve transport of B.

IX-NUCLEATION KINETICS

INTRODUCTION

We have developed in Chapters V and VIII the knowledge needed to derive rates of nucleation and spinodal decomposition. The steady-state nucleation rate is proportional to the product of the equilibrium distribution of nuclei and the rate of attachment of molecules to the nuclei, which for nucleation in condensed phases involves diffusion or diffusion-like jumps across an interface. The time dependence of the nucleation rate follows from a consideration of the mode of growth and shrinkage of embryos. Experimental verification of the theory is very difficult and to the extent that it has been carried out does not contradict the theory. The classical theory needs modification in thin film deposition, because the critical nucleus contains but one atom.

The rate of spinodal decomposition depends on the thermodynamic driving force and diffusion over a penetration distance corresponding to a quarter wave length of a composition wave and can be shown to have a maximum at a particular wave length of the periodic composition wave developed in this mode of decomposition of a metastable host.

1. RATE OF HETEROPHASE NUCLEATION

1.1. Homogeneous nucleation theory and its experimental verification.

One approach to obtaining a relation for the steady-state rate of nucleation assumes that an equilibrium distribution of nuclei in the untrans-formed volume can be maintained despite the draining off of nuclei by their growth. In this approach the steady-state nucleation rate becomes this equilibrium number multiplied by the rate at which nuclei accrete molecules, i.e.

$$J^*_{eq} = \beta^* N \exp[-\Delta G^*/kT] \qquad (9.1)$$

where $N\exp[-\Delta G^*/kT]$ is the equilibrium number of nuclei per untransformed volume and β^* is the number of atoms or molecules within a jump distance of an attachment site about the critical embryo multiplied by the jump frequency toward the embryo. The latter frequency will depend upon the nature of the embryo. For example, in the case of nucleation of a pure solid from a pure liquid the jump frequency will have the form

$$\nu = \nu_0 e^{-Q/kT} \qquad (9.2)$$

where Q is the activation energy for the jump that leads to the attachment of one atom in the liquid to the solid embryo. In the case of formation of a solute rich precipitate Q will be the activation energy for the jump of a solute atom from a position in the lattice to a position on the surface of the critical precipitate embryo. If the latter is coherent with the matrix, then Q is the activation energy for diffusion of the solute in the matrix. Only in the case of nucleation of a condensed phase from a vapor phase will the relation for β^* not contain an activation energy term. In the latter situation, β^* is obtained from the kinetic theory of gases as the collision rate of molecules with the condensed embryo.

Thus, the steady-state rate of heterophase nucleation in condensed phases in the above approach is temperature dependent and of the form

$$J^* = J^\circ \exp[-(Q+ \Delta G^*)/kT] \qquad (9.3)$$

This relation only attempts to describe the steady-state rate of nucleation. However, it is known that the rate of nucleation can be time dependent subsequent to a change in the boundary conditions (e.g. temperature) and neither (9.3) nor its derivation provides an expected theoretical time dependence of the nucleation rate. Thus, an alternate approach to the derivation of (9.3) is required to attempt to describe this time dependence of the nucleation rate.

The classical alternate approach, originally due to Becker and Doring, focusses on the reactions involved in the growth and shrinkage in size of the embryos having n molecules per embryo. This approach yields a system of coupled differential equations of the form

$$[dN_{n,t}/dt] = k^+_{n-1} N_{n-1,t} - [k^-_n N_{n,t} + k^+_n N_{n,t}] + k^-_{n+1} N_{n+1,t} \cdots \qquad (9.4)$$

where k^+_n is the rate of addition of molecules to an embryo of size n, and k^-_n is the rate of loss. The time dependent nucleation rate is given by

$$J_n(t) = k^+_n N_{n,t} - k^-_{n+1} N_{n+1,t}$$

The solution for the time dependent nucleation rate upon substitution and appropriate approximation is

$$J^*(t) \cong J^* e^{-(\tau/t)} \qquad (9.5)$$

where $(\tau \cong -4kT/[\beta^*(\partial^2(\Delta G^\circ)/ \partial n^2)|_{n^*}]$, β^* is the rate at which single atoms impinge upon the critical embryo, ΔG° is the Gibbs free energy of formation of an embryo containing n atoms. The steady state rate of nucleation is given by

$$J^* = Z\beta^* N \exp[-\Delta G^*/kT] \qquad (9.6)$$

where Z is the Zeldovich[2] factor given by

$$Z = [-(1/(2\ kT))(\partial^2(\Delta G^\circ/ \partial n^2)|_{n^*}]^{1/2}$$

The product of the exponential term in (9.6) by N is simply the equilibrium number of heterophase embryos of critical size (see equation 5.6) per unit volume. Z times this number is the actual number of critically sized embryos per unit volume. The Zeldovich factor has a value of about 1/20.

Comparison of equations 9.1 and 9.6 for the steady state nucleation rate reveals that they differ only by the inclusion of the Zeldovich factor in the latter. The practical significance of this difference is difficult to demonstrate. However, the time-dependence of the nucleation rate can be experimentally observed on rapidly changing from one equilibrium distribution of embryos to another by rapid change of the temperature. The time dependence in equation 9.5 is found to be valid in the region near to the approach to the steady-state.

Recently, computer simulation has been used to obtain solutions to the series of coupled equations in (9.4) and these solutions have been used to test the agreement between classical nucleation theory in a condensed system and experiment. A typical result of this confrontation is shown in Figure 9.1 when the experimental treatment is such as to produce sufficiently large embryos where the assumptions that the interface energy and chemical driving force are independent of the cluster size is likely to be valid. However, for treatments likely to produce small nuclei (i.e. n^* less than about 20 molecules) the experimental data diverged from the absolute time dependence predicted by the classical theory, but not by more than a factor of about two. Thus, the classical theory of nucleation is valid for conditions corresponding to the assumptions made in its derivation. We shall show later in this chapter, that we need to be concerned about the latter point when n^* is on the order of unity, as occurs often in deposition of thin films from the vapor phase.

In the recent past doubt had been cast on the prior experimental "verifications" of the classical nucleation theory for nucleation in solids. However, with the definitive, painstaking experimental test performed by LeGoues and Aaronson[5] there is no longer any reasonable basis for doubt that the classical nucleation theory describes homogeneous nucleation in solids accurately.

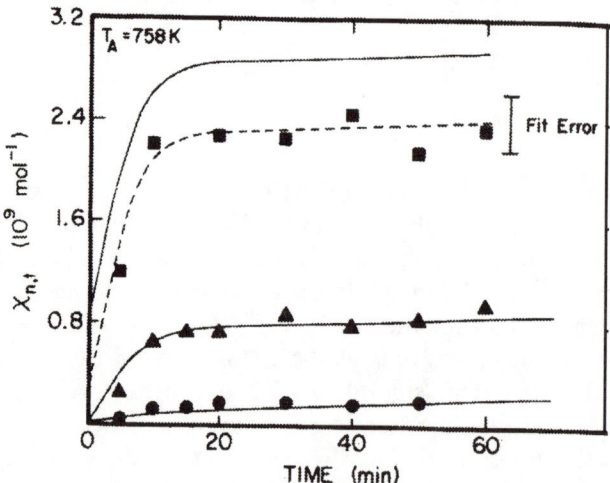

Figure 9.1. Simulation(solid line) and experimental number of nuclei(solid points) for lithium disilicate as a function of time at 758 K after annealing at 713 K for 18h, 724 K for 4.5 h, and 746 K for 45 min. After Kelton and Greer, Phys.Rev.B38, 10089(1988).

1.1.1. Temperature dependence of the steady state nucleation rate

From Chapter V we may obtain the knowledge that ΔG^* is itself temperature dependent and for the case of heterophase nucleation from a metastable parent phase, where the latter is the stable phase above the transition temperature and nucleation is occurring below the transition temperature, ΔG^* decreases as the temperature decreases below the transition temperature. For example, for the case of homogeneous nucleation of spherical precipitates, $\Delta G^* = (16/3)\sigma^3/(g_h - g_l)^2$ and to a first approximation $g_h - g_l = (s_h - s_l)(T_e - T)$. Since both the numerator of (9.6) and the denominator decrease with decreasing temperature, it is to be expected that J^* will reveal a maximum and in fact it does. If we define a time t' as the time required to nucleate a given number, N', of product phase particles per unit volume, then $t' = N'/J^*$. Hence, if J^* reveals a maximum as a function of temperature then t' will reveal a corresponding minimum, as illustrated in Figure 9.2. As shown, this dependence of t' is the same as that found in studies of the rates of transformation, where the time to achieve a given fractional volume of the product phase, t_f, exhibits the so-called C curve behavior in a plot of transformation temperature versus time. Although the latter curves have a

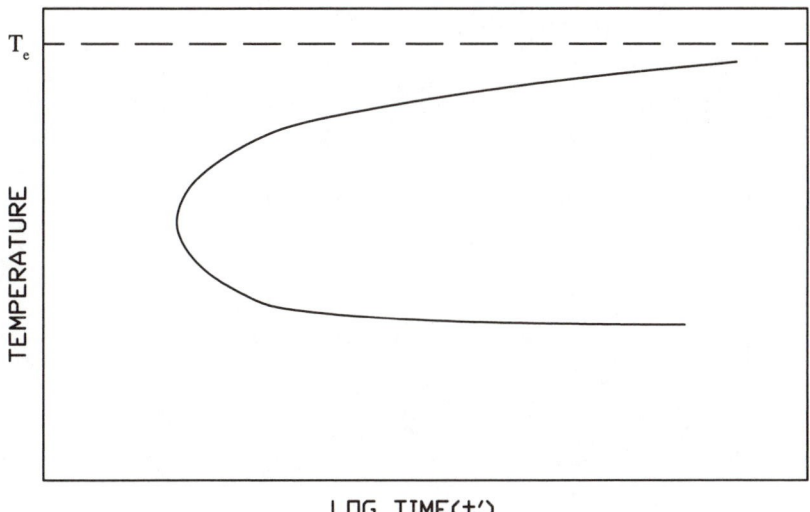

Figure 9.2. Schematic illustration of C-curve behavior in the time to nucleate a given number of stable phase particles.

more complex dependence of t_f on T than of t' on T, due to the effect of impingement of individual growing nuclei, so that $t_f \neq t'$, the presence of a minimum in t_f may or may not be due to the minimum in t'. We shall have more to say on this matter later.

The temperature at which the maximum in J^* or the minimum in t' can be obtained by differentiating J^* with respect to T and setting the derivative equal to zero. The result is that this temperature T^* is defined by

$$d[Q+ \Delta G^*(T^*)]/dT=[Q+ \Delta G^*(T^*)]/T^*$$

Figure 9.3 illustrates a graphical method of solution of this equation.

The existence of a minimum in the time to achieve a certain transformed volume cannot be taken as proof that this minimum correspond to a maximum in the nucleation rate. Another process upon which the transformed volume depends, such as the growth rate, may also exhibit a turning point. For example, suppose that the growth rate can be expressed as a product of the driving force for growth and the mobility for the growth

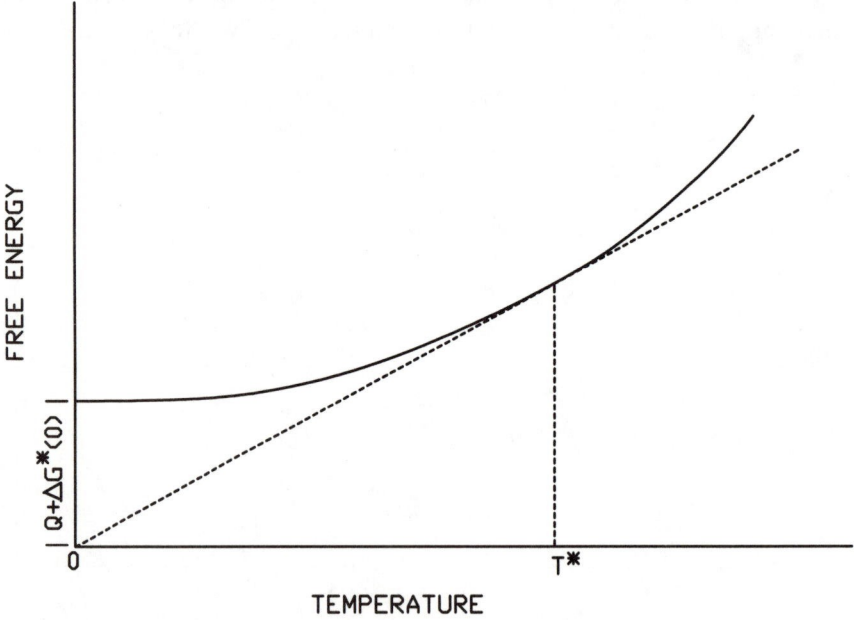

Figure 9.3. Illustrating graphical solution to: $d[Q+ DG^*(T^*)]/dT=[Q+ DG^*(T^*)]/T^*$

process. Also, suppose that the host phase is stable at high temperature while the product phase is stable at low temperature. In this case, as the temperature decreases below that corresponding to equilibrium between the two phases the driving force for growth increases while the mobility for growth decreases. Thus, the growth rate of the stable phase into a metastable host will exhibit a maximum as the temperature decreases below the corresponding equilibrium temperature between the two phases. In general, the temperatures corresponding to the maximum nucleation and growth rates will differ, as illustrate schematically in Figure 9.4. Such separation of nucleation and growth rates has been observed in glass systems.[6] The ease of forming glasses is enhanced when the temperature corresponding to the

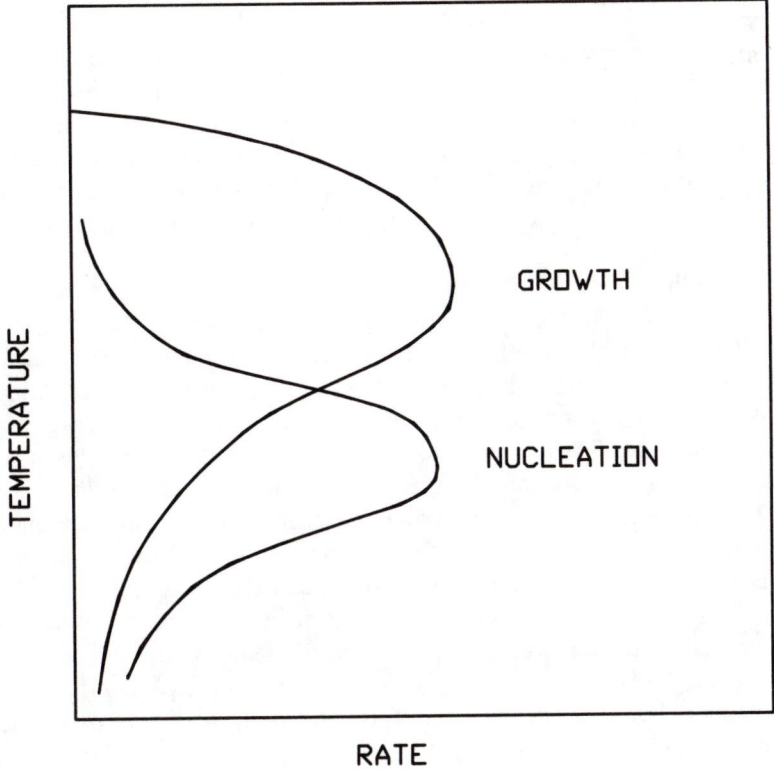

Figure 9.4. Illustrating the different temperatures for the maximum rates of growth and nucleation.

maximum rate of nucleation is less than that corresponding to the maximum in the growth rate.

The expressions given in Chapter V for the work of nucleation apply to sufficiently large embryos where the surface energy and specific free energy of the embryo does not depend on its size. Nucleation of phase transitions is a general phenomenon, not limited to any material and the relations given in this section are applicable to all materials. Specific applications of homogeneous nucleation theory require that accurate calculations of the relevant work of nucleation and the relevant accretion frequency be made. Coherency, strain energy, the need, or the absence of the need, for long range diffusion are factors that affect these parameters and cannot be ignored in any quantitative application of nucleation theory.

1.2. Heterogeneous nucleation

The overwhelming incidence of nucleation in practise is heterogeneous or catalyzed nucleation. Such heterogeneous nucleation occurs at defects within crystals, at interfaces and at any other center of heterogeneity in a parent phase. The equations applicable to homogeneous nucleation for the rate of nucleation are also applicable to heterogeneous nucleation with corrections for the work of nucleation and, where applicable, for the accretion rate.

Where the effect of the defect or interface can be described in terms of a wetting angle then the equations derived in Chapter V (i.e. (5.18)-(5.22)) are probably applicable. However, experiments reveal that grain boundaries are not homogeneous and any grain boundary may contain sites that have more excess energy than that corresponding to the average energy of the boundary. These sites of excess energy seem to be the sites at which heterogeneous nucleation along such boundaries occurs. Further, at least one of the interphase interfaces of heterogeneous nuclei at grain boundaries may be a coherent boundary so that the shape of the nucleus is not that of a lens as assumed in the derivation of the factor $f(\theta)$ in equation (5.18).

The existence of anisotropic interface energies is another factor complicating the quantitative application of the nucleation rate equations. However, the factors mentioned do not affect the principles of the theory of nucleation, but rather the practise of it or the interpretation of experimental nucleation kinetics in terms of it.

1.3. Island nucleation($n^*=1$) in deposition of thin films from the vapor phase.

As indicated above, the classical theory of nucleation is not applicable when the critical nucleus contains about 1 molecule. This situation holds in what is called the "island" regime of thin film deposition from the vapor phase. Let us consider the theory applicable to this mode of nucleation in this section.

The treatment below follows that of Venables et al.[7] When single atoms are mobile on the surface of the substrate onto which atoms from the vapor phase impinge we have

$$dn_1/dt = R - n_1/\tau_a - 2U_1 - \sum_{j=2}^{\infty} U_j \qquad (9.7)$$

$$dn_j/dt = U_{j-1} - U_j \quad (j \geq 2) \qquad (9.8)$$

where n_1, n_j are surface concentrations (number per unit area) and the U_j are the net rate of capture of single atoms by j clusters.

Venables[8] simplified these relations by dividing clusters into 'subcritical', $j \leq i$, and 'stable', $j > i$, and summing all stable clusters via

$$n_x = \sum_{j=i+1}^{\infty} n_j$$

to obtain

$$dn_1 dt = R - n_1/\tau_a - d(n_x w_x)/dt \qquad (9.9)$$

$$dn_j/dt = 0 \quad (2 \leq j \leq i) \qquad (9.10)$$

$$dn_x/dt = U_i - U_c - U_m \qquad (9.11)$$

where R is the rate of impingement of single atoms per unit area. In (9.9) the last term represents the loss of single atoms to n_x stable clusters with an average of w_x per cluster. Equation (9.10) represents detailed balance or equilibrium for the subcritical clusters yielding what is known as the Walton relation

$$\{(n_j/N_o) = (n_1/N_o)^j \sum_m C_j(m) \exp[E_j(m)/kT]$$

where N_o is the number of surface sites per unit area, m the molecular weight of the depositing molecules, C_j are statistical weights determined from configurational and other entropy of these clusters}. The last two terms in equation (9.11) are attempts to deal with coalescence; if stable clusters impinge on each other by growth (U_c) or by mobility across the substrate (U_m), then the number of stable clusters will reduce.

The term U_i in (9.11) is the nucleation rate, J, which can be expressed as

$$J = \sigma_i Dn_1 n_i \qquad (9.12)$$

where D is the single-atom surface diffusion constant, which typically can be written as $\alpha v N_o^{-1} \exp(-E_d/kT)$, σ is a "capture" number that describes the diffusional flow of single atoms to clusters[7]. At the high temperatures where adsorption-desorption equilibrium is quickly established it can be shown that

$$J = \sigma_i D(R\tau_a)^{i+1} C_i N_o^{-(i-1)} \exp(E_i/kT) \qquad (9.13)$$

Thus, measurements of the nucleation rate as a function of the impingement rate R can be used to identify the critical nucleus size i. In island nucleation, as already mentioned i = 1 and as a consequence, the actual rate of deposition is less sensitive to the nucleation rate than to the rate of diffusive capture of single atoms or capture via impingement by stable clusters.To date our knowledge of the theory of nucleation of deposited films has been of little practical use. There is much needed to be learned about the factors that control the microstructure of thin films, their surface roughness, etc.; factors that depend upon the kinetics of deposition. We shall consider some of these factors in a later chapter.

2. RATE OF HOMOPHASE NUCLEATION (SPINODAL DECOMPOSITION)

In Chapter VIII we noted that it is possible to write the chemical diffusivity in the form

$$^oD = (N_2 D_1^* + N_1 D_2^*) C_1 C_2 (d^2 g/dC_2^2)/kT$$

$$^{\circ}D = Md^2g/dC_2^2$$

where g is the molar free energy of the solution. Thus, the chemical diffusivity has the form of a product of a mobility term M and a driving force d^2g/dC_2^2. It is possible for the latter to have a negative value in a metastable solution. This fact is responsible for the so-called "uphill diffusion" with a negative diffusivity[9] that occurs in the spinodal decomposition of a metastable phase.

In Chapter V we noted that the Helmholz free energy F for a solid solution containing gradients of composition is given by

$$F = \int_V [f'(C) + \eta^2 Y(C-C_o)^2/(1-v) + K(\nabla C)^2] dV$$

The change in this free energy due to a change in the composition profile is obtained by taking the differential

$$\delta F = \int_V \{ [(\partial f/\partial C) + 2\eta^2 Y(C-C_o)/(1-v) + (\partial K/\partial C)(\nabla C)^2] \delta C + 2K\nabla C\delta(\nabla C) \} dV$$

Integrating the latter term in the right hand side of this equation by parts yields

$$\delta F = \int_V [(\partial f'/\partial C) + 2\eta^2 Y(C-C_o)/(1-v) - (\partial K/\partial C)(\nabla C)^2 - 2K\nabla^2 C] \delta C dV$$

The term in the square bracket times δC is the change in free energy due to a local change of composition δC. The driving force for change of the composition profile is then just the gradient of the term in the square bracket. We can thus set the flux of component C to be equal to

$$J = -M \text{ grad}[(\partial f'/\partial C) + 2\eta^2 Y(C-C_o)/(1-v) - (\partial K/\partial C)(\delta C)^2 - 2K \nabla^2 C]$$

where M is the mobility coefficient. We now note that

$$-\text{div}J = \partial C/\partial t = \text{div}\{M\text{grad}[\ ]\}$$

Considering only the initial stages of decomposition and neglecting all terms not linear in C yields

$$\partial C/\partial t = M\{(\partial^2 f'/\partial C^2) + 2\eta^2 Y/(1-v)\}\nabla^2 C - 2MK \nabla^4 C$$

The solution of the latter equation is

$$C - C_o = B(\beta, t)\cos(\beta d)$$

where $B(\beta, t)$ obeys

$$\partial B / \partial t = -M\{(\partial^2 f' / \partial C^2) + 2\eta^2 Y / (1-v)\}\beta^2 B - 2MK\beta^4 B$$

according to which

$$B(\beta, t) = B(\beta, 0)\exp[R(\beta)t]$$

where $$R(\beta) = -M\beta^2[(\partial^2 f' / \partial C^2) + 2K\beta^2 + 2\eta^2 Y / (1-v)]$$

The characterisitics of R, called the amplification factor, are that when the term [ ] defined by the term in the square brackets in the latter equation obeys

[ ]>0 then R<0 and the solution is stable with respect to a fluctuation of composition;

[ ]=0 then R=0;

[ ]<0 then R>0 and the solution is unstable with respect to a fluctuation of composition.

In the above it should be recalled that [ ]>0 when $\beta > \beta_c$ or $\lambda < \lambda_c$ and vice versa for reversal of the inequalities. Now for [ ]<0, the larger is R, the larger is B and the larger is the change in the composition achieved in a given time t at a given position. Hence, that fluctuation that yields the maximum value of R will grow fastest. Solving for the maximum R with respect to β yields that

$$\beta_{max} = \beta_c / \sqrt{2}$$

and $$R(\beta_{max}) = 2MK(\beta_{max})^4 = MK(\beta_c)^4$$

Since $(\beta_c)^2 = -(1/2K)[(\partial^2 f' / \partial C^2) + 2\eta^2 Y / (1-v)]$, while the second derivative term in the parenthesis is negative and the second term is positive, then the smaller is the elastic modulus Y, the larger is β_c and R. This dependence on

Y has been substantiated experimentally using observations of the orientation dependence of Y.

Physically, a maximum in the spinodal decomposition rate as a function of the wave length may be explained as follows. As the wave length increases from the critical wave length value the driving force for spinodal decomposition increases, as shown in Figure 9.5. However, in the same variation the distance over which the components have to diffuse increases and hence it will take longer to achieve a given local change in composition. The spinodal decomposition rate is proportional to the quotient of the driving force over the relaxation time associated with a given local composition change. Hence, there is a maximum in the decomposition rate as illustrated schematically in Figure 9.5.

The above classical treatment of spinodal decomposition due to Cahn and Hilliard[10] is applicable to the linear stage of the decomposition process, which occurs early in the decomposition process. The later non-linear stage has been treated by other authors-see Bibliography. Experimentally, the kinetics of the linear stage have been confirmed, but at the time of this writing no non-linear model was in agreement with the corresponding data.

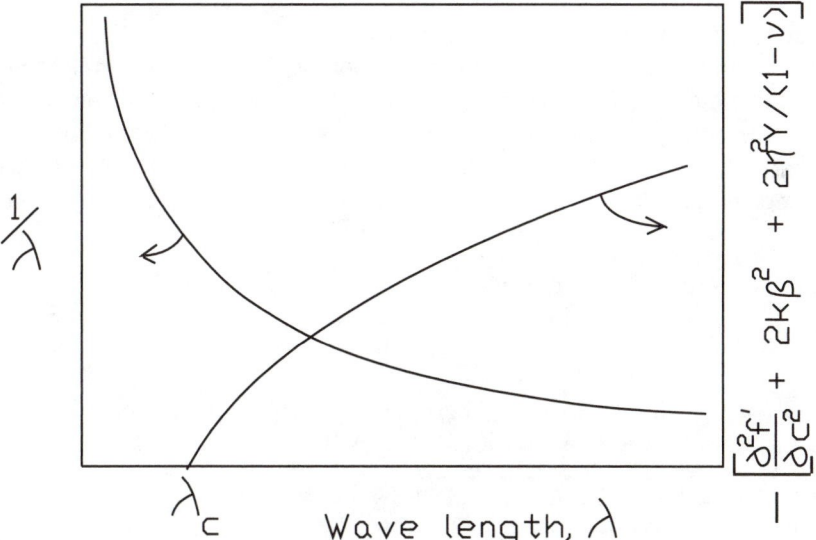

Figure 9.5. Illustrating the dependence on wave length of the inverse diffusion length and the driving force for the diffusion in spinodal decomposition.

3. SUMMARY

We have considered the steady-state rate of heterophase nucleation in a homogeneous medium and found that it is proportional to a product of the equilibrium number of nuclei (critically sized embryos) and the nucleus accretion rate for molecules. A consideration of the rates of growth and shrinkage of embryos, a molecule at a time leads to a system of coupled differential equations. An approximate solution to these equations yields an expression for the proportionality constant between the steady-state rate of nucleation and the product of the equilibrium number of nuclei and the nucleus accretion rate. This solution also gives an expression for the time dependence of the nucleation rate. Experiment confirms this classical model of nucleation.

We have also considered heterogeneous (catalyzed) nucleation and emphasized the need to take into account the anisotropy of the interface energy for many cases of nucleation in the solid state. Also, we have considered the case of nucleation during the deposition onto a substrate from the vapor phase. For conditions that lead to the "island" mode of growth, the critical number of molecules in the nucleus equals unity and, hence, the classical theory of nucleation is not applicable. The nucleation rate is derived for this non-classical case.

Finally, we considered the rate of spinodal decomposition and found that this rate is maximized at a wave length equal to $\sqrt{2}$ times the critical wave length in the linear theory. Experiment verifies the linear theory in its range of application, but not the non-linear models for the later stage of spinodal decomposition.

REFERENCES

1. K.C.Russell, PHASE TRANSFORMATIONS, ASM, Metals Park, Ohio, 1970, p.219.
2. J.B. Zeldovich, Acta Physicochim.,URSS 18, 1(1943).
3. L.Farkas, Zeit.Physik.Chem.125,239(1927).
4. R.Becker and W.Do"ring, Ann.Physik,24 719(1935).
5. F.K.LeGoues and H.I.Aaronson, Acta Met.32,1855(1984).
6. A.H. Ramsden and PJ.F. James, J. Mat.Sci.19, 2844(1984).
7. J.A. Venables, G.D.T. Spiller and M. Hanbucken, Rep.Prog.Phys. 47, 399(1984).
8. J. A. Venables, Phil. Mag.27, 693(1973).

9. P.G. Shewmon, DIFFUSION IN SOLIDS, McGraw-Hill,NY,1963.
10. J.Cahn and J.E.Hilliard, J.Chem.Phys.28,258(1958);31, 688(1959); Acta Met.9, 795(1961);
10, 179(1962).

Bibliography

Summaries of Linear Spinodal Decomposition Stage:

1. J.W.Cahn, Trans. AIME 242,166(1968).
2. J.E. Hilliard, in PHASE TRANSFORMATIONS, ASM, Metals Park, Ohio, 1970, p.497.

Later Spinodal Decomposition Stage:

1. H.E. Cook, Acta Met.18, 297(1970).
2. J.S. Langer in FLUCTUATIONS, INSTABILITIES AND PHASE TRANSITIONS, ed.
T.Riste, Plenum Press, NY, 1975, p.19.
3. B.Ditchek and L.H. Schwartz, Acta Met.28,807(1980).
4. R.D. Doherty in PHYSICAL METALLURGY,3rd edition, eds. R.W.Cahn and P. Haasen, North-
Holland, NY, 1983, p.934.

PROBLEMS

1. Why does the nucleation rate exhibit a maximum as the temperature decreases below the temperature corresponding to equilibrium between the parent and product phases?

2. Why does the rate of spinodal decomposition exhibit a maximum as the wave length of the composition wave increases from the critical value corresponding to the generalized spinodal condition?

3. In the technological process of strengthening an alloy by precipitation in a supersaturated matrix phase, the procedure of quenching the matrix phase from the solutioning temperature to a temperature T_1 and holding for a limited time prior to raising the temperature to T_2 and completing the precipitation process is used. Can you provide a theoretical justification for this procedure? (Hint. The strength of the alloy increases with the population density of the precipitates.)

4. If the condensed product phase were to nucleate from a vapor parent phase would the rate of nucleation exhibit a maximum as a function of temperature? Explain your answer.

5. If you had a measurement of the time for complete spinodal decomposition at one temperature,T_1, a knowledge of the average wave-lengths for the decomposed product at two temperatures, T_1 and T_2, and a knowledge of the activation energy for diffusion, how would you estimate the time for complete spinodal decomposition at the temperature T_2?

6. Design an experiment to yield the interface energy between a product and parent phase from measurements of nucleation rates. Discuss the problems associated with this method.

7. Assume homogeneous nucleation of crystal solids from liquids of the same composition. We compare two cases: nickel versus SiO_2. Assume crystal/melt interface energies of 25 erg/cm^2 in both cases and that the driving force for the phase change is given by $\Delta g=\Delta s(T_M-T)$. For simplicity, we will take $\Delta s=2cal\backslash mol\backslash°C$ and $T_M = 1900K$ for both materials. (Use a handbook to obtain the required specific volumes.) The frequency ν for liquid metals deduced from their viscosity is about $10^9\,T(\#/sec)$. For glasses, including silica, this jump frequency is activation energy dependent with an average activation energy about 14,000 cal/mol over the temperature range of interest, with ν_0 about $10^{12}/sec$. Calculate the number of crystal nuclei produced in 1 cm^3 of each material when the corresponding liquid is quenched from just above 1900K to 300K at the rate of $10^{6}°C/sec$. Hint: use a computer to solve the integral. Assume a nucleus grows with a velocity equal to the product of a lattice parameter, ν and $\Delta g/kT$, where the latter is dimensionless. Calculate the size of the largest crystal particle produced after the quench in each material.

8. In the spinodal decomposition of a supersaturated solution the composition-distance function normal to some crystal plane obeys the relation $C=\delta(t)\sin\beta x$, where $\delta(t)$ is a time dependent amplitude that increases with time and the other variables are defined in the text. If you plot this function for two successive values of the time of annealing you will note that solute is being transported from a low concentration to a region of higher concentration. Provide the basis for this "uphill diffusion" where solute moves up a concentration gradient.

X-KINETICS OF COMPOSITION INVARIANT SOLID/SOLID INTERFACE MIGRATION

INTRODUCTION

Primary recrystallization, secondary recrystallization, grain growth, and A/α (crystalline/amorphous) solid phase epitaxy are some of the phenomena that involve the motion of an interface between phases that have the same composition. We shall consider each of these phenomena from a common viewpoint. However, the mechanism of each phenomenon differs, and needs to be considered separately.

1. DRIVING FORCES FOR RECRYSTALLIZATION AND GRAIN GROWTH.

We consider two classes of rate processes in this chapter. One involves the migration of an interface between crystals of the same phase which, except for the excess energy associated with defects, are stable. The other involves migration of an interface between a crystal of a material and the amorphous phase of the same composition. Both, in phenomenological terms, can be described by the same relation in which the interface velocity is proportional to the product of the driving force and a mobility. However, as will be noted later in the section on mechanisms, there is little to be gained by considering the second process, known as A/α solid phase epitaxy, from this viewpoint, since the driving force does not appear to be a parameter that can be varied and its effect tested, at least at this writing.

The first class comprises the processes that go under the names of recrystallization and grain growth. It is worthwhile to discuss recrystallization and grain growth kinetics from the separate viewpoints of the driving force and mobility. Because of the generally low values of the driving force in these phenomena, it will usually be possible to express the rate of any particular one of these processes in terms of a product of a driving force and a mobility. Let us consider the various driving forces first.

For the sake of illustration of the various phenomena associated with the motion of an interface between identical phases, but of possibly different energy, consider the spherical monocrystal shown in Figure 10.1, that is surrounded by deformed crystals of the same phase in 10.1a, by smaller, but undeformed crystals of the same phase in 10.1b, and by an undeformed crystal of the same phase in 10.1c. Let the free energy per unit volume in the spherical monocrystal and in the crystals of the surrounding grains in 10.1b and 10.1c be g_o, whereas that in the surrounding area of Figure 10.1a is g_e. If g_e is the average energy per unit volume including the energy of grain boundaries in the surrounding volume, then the driving force for the migration of the boundary of the spherical monocrystal into the deformed region of Figure 10.1a is $g_e - g_o - 2\,\sigma V/R$. This is the driving force

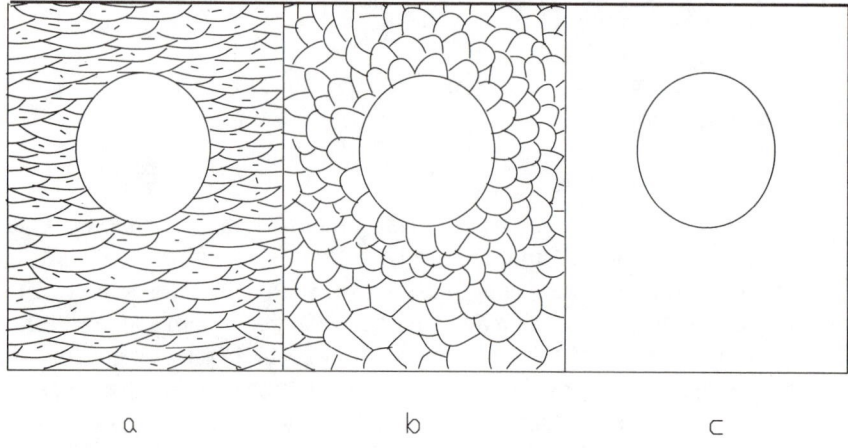

a   b   c

Figure 10.1. Schematic illustration of the initial conditions at the start of a)primary recrystallization, b)secondary recrystallization and c) grain growth. In a) the matrix is deformed. The host of b) is a fine grained polycrystal in which the boundaries are pinned and cannot migrate. The diagram in c) makes the point that during grain growth the driving force for boundary migration is reduction in boundary area.

for the stage of primary recrystallization. In general, $g_e - g_o$ is much larger than $2\sigma V/R$ and is time and temperature dependent. It is made up of the excess energy due to dislocations, grain boundaries and point defects and clusters of same. During the stage of recovery, point defects and dislocations can move so as to either decrease their densities or decrease the total free energy without annihilation. Sufficient energy remains after recovery of sufficiently cold-worked samples as to drive their primary recrystallization into product crystals that have the ground state free energy per unit volume, g_o. In this process, the excess energy due to the excess defects acts to move the interfaces between deformed and undeformed crystals so as to annihilate the former and produce more of the latter.

In the stage corresponding to grain growth, Figure 10.1c, if the radius of the spherical monocrystal increases by dR, the energy associated with the grain boundary increases by $d(4\pi R^2)\sigma = 8\pi R\sigma dR$, which when divided by the volume of the atoms transported across the boundary in its motion by dR, $(4\pi R^2 dR)$, yields $2\sigma/R$ as the free energy increase per unit volume of "transformed" material. Hence, the driving force is $-2\sigma/R$ since the latter is the free energy of the material before "transformation" minus that after "transformation". The same result can be obtained by using a result derived in Chapter IV, which gives the difference in chemical potential between material in the spherical mono-crystal at the interface and material in its surroundings, but at the interface, directly as $2\sigma/R$, with that in the monocrystal being higher than in the surroundings. Thus, the driving force for grain growth is such as to tend to move the grain boundary toward its center of curvature, i.e. to decrease the diameter of the spherical monocrystal. There seems to be a contradiction between the term "grain growth" and the direction in which the grain boundary moves so as to achieve that grain growth. This contradiction can be resolved on considering that in grain growth it is the average grain size that increases, but this average grain size increase is accomplished by the decrease in grain boundary area per unit volume. This decrease in grain boundary area per unit volume corresponds in detail to the motion of individual grain boundary segments toward their centers of curvature. This process is schematically illustrated in Figure 10.2.

The remaining situation of Figure 10.1b corresponds to the process called secondary recrystallization or exaggerated grain growth or abnormal or discontinuous grain growth. In this case, the driving force is the excess energy associated with the grain boundaries of the smaller grains that surround the spherical monocrystal. If the average diameter of these grains is D^*, then the grain boundary area per unit volume associated with these

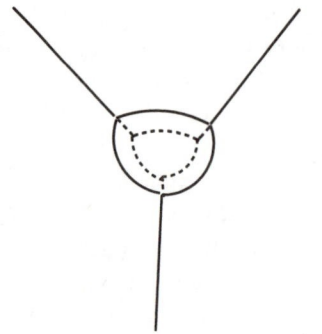

Figure 10.2. Schematic showing decrease in grain boundary area as boundaries move toward their center of curvature.

grains is C/D^*, where C is a constant whose value depends upon the shape of the grains. For cubic shaped grains of the same size C=3; for grains that are equal tetrakadecahedra, the grain boundary area equals $3.35/L$, where L is the separation of square faces. Thus, the driving force for secondary recrystallization is $C\sigma^*/D^* - 2\sigma/R$. Secondary rcrystallization will occur when $D^* < C\sigma^*R/2\sigma$. This condition has to be satisfied during the entire stage of secondary recrystallization If the average grain size of the surrounding grains can increase with time it becomes necessary for R to increase at a greater rate. Because this condition is difficult to satisfy, secondary recrystallization has only been found when the surrounding grain boundaries are prevented from moving. This restraint on boundary motion can be achieved via the use of particles, preferred orientation, or in the case of thin films with grain sizes corresponding to the limit conditioned by the thickness of the film. Let us for the moment accept that these factors can allow us to maintain a constant average grain size in the region surrounding the spherical monocrystal. In the next section we will consider the growth laws for the three stages of growth described above.

In thin films, where the grains extend completely from one surface to the other, there is an additional driving force that can act to cause secondary recrystallization. This driving force is the difference in surface energy between the differently oriented grains. When the surface energy is highly anisotropic then a significant driving force can exist to grow those grains having surfaces with the lowest surface energy at the expense of the other differently oriented grains. The process of producing cube texture oriented Fe-Si transformer sheet takes advantage of this effect.

Most of the data that have been obtained in an attempt to elucidate the phenomenological relations governing recrystallization and grain growth have come from studies on metals. However, observations of these phenomena in ceramics and semiconductors have been found to obey the same relations.

A listing of the magnitudes of the driving forces for the various modes of boundary migration follows

A/α Solid phase epitaxy—————	10^9 N/m^2($\approx$2500 cal/mole)
Primary recrystallization————	10^8 N/m^2($\approx$250 cal/mole)
Grain growth—————————————	10^4 N/m^2($\approx$0.025 cal/mole)
Secondary recrystallization—	10^3 N/m^2($\approx$0.0025 cal/mole)
Electromigration——————————	10^2 N/m^2($\approx$0.00025 cal/mole)

2. GROWTH LAWS FOR PURE MATERIALS.

2.1. Primary recrystallization.

On the assumption that the growth rate dR/dt is given by

$$dR/dt = M(g_e - g_o) = M\Delta g$$

and that the driving force changes with time during the recovery process according to simple second-order kinetics, i.e.

$$\Delta g^{-1} - \Delta g_o^{-1} = k_r t$$

where Δg_o is the driving force corresponding to the start of isothermal annealing at t=0, k_r is a rate constant for the recovery process and t is the time at the annealing temperature, Li[1] has shown that the growth for the primary recrystallized grains obeys

$$(dR/dt)^{-1} = M^{-1}(\Delta g_o^{-1} + k_r t)$$

Figure 10.3 illustrates a case where this law is obeyed experimentally. Incidentally, the fact that the points corresponding to different annealing temperatures are superimposed suggests that the activation energies for the recovery process and the growth of the recrystallized grains are nearly equal in this case.

The phenomenological law of recrystallization is given by the following relation

$$X_v = 1 - \exp(-Bt^k)$$

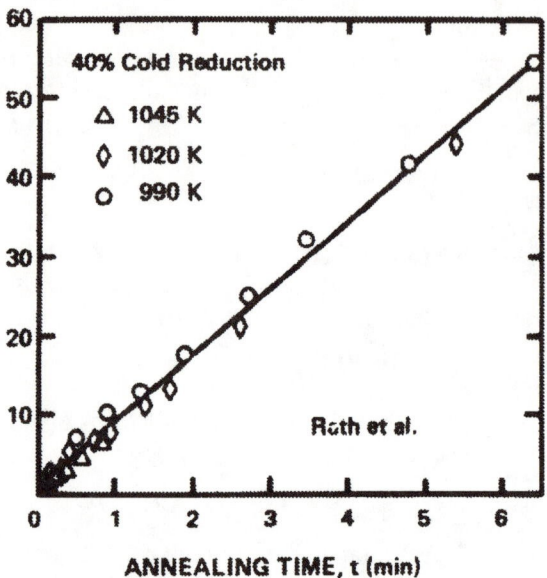

Figure 10.3. Reciprocal growth rate as a function of time for beta titanium. After Rath et al, Met. Trans.10A, 1013(1979).

where X_v is the volume fraction recrystallized, t is time and B and k are constants. Deviations from this law are believed to be due to the simultaneous effect of recovery on the rate of growth, as discussed in the previous paragraph.

2.2. Secondary recrystallization.

There is an incubation period for the onset of secondary recrystallization. After this incubation period the growth rate is independent of time. This is just the behavior expected from the dependence of the growth rate on the driving force discussed above, i.e.

$$dR/dt = M_s[C\sigma^*/2R^* - 2\sigma/R]$$

According to this relation, the incubation period corresponds to the case where R is sufficiently small that the two terms in the bracket are similar in magnitude. When R has grown to the point that the second term in the bracket is small compared to the first term and the first term is constant, then the successive growth rate will be independent of time. Although this relation provides a qualitative explanation of the observations, it has not yet been proven that it is in quantitative agreement with experiment.

2.3. Grain growth.

A law of grain growth can be derived from the relation that

$$dR/dt = M_g[-2\sigma /R]$$

as follows. The rate of change in the total grain boundary area per unit volume

$$d[C/D]/dt = \{d[C/D]/d<R>\}(d<R>/dt)$$

or
$$dD/dt = \{dD/d<R>\}d<R>/dt$$

where D is the average grain diameter and $<R>$ is the average radius of the moving boundaries. Now, it is known that the grain size distribution function is time invariant. This fact implies that the average radius, $<R>$, of the moving boundaries is proportional to D, i.e. $D\sim<R>$. Another giant step is to assume that the velocity corresponding to $d<R>/dt$ is the same as that of a grain having the radius of curvature $<R>$, but of opposite sign. With these assumptions we arrive at the result

$$dD/dt = (k/D)$$

which integrates to give

$$D^2-D_0^{\,2} = 2k(t-t_0)$$

This law is obeyed sometimes. The phenomenological law of grain growth is $D=K_1 t^n$, where n can vary from 1/3 to 1/2.

A more rigorous analysis of grain growth is provided by a Monte Carlo based computer simulation of it.[2] This analysis yields the relation

$$D^m-D_0^{\,m} = 2k(t-t_0) \qquad m=2.4,$$

which is in better agreement with the data. Further, the same computer simulation provides an explanation for this result. Detailed analysis of the computer simulation showed that grains larger than the average execute a random walk in grain size-time space (i.e. grow and shrink in a random fashion). However, grains much smaller than the average all shrink and gradually disappear. Thus, a more appropriate relation for the time dependence of the average grain size is one that incorporates random walk for large grains and curvature driven kinetics for small grains. Louat[3] has given an analysis of grain growth based on random walk of grains in the grain size-time space and Hillert[4] has given another based on curvature driven kinetics.

Computer simulation has clarified another effect on the time dependence of the average grain diameter during grain growth. Anisotropy of grain boundary energies introduces an effect on the time exponent n in the empirical relation $D = Kt^n$. Increasing anisotropy decreases n until it reaches the value 0.25.[5] Thus, although the law governing the motion of an individual boundary may be governed by curvature, other factors have to be taken into account in describing the dependence of average grain diameter on time for an assembly of boundaries.

2.4 A/α solid phase epitaxy.

Since the driving force in solid phase epitaxy for the motion of the amorphous-crystal (α/A) interface between amorphous and crystalline phases having the same composition is independent of time it is expected that the velocity of the interface will be independent of time, as well. This expectation is in agreement with the observations, but this result is not informative of the mechanism applicable in this phenomenon. Other observations will be more useful in this regard and will be discussed in a later section.

3. EFFECT OF DISPERSED PARTICLES ON BOUNDARY MIGRATION

Consider a grain boundary, tending to migrate in the negative y direction, which meets a spherical particle of radius r, as is illustrated in Figure 10.4. The maximum drag force resolved in the y direction due to the particle is $\pi r \sigma$. Given that there are n particles per unit volume, randomly distributed. Hence, a boundary of unit area will intersect a total of 2nr particles (i.e. there are 2nr particle centers in each region r thick about unit area of boundary). But the volume fraction of the particles is $f = 4\pi r^3 n/3$ or $n = 3f/(4\pi r^3)$. Substituting, the number of particles intersecting unit area of grain boundary is then $3f/(2\pi r^2)$. If the boundary is migrating under the influence of its own surface tension then a balance of its driving force, $2\sigma/R$, where R is the radius of curvature of the boundary, and the total drag force, $(\pi r\sigma)(3f/[2\pi r^2])$ yields that

$$R^* = 4r/3f$$

where R^* represents the largest radius of curvature of the moving boundary that can traverse the particles, i.e. there is a limiting grain size above which grain growth in a matrix containing particles is not possible. The precise value of R^* found above, a result originally due to Zener[6], has been questioned, because it appears to be in disagreement with experimental values of R^*. More recent models[4,7] yield improved agreement, but in view of the new insights that computer simulation have provided in the grain growth problem it would be foolhardy to suggest that we have heard the last word in this matter. Nevertheless, it is likely that Zener's concept, for the origin of a critical radius beyond which normal grain growth ceases, is likely to be valid. Thus, one of the means of maintaining an immobile grain boundary array during secondary recrystallization is the use of inert particles deliberately added to the matrix, such as AlN in steels, or via particles precipitated out of a supersaturated matrix at a temperature above that at which the secondary recrystallization is carried out.

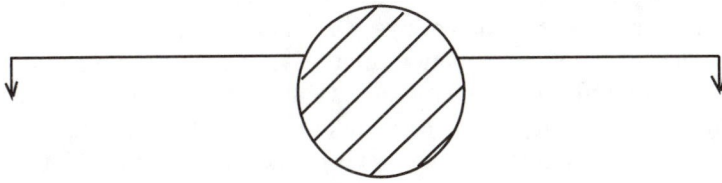

Figure 10.4. Schematic illustration of the increase in grain boundary area as it moves past the particle encompassing it.

4. GRAIN BOUNDARY MOBILITY

Any discussion of grain boundary mobility is faced with the problem, because of the gamut of possible grain boundaries and concomitant grain boundary structures, that there may be a variety of grain boundary mobilities and their dependences upon variables, such as driving force, solute concentration and the like. One may conceive of two extremes of grain boundary behavior. In one extreme atoms may be added with equal probability anywhere to the surface of a growing grain across a grain boundary. In the other, atoms can only be added along kinks at steps on the surface of a growing grain across the grain boundary, where it is likely that these boundary steps are associated with grain boundary dislocations. Indeed, it is likely that both extremes of behavior have been observed as will be noted later. However, there are other possible modes of grain boundary migration. Another that has been recognized is grain boundary migration constrained by the simultaneous glide and climb of grain boundary dislocations. With multiple mechanisms of migration possible for a single grain boundary (i.e. the mechanism may change with the orientation of the boundary for a given misorientation between the bounding grains) it is not difficult to understand that precise rules governing the mobilities of grain boundaries have yet to be developed.

When conditions are such as to bring about a diversity of grain boundary mobilities, then the boundaries with the highest mobility will by their migration tend to produce the largest volume fraction of product. Such product may be either the strain-free grains resulting from primary recrystallization or the abnormally large grains that grow during secondary recrystallization. This phenomenon is responsible for the textures that are produced by these processes. For example, a very common recrystallization texture in fcc metals and alloys is a rotation about a <111> axis by 30 to 40° relative to the deformation texture. As shown in Figure 10.5 the boundaries of grains having a rotation of 40° about <111> relative to the deformation texture move at the maximum rate in deformed aluminum and this is just the rotation of the primary recrystallization texture relative to the deformation texture found for aluminum, as shown in Table 10.1. This recrystallization texture is so common that it has been given the name of the Kronberg-Wilson texture, after those who first noted it. The rapidly moving grain boundaries choke off the growth of the grains with slower moving boundaries and thus those grains having the appropriate relative orientation to the parent matrix to confer maximum mobility to their boundaries act to

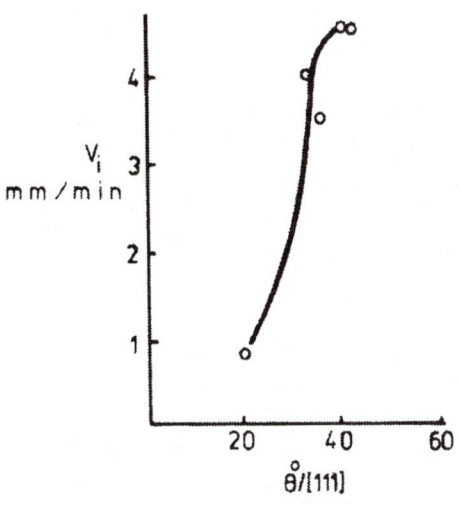

Figure 10.5. Growth rates of new grains into plastically deformed aluminum crystal at 615°C. The [111] axis is always the same one in the deformed crystal. After Liebmann and Lucke, Trans. AIME 209, 427(1956).

consume the parent matrix.

The explanation why the boundaries of those grains having the appropriate orientation move at the maximum velocity is not as yet understood. Originally, significance was placed on the fact that the orientations of the recrystallization textures, such as listed in Table 10.1, are near those for a high degree of coincidence of the atom sites for the two crystal orientations. But there are relative grain orientations that yield higher degrees of coincidence (i.e. a smaller value of Σ than that for the Kronberg-Wilson orientation (7)) as well as lower degrees of coincidence.* No one has noted anything special about the $\Sigma =7$ <111> grain boundaries relative to the other equivalent coincidence boundaries. Further, it appears that a small amount of impurity content in the matrix, but not too much, is necessary for the Kronberg-Wilson texture to be observed. From the work of Aust and his coworkers, it appears that in very pure materials there are many high angle grain boundaries that have mobilities equal to those for the Kronberg-Wilson case

Table 10.1

Material	Rotation angle and axis	Neighboring coincidence orientation
Cu	30° [111]	$\Sigma = 13$
Al	40° [111]	$\Sigma = 7\ 38.2°$ [111]
Zn	30° [0001]	$\Sigma = 13$ [0001]
Fe-Si	27° [110]	$\Sigma = 19\ 26.$ [110]
Al 0.5% Mn . .	43° [661](a)	?
(a) Disorientation corresponding to 153° [335].		

* Σ^{-1} is the fraction of sites belonging to the two lattices of the crystals that bound the grain boundary that are coincident.

but that the effect of a small concentration of impurities is to markedly diminish the mobility of the random high angle boundaries relative to that for the Kronberg-Wilson orientation.(See Figure 10.6.) Thus, one possible explanation of the fact that the Kronberg-Wilson boundaries have higher mobility than the other boundaries is that dilute solute acts to decrease the mobility of all boundaries by a different mechanism than that applicable to the Kronberg-Wilson case. We shall consider later some theories for the effect of solute on the mobility of grain boundaries and return to a consideration of this possible explanation.

The measurements of boundary migration that have been conducted imply that in very pure materials almost all high angle and coincidence boundaries have about the same mobility. The exceptions are the high coincidence coherent twin boundary and the low angle boundaries (except for the simple symmetrical tilt boundaries which can glide). Also, it appears that when the driving forces are high and the solute concentration is high the differences in boundary mobility noted in the previous paragraph disappear.

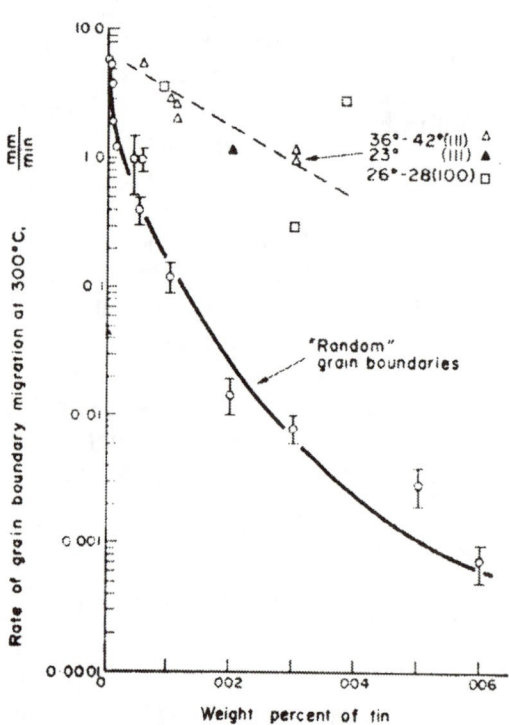

Figure 10.6. Effects of solute concentration on "special" and "random" grain boundaries. After Aust and Rutter, Acta Met.$\underline{6}$, 375(1958).

The mobility of grain boundaries in silicon exhibits an effect not found in metals. In particular, certain dopants, such as P and As, enhance the grain boundary mobility relative to that in pure silicon. It is believed that this effect is a consequence of the effects of these dopants on the concentrations of certain grain boundary dislocation kinks and jogs. (Donors enhance this concentration and acceptors decrease it.[8,9]) In polysilicon, twins comprise the majority of grain boundaries. The migration of twins should be dependent on the climb and glide of grain boundary dislocations containing these kinks and jogs. Thus, because grain boundaries in semiconductors are constrained to minimize the number of dangling bonds and because the electronic states of defects at grain boundaries can be affected by dopants it should be expected that the mobilities of grain boundaries in semiconductors should differ qualitatively from those in metals.

Large angle grain boundaries in ceramics appear to approximate those in metals. There does not appear to be a strong constraint on the types of grain boundaries that can appear in ceramics, except for some particular boundary planes where the geometry forces like ions to be nearest neighbors. Some light on the nature of grain boundaries that appear in polycrystalline materials may be shed by a comparison of observed grain boundary mobilities at a given fraction of the absolute melting point. Table 10.2 compares observed grain boundary mobilities in metals, silicon and ionic crystals.

Table 10.2

Mobilities for grain boundary motion during grain growth and A/α solid phase epitaxy

Material	T/T_M	Mobility (m^3/Ns)	Activation Energy, Q_{bm} (kJ/mol)
Fe	0.62	$1.3\ 10^{-10}$	
Al	0.64	$3.2\ 10^{-11}$	54
MgO	0.63	$8\ 10^{-14}$	275
Si	0.64	$1.8\ 10^{-18}$	230
Si A/α	0.64	$3\ 10^{-16}$	260
Si	0.7	$1.6\ 10^{-16}$	230
Si A/α	0.7	$6\ 10^{-15}$	260
LiF	0.69	$7.5\ 10^{-13}$	150

5. MECHANISMS OF GRAIN BOUNDARY MIGRATION

5.1. Pure materials

As indicated above there are two extremes associated with the motion of grain boundaries. In one mode of boundary migration any site along the boundary is as likely to move from one of the bounding crystals to the other. In the other mode of boundary migration, the jump of an atom from one of the bounding crystals to the other can occur only at certain sites along the grain boundary. Let us consider the first mode at this point. The treatment that follows is that originally given by Turnbull.[5] For an atom to move from a lattice site on one bounding crystal to an adjacent empty lattice site of the neighboring crystal, the unit step is equivalent to that of exchange of an atom with a vacancy in lattice diffusion. Since, every atom along the boundary is assumed to jump at the same rate it is only necessary to relate the boundary velocity to the jump rate of one atom at the boundary. Let the boundary move by the distance d for such a jump. The boundary velocity will then be given by

$$V = d\nu \exp(-\Delta G_{bm}/kT) \, \Delta G/kT$$

where ν is the frequency of vibration of the atoms at the boundary sites, ΔG_{bm} is the free energy of activation for the jump of an atom from one lattice to the other at the boundary in the absence of a driving force and ΔG is the difference in free energy of the system between that when the atom is in the position before the jump and that after the jump.

Measured values of boundary migration velocities in pure metals yield values for the activation energy and preexponential term that are consistent with the above equation. In particular, the activation energy for boundary migration is about half that for self diffusion in metals. Also, the calculated and experimental preexponential term values are in reasonable agreement.

However, it is recognized that the model on which the above equation for the boundary velocity is based is unrealistic in the sense that the kinetics of attachment/disattachment is assumed to be the same for every atom at the boundary. Turnbull's model is more likely to be applicable to the A/α solid phase epitaxy boundary migration. Indeed, reference to Table 10.2 shows that the mobilities of the A/α boundaries are much higher than the grain boundary mobilities in silicon. We must conclude that since the crystalline nature of the grains surrounding the boundary has an effect on mobility of the boundary that Turnbull's model is too simple and not applicable to grain boundary migration.

Other models have since been proposed in an attempt to make more specific the actual motions of the atoms at the boundary in the migration process. In particular, Gleiter[10] has proposed a model based on the assumption that migration occurs by the motion of atoms from kinked lattice steps of close-packed planes in the shrinking crystal, that end at the boundary, to nearby similar steps associated with the growing grain. This model predicts a dependence of the mobility both on grain misorientation and for a fixed grain misorientation on boundary orientation. Some aspects of these predictions are in agreement with experiment (small misorientation) and other aspects disagree with experiment (the dependence near critical misorientations.)

More recently, Smith and coworkers[11] have noted that boundary migration can occur by the migration of DSC lattice dislocations, where such migration is usually limited by diffusion due to the presence of a dislocation climb component during this migration, and by thermally activated atomic shuffles or shears. They have noted that if the sum of the Burgers vectors of these migrating vectors does not equal zero, then there must be a net boundary sliding component accompanying the boundary migration. Evidence that such boundary sliding can occur during boundary migration has been found. However, if the sum of the Burgers vectors equals zero then the boundary must correspond precisely to a low energy structure. The difference between the two situations, other than the absence of sliding in the latter case, is the need for long range diffusion of vacancies in the former and its absence in the latter during boundary migration. It is possible that the difference in mobility of the $\Sigma = 7$ boundary and those of other boundaries is related to the differences just noted for the migration of coincidence and other boundaries. At this writing it does not appear that this matter has been explored in the literature. Indeed, there remains much to be done to advance our knowledge of the detailed mechanism of grain boundary migration in pure materials. However, there are many experiments that support the concept that long range diffusion of vacancies is involved in the boundary migration process. Lucke and coworkers[12] have considered the effects of vacancies on the boundary migration process theoretically, stemming from Lucke's work on the effect of solutes on boundary migration, which will be considered in the next section.

There are certain orientations for which the boundary migrates by a step mechanism, as illustrated in Figure 10.7. In this case, it is obvious that the boundary velocity should not conform to the previous equation. It is likely that in this case, spirals, that are formed when a crystal lattice dislocation meets a grain boundary with its Burgers vector having a component normal to the boundary plane, will provide the means of such stepwise growth, much as they do in the growth of crystals from the vapor

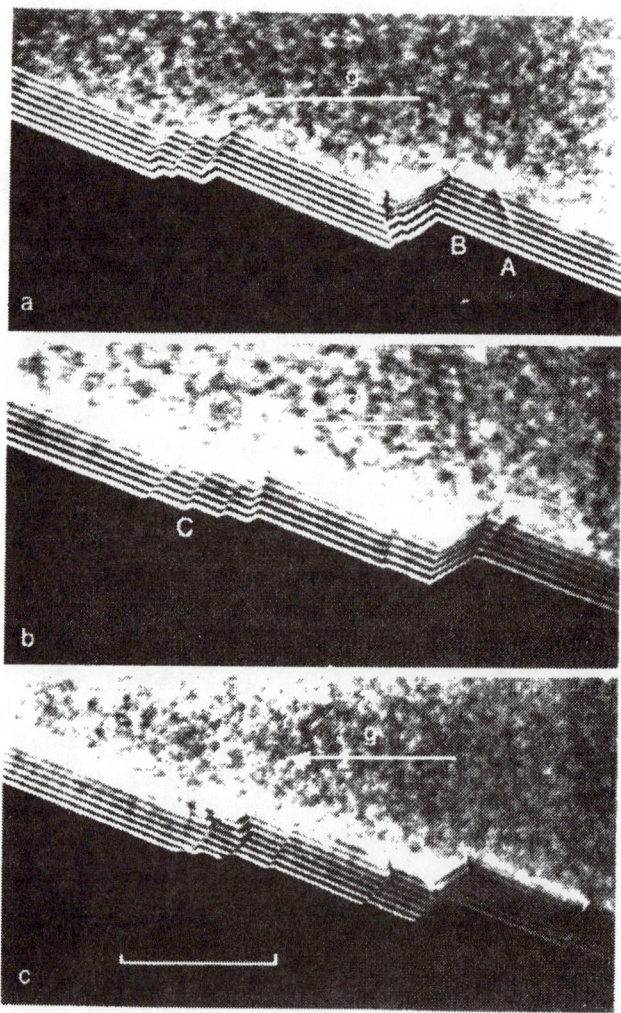

Figure 10.7. Contribution of step motion to growth of an annealing twin in aluminum; the marker represents 100nm. After Smith et al[11].

phase. Accordingly, the equation for the migration velocity becomes

$$V = nv_s h$$

where n the length of step per unit area equals $\Delta G/(20\ h\ \sigma\ d^2)$ in analogy with the crystal growth case[13], v_s is the step velocity and equals $(dv\Delta G/kT)\exp(-\Delta G^*/kT)$, h is the step height, σ is the grain boundary energy, d is the atomic diameter. Substitution then yields

$$V = [(\Delta G)^2 v /(20\ \sigma dkT)]\exp(-\Delta G^*/kT)$$

for the boundary migration velocity. Comparison of the two equations for the boundary velocity shows that the main difference is in the preexponential term, where the latter provides that the boundary velocity depends on the square of the driving force while boundary migration in the absence of spiral steps varies linearly with the driving force. There is evidence for a dependence of boundary velocity on the square of the driving force in grain growth. However, as noted previously, anisotropy of grain boundary energy can also provide the equivalent of a square dependence on the driving force by yielding n=0.25 as the exponent in the empirical growth law.

Recent experiments involving the effect of ion bombardment on grain growth lead to the conclusion that vacancies generated at or near grain boundaries are required for grain boundary migration, at least in Au, Ge and Si.[14] However, comparison of the activation energy for boundary migration and that for vacancy formation yield values of the former that are both smaller and larger than the latter. Much more experimentation will be required to unravel the mechanism of grain boundary migration, which incidentally, need not be the same in all materials.

It is interesting that studies of the effect of ion bombardment on the motion of the A/α interface in solid phase epitaxy in silicon have led to a model for this interface migration.[15] It was proposed that the defect responsible for the motion of the interface is a dangling bond generated in the amorphous phase. If this model is correct, then it can not be applied to grain boundary migration. However, another candidate for this defect, examined in reference 14, was a combination of vacancies and interstitials in the crystalline phase. It was not rejected for a valid reason; the dangling bond concept was more esthetically satisfying to the investigator. Since the combination of vacancy and interstitial is also a candidate defect that could explain the observations on the effect of ion bombardment on the migration of grain boundaries[14] it would appear to this author to be a more esthetically satisfying choice. Hence, the question concerning the identity of the defect responsible for interface migration is still an open one.

5.2. Effect of interface roughness on migration mechanism.

Since singular interfaces can migrate only as a consequence of the lateral motion of steps, whereas a rough interface may migrate via a continuous mode (as per Turnbull) or via a stepwise growth, it is desirable to have a criterion which can distinguish between these modes of growth. Two procedures have been used to devise such a criterion. One is based on thermodynamic concepts and is due to Cahn[16.] The other is based on statistical concepts and is due to Jackson[17]. Although the former was developed for solid state transformations and the latter for liquid/solid interfaces there is no reason why they should not apply to incoherent interfaces between phases having constant compositions, regardless of the nature of the phases.

According to Cahn, if the interface can attain a metastable equilibrium configuration during growth, in presence of a driving force, then stepwise growth is required when the driving force is less than some critical value. On the other hand, if the local displacements of an interface in its motion lead to no local minima in the free energy of the interface then the interface configuration can change in the course of growth. This condition is fulfilled for all interfaces above a critical value of the driving force. The need for consideration of the magnitude of the driving force in evaluating the operating mode of growth (i.e. stepwise or continuous) is obvious for the case of a singular interface near a cusp orientation. In this case, unless the interface can be reoriented away from the cusp orientation the interface will remain singular and grow by step propagation. The driving force needs to be sufficiently large to accomplish this reorientation for continuous growth to be possible. On the other hand, it is not obvious why non-singular interfaces cannot grow in a continuous manner. Yet, according to Cahn, they may or may not, depending upon the criterion that a lateral growth step mechanism is required whenever the interface is able to attain a metastable condition in the presence of a driving force.

According to Jackson[17a], the interface can be taken to correspond to a mixture of empty and filled sites at the interface. With N the total number of sites per unit area, the fraction of filled sites is taken to equal N_A/N and $(1-N_A/N)$ is the number of unfilled sites. The change in free energy associated with the filling of such sites is

$$\Delta G/NkT_e = \alpha N_A[(N-N_A)/N^2] - \ln\{N/(N-N_A)\} - N_A/N\ln[(N-N_A)/N_A]$$

where $\alpha = L/kT_e$, L is the heat of the transformation and T_e is the equilibrium temperature of the transformation. This equation predicts that for $\alpha < 2$, the interface is rough and growth is continuous, while for $\alpha > 2$, the interface is smooth and growth occurs in a stepwise mode. Regardless of the validity of the derivation, there is evidence to support this separation between the two modes of growth in the liquid/solid transition at about $\alpha = 2$.

Jackson[17b] has shown that α plays a role in the rate of migration of the solid/liquid interface in a pure material. He derives a net growth rate, given by

$$V = v'J\alpha f(h,k)(\Delta T/T)\exp(-Q/RT)$$

where v' is the atomic volume, $\alpha = L/kT_e$ as before, f(h,k) is a geometric factor that is dependent on α and the Miller indices, $\Delta T = T - T_e$, Q is the activation energy for diffusion in the liquid and J is a frequency factor. Jackson found that for the orientation of the fastest growing edges (faces), f(h,k) is independent of α. Also, he found that for small α, f(h,k) is nearly independent of the crystal indices. However, at high α, where there is a large difference between the f(h,k) values for different orientations, the fastest growing edges (faces) disappear early in the growth leaving behind the slowest growing edges (faces). The implication is that at small values of α, growth is almost isotropic (non-faceted), whereas, for high α, growth is faceted; a conclusion that is in agreement with Jackson's first model.

5.3. Effect of solute on boundary mobility

It is well known that boundary mobility can be very sensitive to the presence of solute. Indeed, only a few ppm of solute can change the boundary mobility by orders of magnitude, as shown in Figure 10.6. The theory of the effect of solute on grain boundary mobility has been explored by Lucke and coworkers[18] and by Cahn[19]. Their theories based on the existence of a distance-dependent interaction energy between the solute and the grain boundary, E(x), yield different relations between the boundary velocity and the driving force. For the low velocity limit, this relation takes the form

$$v = P/[M^{-1} + \alpha C]$$

where M is the mobility, P is the driving force, C is the bulk solute concentration and

$$\alpha = 4N_V kT \int_{-\infty}^{+\infty} \{[\sinh^2 (E(x)/2kT)] / D(x)\} dx$$

where $D(x)$ is the solute diffusivity, which is a function of the distance, x, between solute and boundary. At the high velocity limit, the boundary velocity becomes

$$v = MP(1 - \alpha C/(\beta^2 P^2 M))$$

where

$$\alpha/\beta^2 = (N_V/kT) \int_{-\infty}^{\infty} (\partial E/\partial x)^2 D(x) dx$$

The relation applicable to both regimes is

$$P = M^{-1}v + \alpha Cv/(1 + \beta^2 v^2)$$

There are data which support some of the concepts associated with these theories. In particular, in the low velocity region it is predicted that the reciprocal of the boundary velocity should depend linearly on the solute concentration and this functional relationship has been found in the work of Gordon and Vandermeer[20] in dilute aluminum alloys. Further, the relation between velocity and driving force given by the last equation has been verified qualitatively in the experiments of Sun and Bauer[21] and Drolet and Galibois[22]. However, some aspects of the results of the latter workers are not in agreement with the theory, as noted by Simpson et al[23]. There is little doubt that, as assumed in the theory, the solute drag on the grain boundaries does involve the lattice diffusion of the solute, as indicated by the fact that the activation energy for boundary migration in the low temperature and low velocity regimes approximates that for self-diffusion.

The models of solute drag have also been applied to a study of the effect of solute on boundary migration in non-metallic substances.[24]

6. SUMMARY

We have shown that the phenomena of A/α solid phase epitaxy, primary recrystallization, secondary recrystallization and grain growth can be described in common terms. The driving forces for these processes decrease in the order listed. The laws governing grain growth in each of these regimes have been discussed. We found that a distribution of insoluble particles can lead to the cessation of growth. The mobilities of the grain boundaries were then discussed and mechanisms for their migration were suggested. The large effect of solute on these mobilities was then studied.

REFERENCES

1. J.C.M. Li in RECRYSTALLIZATION,GRAIN GROWTH AND TEXTURES,ed. H. Margolin, ASM, Metals Park, Ohio, 1966, p.45.

2. M.P. Anderson, D.J. Srolovitz, G.S. Grest and P.S. Sahni, Acta Met. $\underline{32}$, 783(1984).

3. N.P. Louat, Acta Met.$\underline{22}$, 721(1974).

4. M. Hillert, Acta Met.$\underline{13}$, 227(1965).

5. D. Turnbull, Trans. Met. Soc. AIME $\underline{191}$,661(1951).

6. C. Zener, in C.S. Smith, Trans.A.I.M.E.$\underline{175}$, 47(1948).

7. P.R. Rios, Acta Met.$\underline{35}$, 2805(1987).

8. P.B. Hirsch, J. Physique, Colloq.$\underline{C6}$, 117(1979).

9. D.A. Smith, T.Y. Tan and C. Fontaine in Advances in Ceramics, vol.6, eds. M.F. Yan and A.H. Heuer, American Ceramics Society, 1983.

10. H. Gleiter, Acta Met.$\underline{17}$, 565,853(1969).

11. D.A. Smith, C.M.F. Rae and C.R.M. Grovenor in GRAIN BOUNDARY STRUCTURE AND KINETICS, ASM, Metals Park, Ohio,1980, p.337.

12. Y.Estrin and K. Lucke, Acta Met. $\underline{30}$, 983(1982).

13. W.K. Burton, N. Cabrera and F.C. Frank, Phil.Trans.Roy.Soc. $\underline{A243}$,299(1950-1).

14. H.A. Atwater, C.V. Thompson and H.I. Smith, J. Appl. Phys.$\underline{64}$, 2337(1988).

15. K.A. Jackson, J. Mater. Res.$\underline{3}$, 1218(1988).

16. J.W. Cahn, Acta Met.$\underline{8}$, 556(1960).

17. a)K.A. Jackson in LIQUID METALS AND SOLIDIFICATION. American Society for Metals, 1958; b) J. Cryst. Growth, 3/4, 507(1968) .

18. K. Lucke and K. Detert, Acta Met.$\underline{5}$,628(1957); K.Lucke and H.P. Stuwe in RECOVERY AND RECRYSTALLIZATION OF METALS, ed L. Himmel, Wiley-Interscience,1962,p.171;K.Lucke and H.P.Stuwe,Acta Met.$\underline{19}$, 1087(1971).

19. J.W. Cahn, Acta Met.$\underline{10}$, 789(1962).

20. P. Gordon and R.A. Vandermeer, Trans. Met. Soc. AIME $\underline{215}$, 577(1959);$\underline{224}$, 917(1962);in RECRYSTALLIZATION, GRAIN GROWTH AND TEXTURES,ed H. Margolin, ASM, Metals Park, Ohio, 1966, p.205.

21. R.C. Sun and C.L. Bauer, Acta Met. $\underline{18}$, 635, 639(1970).

22. J.P. Drolet and A. Galibois, Met Trans$\underline{2}$, 53(1971).

23. C.J. Simpson, W.C. Winegard and K.T. Aust, in GRAIN BOUNDARY STRUCTURE AND PROPERTIES, eds. G.A. Chadwick and D.A. Smith, Academic Press, NY, 1976, p.201.

24. A.M. Glaeser, H.K. Bowen and R.M. Cannon, J. Am. Cer. Soc.$\underline{69}$, 119(1986).

BIBLIOGRAPHY

R. W. Cahn in PHYSICAL METALLURGY, eds R.W. Cahn and P. Haasen, North-Holland, NY, 1983, vol.2, p.1595.

H. Hu in Metallurgical Treatises, eds. J.K. Tien and J.F. Elleiott, TMS-AIME, Warrendale, Pa., 1981, p.385.

E.Nes, N.Ryum and O. Hunderi, Acta Met.<u>33</u>, 11(1985). This paper is a good summary of theories on the effect of particles on boundary migration.

PROBLEMS.

1. Distinguish in terms of driving force between primary recrystallization, secondary recrystallization and grain growth.

2. Why do grain boundaries migrate towards their center of curvature in grain growth and not in the recrystallization processes?

3. Why is it necessary to be able to achieve secondary recrystallization to immobilize the as-primary-recrystallized grain boundaries?

4. Why do low angle boundaries have low mobilities for migration?

5. Provide a reason for the fact that the activation energy for high angle boundary migration is less than that for diffusion along these boundaries.

6. What would you expect the grain boundary diffusivity for a coincidence boundary to be as compared to that for a boundary inbetween coincidence orientations? Substitute the words "migration mobility" in place of "diffusivity" in the above and answer the question again.(Hint, the answer to the latter question depends upon the excess volume at the grain boundary.)

7. Why does the presence of solute exert such a large effect upon the velocity of grain boundary migration?

8. Why is the effect of impurities on grain boundary velocity small at high values of the velocity or driving force?

9. Why does the impurity drag on grain boundaries exhibit a maximum as the velocity of g.b. migration increases? (Hint, at high velocity how does the concentration of solute at the grain boundary compare to that at low velocity?)

XI-GROWTH OF PHASES
(DIFFUSION OR REACTION CONTROL).

INTRODUCTION

Most of the processes studied in this chapter involve transport of matter across an interface while a chemical reaction also takes place at this interface. The interaction between the diffusion and chemical reaction is of great interest and we develop the coupling between the two processes. Also, we consider processes that involve matter transport to an interface with a divergence of this transport at the interface. In all the cases we treat in this chapter, the interface is either planar, or acts equivalently to a planar interface in the sense that the interface shape is stable during the process. In many of these processes, non-equilibrium thermodynamics contributes a formalism that is often useful in defining the relations between parameters.

1. A/B DIFFUSION COUPLE-NO INTERMEDIATE PHASE.

1.1. Components have equal molar volumes.

Suppose that a semi-infinite element A is placed into intimate contact with another semi-infinite element B and then is brought to a temperature T at which diffusion occurs for a finite time. If the corresponding binary phase diagram is the type illustrated in Fig.11.1a, then a plot of the composition along the diffusion couple would appear as shown in Fig. 11.1b.

It is often asked why does not a two phase region appear between the

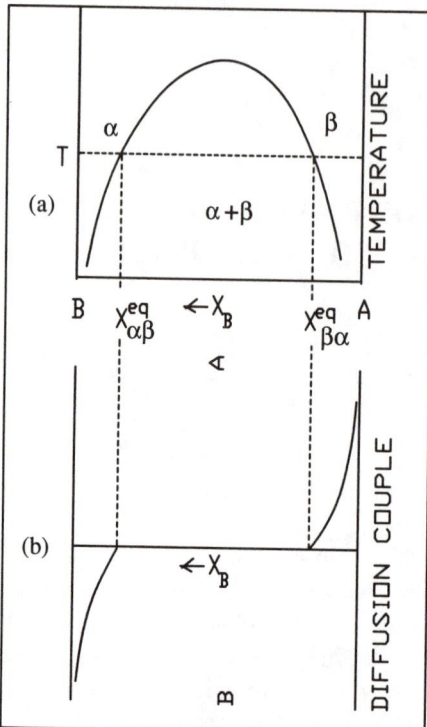

Figure 11.1. Illustrating compositions at interface of diffusion couple relative to phase boundary compositions in phase diagram.

two terminal solid solutions. The answer to this question is simple. Suppose that a two phase region was produced between the two terminal solid solutions having the respective equilibrium compositions of the corresponding terminal solid solutions. Thus, in phase alpha, the concentration of A(B) is constant at that equilibrium value given by the phase diagram. Similarly, for phase beta. Since the concentration gradients in the two phase region equal zero, there is no transport of A or B across the two phase region! Consequently, only at the interfaces of the two phase region with the terminal solid solutions will the transport of atoms across the interface be possible. In particular, at the interface between the B poor solid solution (alpha) and the two phase region, B atoms diffuse from the interface into the alpha solution. These B atoms cannot be provided by diffusion across the two phase region, as just demonstrated. Hence, they can only be provided by the transformation of B rich beta phase at the interface to the B poor alpha phase, thereby causing the alpha/two phase interface to move in the direction of consumption of the two phase region. Similarly, at the A poor beta/two phase interface, the diffusion of A into the beta phase requires the conversion of the A rich alpha phase to the A poor beta phase and the motion of the beta/two phase interface in the direction of consumption of the two phase region. Consequently, the two phase region will disappear—it is not stable under the boundary conditions.

The point so laboriously made above can be made more simply using a plot of the chemical potentials of the components in the diffusion couple, as shown in Fig. 11.2. Contrary to the concentrations, which exhibit a discontinuity at the alpha/beta interface, the chemical potentials are

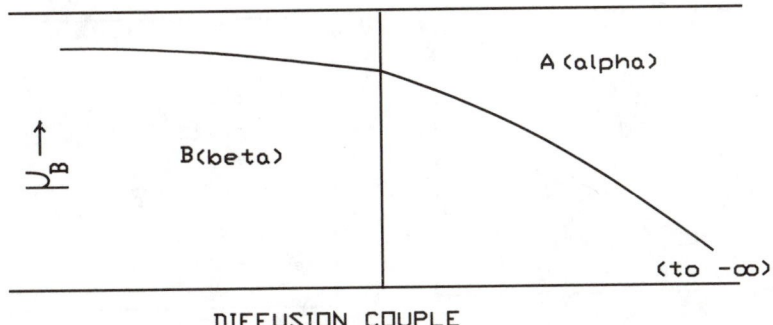

DIFFUSION COUPLE

Figure 11.2. Illustrating how the chemical potential of species B varies continuously along the diffusion couple in the absence of interface migration.

continuous, as shown. If a two phase region were to exist between the terminal solid solutions, it is now obvious that the chemical potentials would have to be constant in this two phase region. Thus, there is no driving force for transport of matter across the two phase region.

1.2. Components have unequal partial molar volumes.

The situation described in the previous paragraph corresponds to the case that local equilibrium holds at the α/β interface, that the latter is incoherent, and that the partial molar volumes of the diffusing species are equal. In the event that the partial molar volumes of the diffusing species differ then the gradients in composition will develop gradients in stress, which will affect the compositions along the diffusion couple. In this case, the diffusion potential replaces the chemical potential in the phenomenological relations for diffusion. Also, it is necessary to incorporate the effect of stress gradients on the diffusive fluxes. This problem has been considered by Larche and Cahn[1], who have shown, in the absence of a stress normal to the interface, that interface compositions can be determined using free energy-composition curves. Figure 11.3 illustrates the construction and results. The shifted Helmholtz free energy for the phase i is obtained by adding to that corresponding to the absence of stress the elastic energy per mole, which is just a function of the local composition and is given by

$$f_{el} = V_o E\eta^2(c^i - c_o^i)^2/(1-\nu)$$

where V_o is the molar volume at composition c_o far from the interface, E is Young's modulus, $\eta = (V_1 - V_2)/(3V_o)$ and ν is Poisson's ratio. The interface

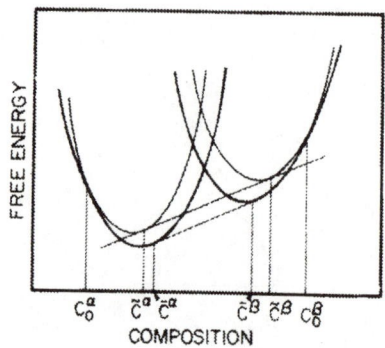

Figure 11.3. The unstressed free energies are shifted by f_{el} upwards. The common tangent gives the coexisting compositions.

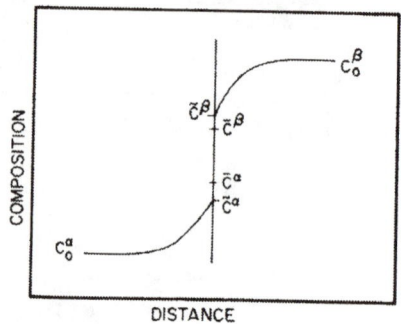

Figure 11.4. Compositions in a diffusion couple at incoherent interface. See Figure 11.3 for definition of compositions. Both figures after Larche and Cahn[1].

compositions for $\eta \neq 0$ are shown relative to those for $\eta = 0$ in Figure 11.4.

2. A/B DIFFUSION COUPLE—WITH INTERMEDIATE PHASE, INCOHERENT INTERFACES AND AT CONSTANT MOLAR VOLUME.

2.1. One intermediate phase.

Suppose now that the phase diagram for the A-B system exhibits one stable intermediate phase at the temperature T, as shown in Fig.11.5. What will occur after a diffusion anneal of the diffusion couple formed from A and B? The situation is more complex than that discussed above. Instead of only diffusion there is now an additional reaction-that of the transformation of crystal structures, which occurs at two interfaces: that of the interface between alpha and gamma and that of the interface between beta and gamma. This transformation reaction is called an interface reaction. Normally, for a kinetic reaction to take place, the driving force for that reaction must be finite. For example, diffusion requires a finite value of the chemical potential gradient to procede. In the case of the interface reaction, there must be a finite value of the difference in free energy between reactants and

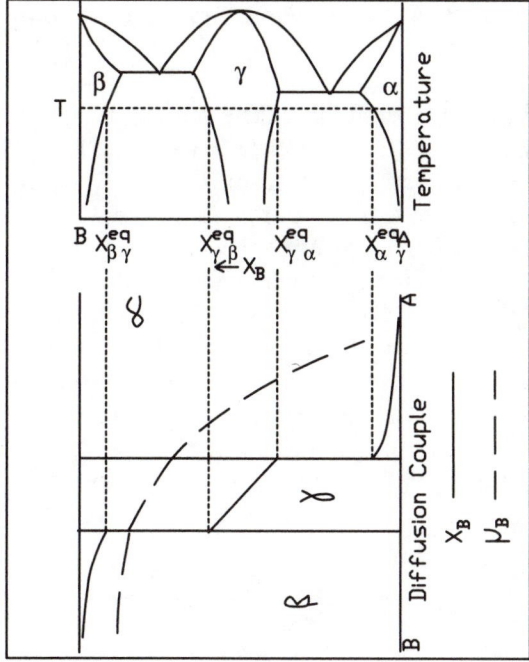

Figure 11.5. Interface compositions and solute chemical potential in a diffusion couple with intermediate phase under diffusion controlled conditions.

products to cause it to move and thereby transform one crystal structure to another of a different composition. If the chemical potentials of the species were continuous across the interfaces, as shown in Fig.11.5, then it is possible for species to be transported across these interfaces, but the interfaces themselves would not move, i.e. there would be no transformations and the thickness of the gamma phase would be constant. For transformations and interface motion to be possible, there must be a discontinuity in the chemical potentials at the interface, as shown in Fig.11.6, so that the difference in free energy between reactants and products is finite. In non-equilibrium thermodynamics, this driving force for the chemical reaction at the interface is called the chemical affinity.

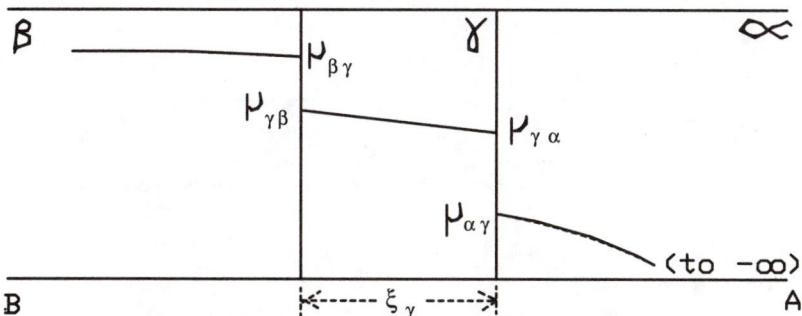

Figure 11.6. Chemical potential of solute as a function of distance along diffusion couple with moving boundaries.

2.1.1. Growth of gamma phase under condition that terminal solubilities are very small:

In this section we will assume: either that the terminal phases are saturated; or that the terminal phases alpha and beta have such small solubilities for the respective minor components that diffusion in alpha and beta can be neglected with respect to diffusion in gamma. We will also assume that the difference in specific free energy between the pure component and that for the terminal phase, at the composition in equilibrium with the intermediate phase, is negligible, as shown in Figure 11.7. Further, we will assume that in gamma the diffusivity of one component is much faster than that of the other component. These assumptions act not only to simplify the algebra, but, in fact, are probably good approximations for the growth of some silicide intermediate phases from their terminal phases.

Conservation of B at the gamma/beta interface yields the following equation:

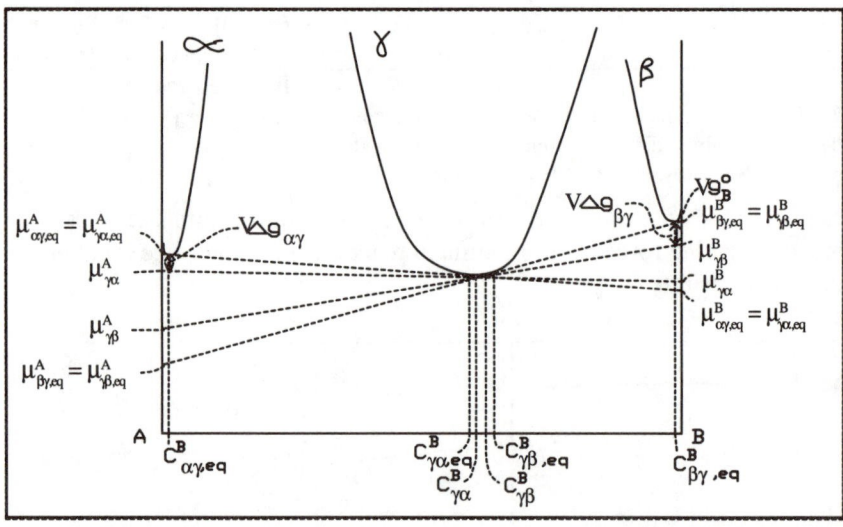

Figure 11.7. Defining chemical potentials, compositions and driving forces for growth of the intermediate gamma phase at the expense of the terminal alpha and beta phases. Note that the deviations from the local equilibrium compositions occur in the gamma phase and not in the terminal phases.

$$(C^B_{\beta\gamma} - C^B_{\gamma\beta})(d\xi/dt)_{\beta\gamma} = D^B_\gamma (dC^B_\gamma/dy)_{\gamma\beta} \qquad (1)$$

$C = X/V$, where X is the atom fraction and V the molar volume and we assume that all three phases have the same molar volume. Similarly, conservation of B at the alpha/gamma interface yields:

$$(C^B_{\gamma\alpha} - C^B_{\alpha\gamma})(d\xi/dt)_{\alpha\gamma} = - D^B_\gamma (dC^B_\gamma/dy)_{\alpha\gamma} \qquad (2)$$

Let us assume that within gamma the gradient of composition of B is independent of distance. Thus,

$$(dC^B_\gamma/dy)_{\gamma\beta} = (dC^B_\gamma/dy)_{\alpha\gamma} = J_B/(-D^B_\gamma) = (C^B_{\gamma\beta} - C^B_{\gamma\alpha})/\xi_\gamma \qquad (3)$$

which states that the flux of B is everywheres constant in gamma including at the interfaces.

But, if the reactions at the interfaces are homogeneous along the interface (i.e. the interface does not move via a step or ledge mechanism), then they can be taken to be proportional to the difference in free energy between reactants and products. Hence,

$$(d\xi/dt)_{\beta\gamma} = L_{\beta\gamma}(\Delta g_{\beta\gamma}); \quad -(d\xi/dt)_{\alpha\gamma} = L_{\alpha\gamma}(\Delta g_{\alpha\gamma})$$

The assumption we have made, that diffusion can be neglected in alpha and beta, makes the calculation of the difference in free energy between reactants and products particularly simple. We note that in a motion of the interface into either terminal phase, the constituents transported across unit area of interface correspond simply to the product of the composition of the terminal phase, at the interface, by the distance moved by the interface. Thus, for unit motion of the interface, the difference in free energy across the beta/gamma interface will be

$$\Delta g_{\beta\gamma} = (X^B_{\beta\gamma}/V)(\mu^B_{\beta\gamma} - \mu^B_{\gamma\beta}) + [(1 - X^B_{\beta\gamma})/V](\mu^A_{\beta\gamma} - \mu^A_{\gamma\beta})$$

Also, that across the alpha/gamma interface will be

$$\Delta g_{\alpha\gamma} = (X^B_{\alpha\gamma}/V)(\mu^B_{\alpha\gamma} - \mu^B_{\gamma\alpha}) + [(1 - X^B_{\alpha\gamma})/V](\mu^A_{\alpha\gamma} - \mu^A_{\gamma\alpha})$$

In both cases, we make the further reasonable assumptions: in the gamma phase the interface concentrations are not those corresponding to equilibrium, whereas in the terminal phases the interface compositions do correspond to the respective equilibrium concentrations; and the deviations of the chemical potentials from the equilibrium values at the interfaces are small enough that we need only expand to first order in the difference in interface compositions from the respective equilibrium values. Figure 11.7 illustrates the various terms that contribute to the driving forces for interface motion. It shows that small deviations from the equilibrium interface compositions in gamma can yield large positive driving forces for motion of the interfaces into the terminal phases, whereas with equilibrium interface compositions in gamma and off equilibrium composition at the interface in the terminal phases, the possible driving force for motion of the interfaces into the terminal phases are negative. Hence, for the approximations made the driving forces become

$$\Delta g_{\beta\gamma} = g_\beta^o - (1/V)(X_{\beta\gamma}^B \mu_{\gamma\beta}^B + X_{\beta\gamma}^A \mu_{\gamma\beta}^A)$$

and

$$\Delta g_{\alpha\gamma} = g_\alpha^o - (1/V)(X_{\alpha\gamma}^B \mu_{\gamma\alpha}^B + X_{\alpha\gamma}^A \mu_{\gamma\alpha}^A)$$

We now expand the chemical potentials in the gamma phase about their interface equilibrium values to obtain

$$\mu_{\gamma\beta}^B = \mu_{\gamma\beta,eq}^B + (\partial\mu_\gamma^B/\partial X_\gamma^B)(X_{\gamma\beta}^B - X_{\gamma\beta,eq}^B) + ...$$

$$\mu_{\gamma\alpha}^A = \mu_{\gamma\alpha,eq}^A + (\partial\mu_\gamma^A/\partial X_\gamma^A)(X_{\gamma\alpha}^A - X_{\gamma\alpha,eq}^A) + ...$$

Since, $\mu_{\beta\gamma,eq}^B = \mu_{\gamma\beta,eq}^B$; $\mu_{\alpha\gamma,eq}^A = \mu_{\gamma\alpha,eq}^A$; $Vg_\beta^o \approx X_{\beta\gamma}^B \mu_{\beta\gamma,eq}^B + X_{\beta\gamma}^A \mu_{\beta\gamma,eq}^A$

and $Vg_\alpha^o \approx X_{\alpha\gamma}^B \mu_{\alpha\gamma,eq}^B + X_{\alpha\gamma}^A \mu_{\alpha\gamma,eq.}^A$

and by the Gibbs-Duhem relation $(\partial\mu^A/\partial X^A) = (X^B/X^A)(\partial\mu^B/\partial X^B)$

then, substitution yields

$$(d\xi/dt)_{\beta\gamma} = L_{\beta\gamma}(\partial\mu_\gamma^B/\partial X_\gamma^B)[X_{\beta\gamma}^B - (X_{\beta\gamma}^A X_{\gamma\beta}^B/X_{\gamma\beta}^A)](C_{\gamma\beta,eq}^B - C_{\gamma\beta}^B) \quad (4)$$

$$= M_{\beta\gamma}(C_{\gamma\beta,eq}^B - C_{\gamma\beta}^B)$$

$$(d\xi/dt)_{\alpha\gamma} = L_{\alpha\gamma}\,(\,\partial\mu^A_\gamma/\partial X^A_\gamma)\,[X^A_{\alpha\gamma} - (\,X^B_{\alpha\gamma}\,X^A_{\gamma\alpha}/\,X^B_{\gamma\alpha})](C^B_{\gamma\alpha,eq} - C^B_{\gamma\alpha})\quad(5)$$

$$= M_{\alpha\gamma}\,(C^B_{\gamma\alpha,eq} - C^B_{\gamma\alpha})$$

Let us now solve for $C^B_{\gamma\beta}$ and $C^B_{\gamma\alpha}$ and substitute into (3)

$$C^B_{\gamma\beta} = C^B_{\gamma\beta,eq} - (d\xi/dt)_{\beta\gamma}\,/\,M_{\beta\gamma}\;;\;\;C^B_{\gamma\alpha} = C^B_{\gamma\alpha,eq} - (d\xi/dt)_{\alpha\gamma}/\,M_{\alpha\gamma}$$

$$J_B = - (D_\gamma/\xi_\gamma)[C^B_{\gamma\beta,eq} - C^B_{\gamma\alpha,eq} - (d\xi/dt)_{\beta\gamma}\,/\,M_{\beta\gamma} + (d\xi/dt)_{\alpha\gamma}/\,M_{\alpha\gamma}]$$

Now replace $(d\xi/dt)_{\beta\gamma}$ from (1) and $(d\xi/dt)_{\alpha\gamma}$ from (2) to obtain

$$J_b = - (D_\gamma/\xi_\gamma)[\Delta C^B_{\gamma,eq} + J_B/\{(C^B_{\beta\gamma} - C^B_{\gamma\beta})M_{\beta\gamma}\} + J_B/\{(C^B_{\gamma\alpha} - C^B_{\alpha\gamma})\,M_{\alpha\gamma}\}]$$

Solving for J_B yields

$$J_B = - \Delta C^B_{\gamma,eq}/[(\xi_\gamma/\,D^B_\gamma) + 1/\{(C^B_{\beta\gamma} - C^B_{\gamma\beta})M_{\beta\gamma}\} + 1/\{(C^B_{\gamma\alpha} - C^B_{\alpha\gamma})\,M_{\alpha\gamma}\}]$$

We now relate the growth rate of the intermediate phase gamma to J_B

$$(d\xi/dt)_\gamma = (d\xi/dt)_{\beta\gamma} - (d\xi/dt)_{\alpha\gamma} = -J_B/(C^B_{\beta\gamma} - C^B_{\gamma\beta}) - J_B/(C^B_{\gamma\alpha} - C^B_{\alpha\gamma})$$

$$= -J_B\,\{1/(C^B_{\beta\gamma} - C^B_{\gamma\beta}) + 1/(C^B_{\gamma\alpha} - C^B_{\alpha\gamma})\}$$

Therefore,

$$(d\xi/dt)_\gamma = \frac{\Delta C^B_{\gamma,eq}\,\{1/(C^B_{\beta\gamma} - C^B_{\gamma\beta}) + 1/(C^B_{\gamma\alpha} - C^B_{\alpha\gamma})\}}{(\xi_\gamma/\,D^B_\gamma) + 1/\{(C^B_{\beta\gamma} - C^B_{\gamma\beta})M_{\beta\gamma}\} + 1/\{(C^B_{\gamma\alpha} - C^B_{\alpha\gamma})\,M_{\alpha\gamma}\}}\quad(6)$$

Let us now examine the behavior of (6). When the thickness of the gamma phase, ξ_γ is small near the start of the growth process, then the last two terms in the denominator can exceed the term containing the thickness ξ_γ. One of these two remaining terms usually will be much larger than the other, but in any case in this regime of behavior it is apparent that the growth rate will be nearly independent of the time. The slowest process in this regime is obviously that for interface migration and, hence, the regime is called interface or reaction controlled. When ξ_γ becomes larger than a critical value given by

$$\xi_\gamma^* = D_\gamma^B \left[1/\{ (C_{\beta\gamma}^B - C_{\gamma\beta}^B) M_{\beta\gamma} \} + 1/\{ (C_{\gamma\alpha}^B - C_{\alpha\gamma}^B) M_{\alpha\gamma} \} \right]$$

then the growth rate of the gamma phase depends upon the thickness of the gamma phase. Because the term containing the thickness is the term depending upon diffusion of the most rapid species in the gamma phase, this regime of behavior, in which the thickness depends parabolically on time, is termed diffusion controlled. These two regimes are illustrated in Figure 11.8. The literature abounds with examples of reaction controlled growth of intermediate phases. For example, SiO_2 first grows linearly with time then parabolically with time in the oxidation of silicon.* Many thin film silicides grown by reaction between a thin metal film and a silicon substrate exhibit the same behavior. Indeed, this mode of growth is common in non-metallic

---

* Doremus (J. Appl. Phys.<u>66</u>,4441(1989)) has argued that the linear regime in the growth of SiO_2 is not a manifestation of reaction controlled growth.

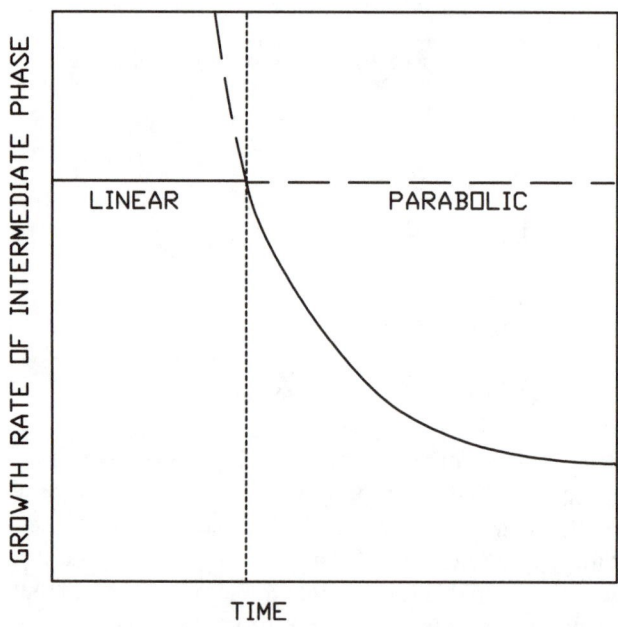

TIME

Figure 11.8. Illustration of the behavior of equation (6) in this chapter. The initial stage yields a linear dependence of the intermediate phase thickness ξ as a function of the time, whereas the second stage yields a parabolic dependence of ξ on time. Notice that the two processes are coupled in series and hence that the slower controls the observed rate of growth.

systems.

It is easy to show that for the case of diffusion controlled growth of the intermediate phase A_nB_m in the type of system considered above, that the rate constant k in the relation $\xi^2 = 2kt$ is nearly given by

$$k = (n+m)(D_A/n + D_B/m)|\Delta G|/RT.$$

(It is merely necessary to show that for the case where the terminal solutions are dilute at the phase boundaries and for the intermediate phase A_nB_m, that the difference in chemical potentials in equilibrium at the phase boundaries of a given species is related to the free energy for the reaction $|\Delta G|$ by the lever law (similar triangles), that $d\xi/dt = \Sigma c_i v_i$ and to use for the flux of a given species through the intermediate phase the Nernst-Einstein relation $c_i v_i = F_i D_i/RT$.) This is left as a problem for the student to demonstrate.

A more extended analysis of the growth rate for the intermediate phase under <u>diffusion controlled conditions</u> has been given by Kidson[2]. See also the articles in the Bibliography for additional treatments of diffusion controlled intermediate phase growth or what is sometimes called "reaction diffusion".

2.2. Many possible intermediate phases.

Of course, when many intermediate phases are stable at the diffusion temperature then a description of the kinetics is more complicated than given in the previous section.[2] In semi-infinite bulk diffusion couples all the intermediate phases will ultimately form and grow. However, in thin film/bulk couples, as in the formation of silicides by reaction between a deposited thin metal film and a silicon substrate, it is usually possible for only one intermediate phase to form during the time period it takes to exhaust the thin film source. A further reaction between this intermediate phase and the bulk substrate can occur, but, generally, the kinetics involved in the various possible phase transformations vary so greatly that it is necessary to raise the temperature to initiate a new reaction in a reasonable time period.

One question that arises in a consideration of the thin film/bulk couples is what condition governs the choice of the intermediate phase that appears first on increasing the diffusion temperature. One "answer" is that the phase that appears first as the diffusion temperature is raised is the one that consumes the thin film material the fastest. The reader may recall our discussion of competing rate processes in Chapter VII. However, this "answer" does not allow us to predict which possible phase will appear first, without a detailed knowledge of the various diffusivities, nucleation kinetics, and reaction rates for all the possible reactions. Nevertheless, it is possible to investigate some of the factors that affect the relative kinetics as follows.

We can ask the question, "Suppose that somehow two intermediate phases formed with planar interfaces between two terminal phases, one a thin film and the other a substrate, that composed the original diffusion couple. What are the conditions that govern whether these two phases can grow or one will shrink at the expense of the other's growth?" To answer this question, under the assumption that diffusion controlled conditions apply to both intermediate phases, we set up the equations for conservation of matter at the interfaces and for diffusion through the intermediate phases. We assume, for simplicity, that the solubilities for the corresponding solute in the terminal phases are negligible, that the diffusivity of one species overwhelms that for the other species in both intermediate phases($D_2 >> D_1$), and that the solubility ranges of the intermediate phases are very small. Figure 11.9 illustrates both the diffusion couple and the free energy composition diagram for this system and defines the various quantities to be used. We focus on each interface and set down the matter conservation relations that act to define the velocities of the interfaces. Thus, for the interface between the film and the phase α we can write with c, corresponding to the concentration per unit volume of component 2, and μ, the chemical potential of this component

$$(c^M - c^\alpha)v_1 = -(c^\alpha/RT)(\mu^{\alpha M} - \mu^{\alpha\beta})D^\alpha/h_\alpha$$

Similarly, for the interface between the substrate and phase β

$$(c^\beta - c^X)v_3 = (c^\beta/RT)(\mu^{\beta\alpha} - \mu^{\beta X})D^\beta/h_\beta$$

and for that between phases α and β

$$(c^\alpha - c^\beta)v_2 = (c^\alpha/RT)(\mu^{\alpha M} - \mu^{\alpha M})D^\alpha/h_\alpha - (c^\beta/RT)(\mu^{\beta\alpha} - \mu^{\beta X})D^\beta/h_\beta$$

$$= -(c^M - c^\alpha)v_1 - (c^\beta - c^X)v_3$$

Now, the criterion for growth of any intermediate phase is $v_{n+1} - v_n > 0$.

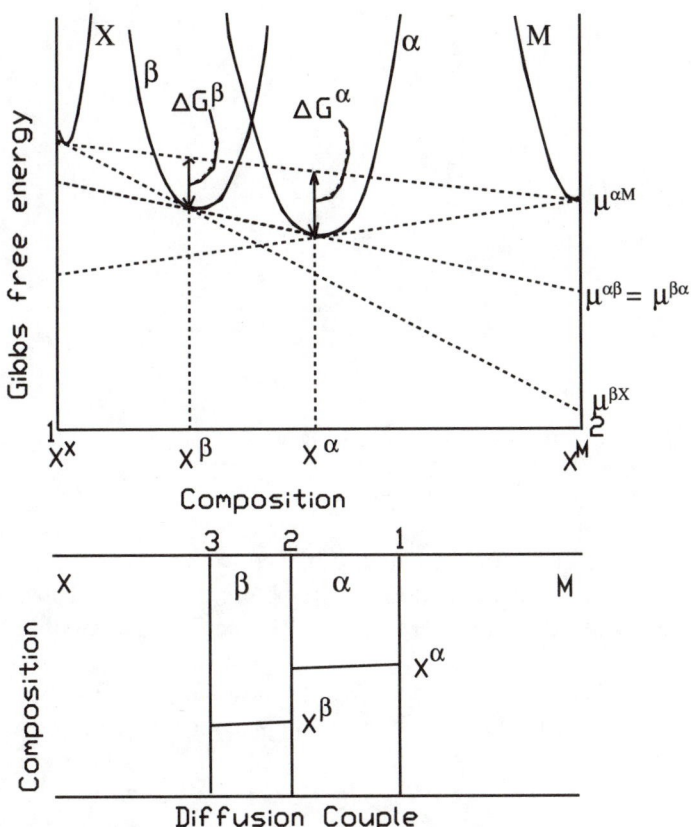

Figure 11.9. Defining chemical potentials and average compositions of phases for analysis of relative growth or disappearance of intermediate phases.

Thus, to determine whether the phase alpha grows or shrinks we need to evaluate whether $(v_{n+1} - v_n)$ are greater or smaller than zero, respectively. Substituting and collecting terms we arrive at the relations governing growth or shrinkage given by

$$dh_\alpha/dt = A/h_\alpha - B/h_\beta$$

$$dh_\beta/dt = C/h_\beta - D/h_\alpha$$

where $A = D^\alpha(C^\alpha/RT)(C^M - C^\beta) \dfrac{(1-X^\beta)\Delta G^\alpha - (1-X^\alpha)\Delta G^\beta}{(C^\alpha-C^\beta)(C^M-C^\beta)(X^\alpha-X^\beta)}$

$$B = D^\beta(C^\beta/RT) \frac{(1-X^\beta)(X^\alpha\Delta G^\beta - X^\beta\Delta G^\alpha)}{(C^\alpha-C^\beta)X^\beta(X^\alpha-X^\beta)}$$

$$C = D^\beta(C^\beta/RT) \frac{(1-X^\beta)[X^\alpha\Delta G^\beta/X^\beta - \Delta G^\alpha](C^\alpha-C^X)}{(C^\beta-C^X)(X^\alpha-X^\beta)(C^\alpha-C^\beta)}$$

and $D = D^\alpha(C^\alpha/RT) \dfrac{(1-X^\beta)\Delta G^\alpha - (1-X^\alpha)\Delta G^\beta}{(C^\alpha-C^\beta)(X^\alpha-X^\beta)}$

Thus, depending upon the values of the constants, these equations can lead to either growth of both intermediate phases , or to the growth of one and the shrinkage of the other. Figure 11.10 describes the possible behavior in the h_a, h_b space.

    For silicides we simply do not have enough information to evaluate A,B,C and D. Nor do we have enough data to determine whether the analysis provides an answer to the question concerning the sequential nature of appearance of the intermediate phases in thin film couples.

    In all of the above we have implicitly, or explicitly, assumed that the interfaces were incoherent so that vacancies and interstitials could be created or destroyed at the interface. If the interface is coherent or partially coherent

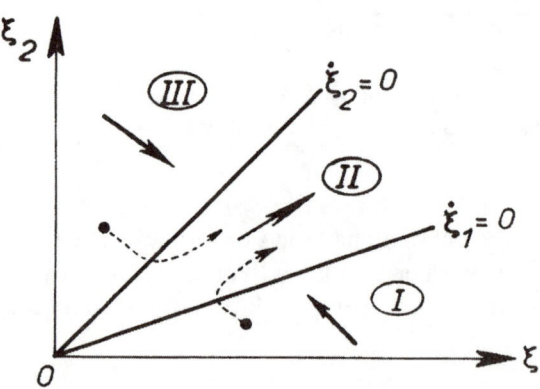

Figure 11.10. Read $\xi = h$. In domain I phase α regresses and phase β grows. In domain II phase α grows and phase β grows. In domain III phase α grows and phase β regresses.

then we must be concerned with this fact. A divergence of vacancy(interstitial) flux at the interfaces will develop stresses or voids or generate climb type dislocations.

2.3. Amorphous intermediate phase

One of the recent interesting discoveries is that metastable amorphous phases can form as intermediate phases in some diffusion couples.[3] It is noted that the systems that exhibit this behavior are characterized by having large negative free energies of formation for the stable intermediate phases and an appreciable difference in atomic radii between the components. In this case, it is conceivable that an intermediate composition amorphous phase will be stable relative to the stable mixture of the terminal phases, although it will be metastable relative to the stable intermediate phases. Thus, following the expectation that the fastest reaction between the two terminal phases will be the one to appear first and the possibility that diffusion through the amorphous phase can be rapid for the smaller species, if it is small enough to diffuse through the interstitial space in the amorphous phase, then the observed reaction of two crystalline terminal phases to yield an intermediate amorphous phase is at least plausible. Indeed, it is known that one component, the smaller one, diffuses anomalously fast through the amorphous phase. Also, it is known that provision of nucleation barriers to the formation of the amorphous phase, such as by making one of the terminal phase thin films monocrystalline, slows down the reaction sufficiently that the amorphous phase does not form in systems where it does form when the reactant thin films are polycrystalline. Obviously, this field is ripe for the adventurous explorer.

3. GROWTH FROM SUPERSATURATED SOLUTION.

Another mode of growth that involves both diffusion and a chemical reaction in the kinetics of formation of the growing phase is that of precipitation from a supersaturated solution. An example of a process that uses this mode of synthesis to produce a desired material is liquid phase epitaxy. This process also occurs in the precipitation of phases in the aging of supersaturated alloy solid solutions to produce distributions of precipitates in alloy systems. As shown in Figure 11.11 the supersaturated solution is generally produced by cooling a solution from a temperature at which it is stable to another at which it is supersaturated. We will assume that no diffusion occurs in the precipitated phase, beta, but does occur in the parent phase alpha and that the composition of beta is that in equilibrium with alpha. Also, we will consider the case of growth by the motion of a planar interface separating the two phases. Hence, the plot of concentration of the solute species against distance is schematically as illustrated in Figure 11.12. We will also assume that the solute species is the more mobile of the two species and that the temperature distribution does not affect the kinetics.

The change in free energy due to the transport of matter across the interphase interface per unit area of interface per unit motion of interface is given by

$$dG/d\xi = \lambda(c^s - c^\alpha_e) \tag{7}$$

where c^s is the concentration of the solute at the interface and c^α_e is that given by the phase diagram for local equilibrium between parent and product phases. Also,

$$\lambda = \Sigma_{i=1}^n c^\beta_i \, \partial\mu^\alpha_i / \partial c^\alpha_i. \tag{8}$$

We shall assume linear kinetics (the interface is rough and growth does not involve lateral motion of ledges), i.e.

$$d\xi/dt = M_{\alpha\beta}(dG/d\xi) \tag{9}$$

Because it has been assumed that only the solute component is diffusing, conservation of this component at the interphase interface yields

$$(c^\beta - c^s)(d\xi/dt) = D^\alpha(\partial c^\alpha/\partial x)_{x=\xi} \tag{10}$$

Now, diffusion in the supersaturated phase alpha must satisfy Fick's 2nd law and the boundary conditions or

$$\partial c^\alpha/\partial t = D^\alpha \partial^2 c^\alpha/\partial x^2 \tag{11}$$

with the boundary conditions being

$$c^\alpha = c^\alpha_o \text{ at } x = \infty; \quad (\partial c^\alpha/\partial x)_{x=\infty} = 0$$

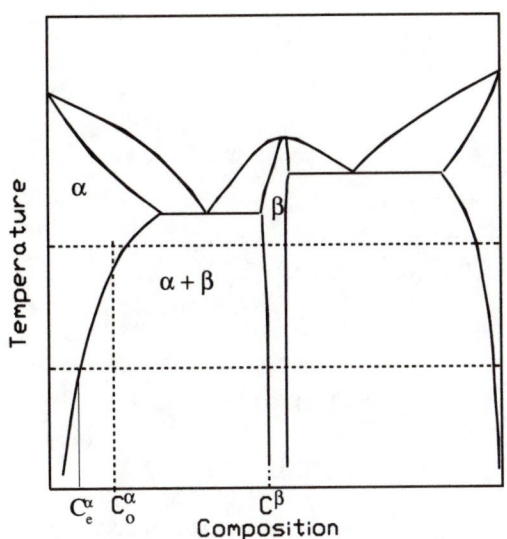

Figure 11.11. Phase diagram illustrating the compostions for local equilibrium at the interface and that in the supersaturated solution.

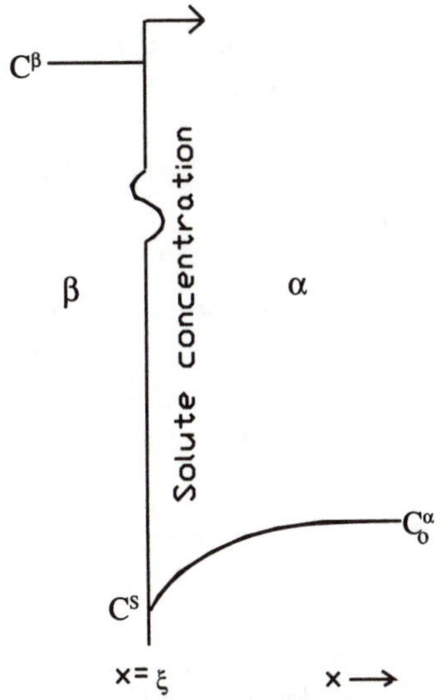

Figure 11.12. Showing the interface between the precipitate β and the supersaturated α phase and the concentration profile in the latter..

and
$$c^\alpha = c^s \text{ at } x = \xi; \text{ for all } t > 0. \qquad (12)$$

Equations (11) and (12) are satisfied by

$$c^\alpha = A + \text{Berfc}\{x/[2(D^\alpha t)^{1/2}]\}$$

Substitution of this solution in (12) yields

$$c_0^\alpha = A + \text{Berfc}(\infty) = A + B; \quad c^s = A + \text{Berfc}\{\xi/[2(D^\alpha t)^{1/2}]\}$$

Hence, $c^s = c_0^\alpha - B + \text{Berfc}\{\xi/[2(D^\alpha t)^{1/2}]\}$

or
$$c_0^\alpha - c^s = B(1 - \text{erfc}\{\xi/[2(D^\alpha t)^{1/2}]\})$$

But, $\partial c^\alpha/\partial x = B\exp(-x^2/4D^\alpha t)/[\pi(D^\alpha t)^{1/2}]$

(because $(\partial/\partial x)(\int_0^u \exp(-\eta^2)d\eta) = (du/dx)\exp(-x^2/4D^\alpha t)$, where $u = x/[2(D^\alpha t)^{1/2}])$.

Substitution for B yields

$$\partial c^\alpha/\partial x = \exp(-x^2/4D^\alpha t)\frac{(c_0^\alpha - c^S)/[\pi(D^\alpha t)^{1/2}]}{(1 - \mathrm{erfc}\{\xi/[2(D^\alpha t)^{1/2}]\})}$$

We now substitute this relation into (10) to obtain

$$(c^\beta - c^S)(d\xi/dt) = D^\alpha \exp(-\xi^2/4D^\alpha t)\frac{(c_0^\alpha - c^S)/[\pi(D^\alpha t)^{1/2}]}{(1 - \mathrm{erfc}\{\xi/[2(D^\alpha t)^{1/2}]\})} \qquad (13)$$

Now, let $\xi = j(4D^\alpha t)^{1/2}$. Hence $d\xi/dt = (j/2)(D^\alpha/t)^{1/2}$. We substitute for $d\xi/dt$ in (13) to obtain the relation for j, as follows

$$(c^\beta - c^S)\,j = 2(c_0^\alpha - c^S)\exp(-j^2)/[\sqrt{\pi}\,(1 - \mathrm{erfc}(j)\,)] \qquad (14)$$

Since $(c_0^\alpha - c^S) \ll (c^\beta - c^S)$ an analytical approximation for j is

$$j = 2(c_0^\alpha - c^S)/[\sqrt{\pi}\,(c^\beta - c^S)] \qquad (15)$$

We equate the expressions for $d\xi/dt$ in (9) and (10) and substitute for j in $(\partial c^\alpha/\partial x)_{x=\xi}$ and for the latter in (10) to obtain the following relation for small values of j

$$M_{\alpha\beta}\lambda(c^S - c_e^\alpha) = 2D^\alpha(c_0^\alpha - c^S)^2/[\pi(c^\beta - c^S)^2\xi] \qquad (16)$$

We note that $c^S \ll c^\beta$, which allows us to obtain the following approximate relation for c^S

$$K(c^S - c_e^\alpha) \underset{0}{=} (c_0^\alpha - c^S)^2 \qquad (17)$$

where $K = [M_{\alpha\beta}\lambda\xi\pi(c^\beta)^2/(2D^\alpha)]$

Thus, for $K \longrightarrow 0$, $c^S \longrightarrow c_0^\alpha$ and for $K \longrightarrow \infty$, $c^S \longrightarrow c_e^\alpha$. Thus, diffusion control holds at large values of the thickness ξ and the ratio of $M_{\alpha\beta}/D^\alpha$, whereas for small values of $M_{\alpha\beta}/D^{\alpha}$ interface reaction control holds. Interface reaction control is more common in non-metallic systems than in alloys.

When the parent supersaturated solution is liquid then it is necessary to take convection into account. One way to do this is to assume that a laminar layer develops in the convecting liquid in contact with the crystal being grown. Diffusion then occurs through this constant laminar layer thickness with concentrations at both interfaces of the laminar layer being maintained constant. Thus, the growth rate is independent of time and can be evaluated with a knowledge of the laminar layer thickness, the interface concentrations and the diffusivity in the liquid. Usually, these parameters are not known and must be derived from measurements.

4. SOLIDIFICATION.

4.1. Pure materials.

In pure materials the extraction of the heat of fusion usually controls the rate of solidification, except for those materials which have such slow crystallization rates that the latter control. Examples of the latter materials are SiO_2, P_2O_5 and GeO_2 and many polymers. Doremus[4] has shown how the crystallization velocity may be estimated on the assumption that the free energy dissipated per unit jump of a molecule across the solid/liquid interface in the solidification step equals the work to move the molecule through a viscous liquid at the solidification velocity. Accordingly, the crystallization velocity is then

$$v_C = \Delta H_f(T_M - T)/(3\pi\eta T_M\lambda^2)$$

where ΔH_f is the heat of fusion, T_M the melting point, η is the viscosity, and λ is the jump distance for a molecule in the liquid at the solid/liquid interface needed to join the solid.

The shape of the solid/liquid interface is controlled by the factor α described in section 5.2 of Chapter X (i.e., $\alpha = \Delta S_f/R$). For α values less than 2 the interface tends to be rough and non-faceted. Metals have α values less than 2. Most organic compounds have α values > 2 and grow facets, while polymers have large α values and grow either faceted or as spherulites. The development of dendrites during growth will be considered in the next chapter.

Although not apparent from the above relation between crystallization velocity and undercooling, it is known empirically that the coefficient is a function of the interface orientation for faceted crystallization. Thus, this relation must be considered to be a rough approximation only.

The generation of defects in the solid at the solid/liquid interface has technological significance, especially in the growth of large single crystal ingots of semiconductor grade silicon and gallium arsenide. This technology is outside the scope of this book.

4.2. Multicomponent systems.

4.2.1. Local equilibrium at solid/liquid interface.

In the process of solidification of a melt, having a phase diagram, such as shown in Figure 11.13, with a planar solification front, the concentrations in the melt and solid proceed with time as illustrated in Figure 11.14, on the assumption of no convection, no diffusion in the solid phase, local equilibrium at the solid/liquid interface and a constant velocity for the solidification front. Contrary to the cases treated in sections 2 and 3, the driving force for interface motion does not derive solely from chemical potential differences across the interface, but does depend primarily on the dissipation of the heat of fusion via the appropriate temperature gradient at the interface. When steady state is reached the concentration in the liquid with respect to the solidification front is given by

$$c_L/c_o = 1 + [(1 - k)/k]\exp(-Rx/D)$$

where k is the distribution coefficient (c_S/c_L), R is the solid/liquid interface velocity and x is distance measured from the solid-liquid interface. We can now calculate a liquidus temperature-distance curve corresponding to the composition $c_L(x)$ function. If the liquidus and solidus phase boundaries are straight lines then the equation for $T_L(x)$ is

$$T_L = T_o - mc_o\{1 + [(1 - k)/k]\exp(-Rx/D)\}$$

where m is the slope of the liquidus (dT_L/dc). A plot of a $T_L(x)$ function is

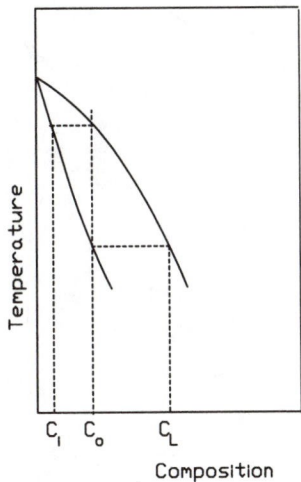

Figure 11.13. Solidus-liquidus-boundaries in phase diagram defining compositions.

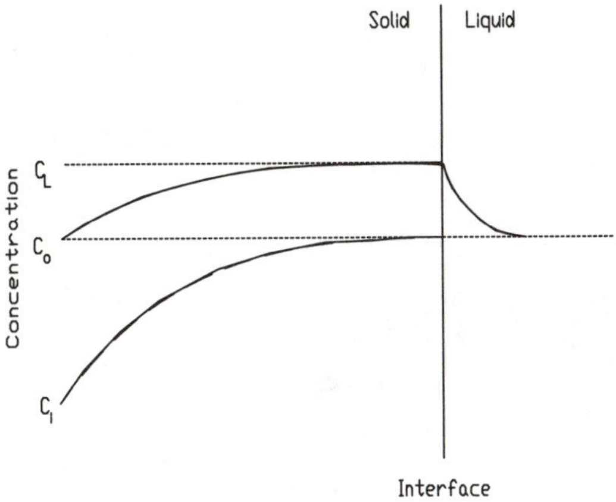

Figure 11.14. Composition-distance plots relative to solid-liquid interface moving at constant velocity. Refer to Figure 11.13 for definitions of compositions relative to phase diagram.

shown in Figure 11.15. Suppose now that the actual temperature in the liquid is as shown by the dashed line in this figure. In this case there is a zone ahead of the solid/liquid interface in which the actual temperature is less than the liquidus temperature corresponding to the concentration in the liquid at that position. This zone has been called the constitutionally supercooled zone. We will show in the next chapter that any supercooled zone leads to a tendency towards morphological instability of the interface, because any part of the interface that moves ahead of the planar front can get rid of heat and solute laterally and thus move at a faster velocity than the planar front.

At this point it is appropriate to point out that Pfann[5] helped initiate the revolution that brought materials science to its present state of development through his recognition of the role of the liquid/solid solute distribution coefficient in his invention of the process of zone melting. He recognized that the purification of the solid relative to the liquid in the first part to freeze, as illustrated in Figure 11.13, could be used to achieve the purification of semiconductors needed to make transistors a reality. His scheme was to cause a liquid zone, convectively mixed, to travel through a solid bar. Simple, but effective.

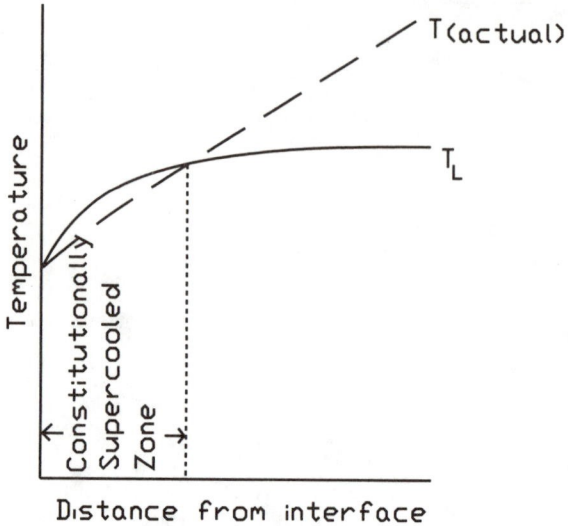

Figure 11.15. Showing the actual temperature in the liquid phase ahead of the solid-liquid interface(dashed line). Also, the liquidus temperature corresponding to the liquidus composition-distance plot in Figure 11.14 is shown as the solid line. The region where the actual temperature is less than the liquidus temperature is the constitutionally supercooled zone.

4.2.2. Non-equilibrium liquid/solid interface conditions.

It is possible with modern methods of heating very thin surface layers and with rapid solidification techniques to achieve, in semiconductors and even in metals, crystallization rates that are not limited by the rate of heat removal. Under these conditions, it is no longer valid to assume that local equilibrium exists at the solid/liquid interface. If we use the viscosity to estimate a maximum crystallization velocity for metals we arrive at a magnitude of about 1 m/sec. It is possible for the solid/liquid interface velocity to exceed this value experimentally. Thus, we need a new concept to determine the maximum crystallization velocity. Turnbull[6] has suggested that the solid/liquid interface can move by essentially a diffusionless mode in these materials, limited in velocity only by the velocity of sound. With this concept it is then easy to comprehend how solute may be trapped at concentrations greatly exceeding those corresponding to local equilibrium at the interface.

Aziz[7] has treated the problem of calculating the effective distribution coefficient in rapid solidification. He assumes that there are two processes occurring at the interface: solvent atoms joining on to the solid phase via diffusionless shifting of positions and dragging along with them solute atoms in this process and concurrently the loss of solute atoms in the solid, at the solid/liquid interface, by thermally activated jumps into the adjoining liquid. This model leads to the following expression for the distribution coefficient when the mode of interface migration involves the lateral motion of steps

$$k(v,T) = k_e(T) + [1 - k_e(T)]\exp(-v_D/v)$$

and when the process is continuous, as at a rough interface, the distribution coefficient is given by

$$k = [(v/v_D) + k_e(T)]/[(v/v_D) + 1]$$

where $k_e(T)$ is the equilibrium distribution coefficient at the temperature T of the interface, and v_D is the maximum crystallization velocity corresponding to diffusion limited motion of atoms at the interface given by either the relation in section 4.1 or by D_i / λ, where D_i is the diffusivity for the jump across the interface and λ is the interatomic distance. Thus, for large v/v_D, k can approach unity. The trapping of supersaturations of solute from the liquid is important technologically in the semiconductor area where it is sometimes desired to develop large supersaturations of desired dopants in electrically active substitutional lattice sites.

Caroli et al[8], using non-equilibrium thermodynamics, have considered, for linear interface kinetics (rough and not stepped interfaces), the general conditions which can and cannot yield solute trapping in the solid. Further, they show, contrary to previous assertion[9], that the Onsager reciprocal relations are applicable to all the representations of the linear interface kinetics. (Interface motion requiring nucleation, or lateral motions of steps emanating from screw dislocations does not correspond to linearizable kinetics.) One result is that solute trapping is possible only in systems with positive values of the off-diagonal (cross) conductance coefficients in the linear relations between generalized forces and fluxes. They generalized this result to far from equilibrium solidification and suggest that solute trapping may result from interactions between unlike atoms such that the freezing-in of one constituent leads to the freezing-in of interacting constituents.

The non-equilibrium thermodynamic treatment of this problem by Caroli et al[8] is so instructive that it is reproduced below. The mass currents are defined by

$$j_i = \rho_i^{L,S}(v_i^{L,S} - v_s) \cdot n$$

where n is the unit vector normal to the interface pointing into the liquid, v_s is the interface velocity, and $v_i^{L,S}$ the velocity of species i and ρ_i the mass per unit volume in the L (or S) phase. The currents j_i thus correspond to the flux of i through the interface. The linear relations between these currents and their corresponding driving forces are for a binary system

$$j_A = L_{AA}\delta\mu_A + L_{AB}\delta\mu_B$$

$$j_B = L_{BA}\delta\mu_A + L_{BB}\delta\mu_B$$

$$\text{where } \delta\mu = \mu^S - \mu^L$$

These linear kinetic coefficients satisfy the Casimir-Onsager relations

$$L_{AB} = L_{BA}$$

$$L_{ii} \geq 0; \quad L_{AA}L_{BB} - L_{AB}^2 \geq 0.$$

Other currents, which are linear combinations of j_A and j_B can be devised, which are useful. In particular, the total mass current

$$J = j_A + j_B$$

and the diffusion current in either of the two phases e = L,S

$$J_c^e = (1 - C_e)j_B - C_e j_A$$

where

$$C_e = \rho_B^e/(\rho_A^e + \rho_B^e)$$

is the mass concentration of the solute species B in the e phase.

Caroli et al showed that the linearized flux relations based on these currents that yield the same interface entropy production also obey the

Onsager conditions, contrary to the previous assertion of Baker and Cahn[9].

For the case that diffusion in the solid can be neglected, $J_c^s = 0$ provides a constraint on the relative values of m_A and m_B given by

$$\delta\mu_B = -\frac{(1 - C_S)L_{AB} - C_S L_{AA}}{(1 - C_S)L_{BB} - C_S L_{AB}} \, \delta\mu_A$$

Also, this constraint leads to

$$j_A = (1 - C_S)J; \quad j_B = C_S J$$

$$j_L = (C_S - C_L) J$$

Also, if $L_{AB} = 0$, then

$$J = j_A/(1 - C_S) = L_{AA} \, \delta\mu_A/(1 - C_S) = j_B/C_S = L_{BB} \, \delta\mu_B/C_S.$$

Since $L_{AA} > 0$ and $L_{BB} > 0$, and for the case of solidification, for which $J < 0$, then $\delta\mu_A < 0$ and $\delta\mu_B < 0$. This result implies that solute trapping is impossible because for solute trapping $\delta\mu_B > 0$.

Consequently, we have arrived at the conclusion that solute trapping requires $L_{AB} \neq 0$ and, as we show below, it requires that $L_{AB} > 0$. Indeed, the condition for solute trapping is $\delta\mu_B/J < 0$. But,

$$J = j_B/C_S = L_{AB} \, \delta\mu_A + L_{BB} \, \delta\mu_B$$

$$= -\{(L_{AA}L_{BB} - L_{AB}^2)C_S/[(1 - C_S)L_{AB} - C_S L_{AA}]$$

Thus, $\delta\mu_B/J < 0$ requires that

$$[C_S L_{AA} - (1 - C_S)L_{AB}]/\{(L_{AA}L_{BB} - L_{AB}^2)C_S\} < 0.$$

Since $(L_{AA}L_{BB} - L_{AB}^2) \geq 0$ and $L_{AA} \geq 0$, by the Casimir-Onsager conditions then

$$L_{AB} > [C_S/(1 - C_S)]L_{AA} > 0$$

is the condition that satisfies solute trapping.

We hasten to add that the above model assumes near equilibrium conditions, where the interface velocity does not depend upon the flux of heat. It provides an analysis of the solute trapping effect, however, that is of help in analyzing solute trapping models applicable to far from equilibrium conditions.

It is clear that the treatment of the solidification problem by Caroli et al is applicable to other growth processes involving a reaction at an interface and diffusion to or from it. We have neglected cross-coupling effects in such processes examined in this chapter. The formal treatment is given in the second paper by these authors listed in reference 8.

5. PHYSICAL VAPOR DEPOSITION(PVD).

At the outset we need to distinguish between two regimes of physical vapor deposition. In one, the growth rate is given by the impingement rate and occurs at high supersaturation ratios and low substrate temperatures, where the evaporation rate is much smaller than the impingement rate. This is the situation normally encountered in PVD. In the other regime, the supersaturation is low and deposition is accomplished at near equilibrium conditions. This regime is sometimes used in the production of defect-free, monocrystalline, epitaxial silicon thin films. We will explore the kinetics associated with the latter regime.

Many stages are involved before a molecule incident upon a surface is incorporated into the lattice of the surface material. The possible stages involved in this process are as follows:

1. Adsorption onto the surface.
2. Diffusion along the surface.
3. Adsorption along a surface step. (See Figure 11.16 for definition of a surface step and kink along the step.)
4. Diffusion along the step.
5. Incorporation into the lattice at a kink.

Let us define certain concepts necessary to arrive at the growth rate. First, since along vicinal surfaces atoms can only be incorporated into the lattice at kink sites, which exist along surface steps, and such incorporation leads to the lateral motion of such steps, the growth rate is given by

$$G = hv/d_s$$

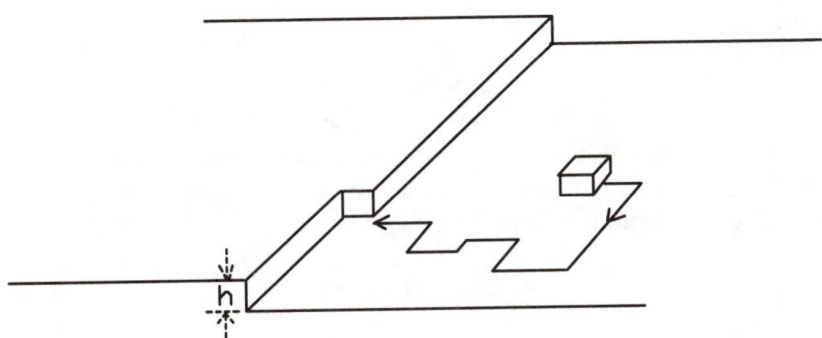

Figure 11.16. Showing a kink at a step on a crystal surface, diffusion of a condensed molecule along the surface to the kink where it is incorporated into the crystal.

where h is the height of the surface step, v is the velocity of the step normal to itself along the surface and d_s is the distance between such parallel steps. We will assume that the surface consists of parallel steps. (Indeed, in the epitaxial growth of silicon films the substrate is chosen to have an off-axis orientation to produce an arrangement where the surface consists of parallel surface steps. i.e. $d_s = h/\theta$, where θ is the angle between the surface normal and the low index plane normal.)

To get at the lateral velocity of these steps let us note that the impingement rate I given by gas kinetic theory is

$$I = P/(2\pi mkT)^{1/2}$$

where m is the mass of the molecule, P is the pressure and T is the temperature of the gas. At equilibrium, the impingement rate is equal to the evaporation rate. The latter, far from the steps, is given by

$$E = - \, dn_s/dt = n_s v \exp(- \, g_{des}/kT) = n_s/\tau$$

where v is the frequency of attempts at the surface for a surface admolecule to evaporate and g_{des} is the difference in free energy between an admolecule and an evaporated one. Also, it is apparent that

$$\tau = \exp(g_{des}/kT)/v.$$

Thus,

$$n_S = P\tau/(2\pi mkT)^{1/2}$$

Equilibrium exists also near the kink sites for the reaction betwen surface molecules and incorporation of same at kink sites. This reaction yields for the surface concentration near the kink sites

$$n_{Se} = n_o \exp(-g_k/kT)$$

where g_k is the difference in free energy between an admolecule and one at a kink site and n_o is the number of molecular sites per unit area on the surface where adsorption can take place. Thus, a surface concentration profile of admolecules is set up between the region near the steps and far from them.

Let us consider now the diffusion of molecules from the surface to the kinks which are sinks for these molecules. Fick's law gives

$$J_S = -D_S(\partial n_S(x)/\partial x)$$

Conservation of matter requires that

$$dJ_S/dx = I - E = (n_S - n_S(x))/\tau$$

$$\text{or} \quad -D_S(\partial^2 n_S/\partial x^2) = (n_S - n_S(x))/\tau$$

We define a diffusion distance, x_S by

$$x_S^2 = D_S\tau$$

Hence,

$$x_S^2 (\partial^2 n_S/\partial x^2) = n_S(x) - n_S.$$

For the boundary conditions, $n_S(x) = n_{Se}$ at $x = 0$ at the step and $n_S(x) = n_S$ far from it, this equation has the solution

$$n_S(x) = n_S - (n_S - n_{Se})\exp(-x/x_S)$$

where x is considered positive on both sides of the step. Now, the step velocity is related to the flux by

$$vn_o = -2J_S|_{x=0}$$

and the latter is given by

$$J_S|_{x=0} = -D_S(\partial n_s(x)/\partial x)_{x=0} = -D_S(n_s - n_{se})/x_S$$

$$= -D_S(P - P_e)\tau/[x_S(2\pi mkT)^{1/2}] = -x_S(P - P_e)/(2\pi mkT)^{1/2}$$

Thus, $v = 2x_S(P - P_e)/[n_o(2\pi mkT)^{1/2}] = 2x_S I_e(S-1)/n_o$, where S is the supersaturation ratio, i.e. $S=P/P_e$.

We have neglected the effect of step velocity on the concentration $n_s(x)$. (See Voigtlander et al, Appl. Phys.A38,(1985) for a treatment that takes this effect into account.)
        We have assumed implictly in the above derivation that the concentration at the steps is given by the equilibrium concentration of molecules, and that the diffusion distance x_S is >> than the distance between kinks along a step. If the distance between steps is less than the diffusion distance then another solution is found. In this case the boundary condition is that at the midpoint between the steps $dn_s(x)/dx = 0$. The solution is

$$(n_s - n_s(x))/(n_s - n_{se}) = \cosh[(d_s - 2x)/2x_S]/\cosh(d_s/2x_S)$$

Taking the derivative of $n_s(x)$ and substituting into Fick's law we have

$$J_S(x=0) = -x_S(n_s - n_{se})\tanh(d_s/2x_S)/(2x_S)$$

The step velocity is then

$$v = x_S(S - 1)I_e\tanh(d_s/2x_S)/n_o$$

where S, the supersaturation ratio, equals $n_s/n_{Se} = P/P_e$. The growth rate is related to the step velocity by

$$G = hv/d_s$$

Thus, the growth rate of a surface having parallel steps will be linearly related to the supersaturation ratio. A more complicated step arrangement on the surface, such as a series of spiral steps about an emerging screw dislocation, will yield a more complicated relation between the growth rate and the supersaturation ratio.

6. CHEMICAL VAPOR DEPOSITION(CVD).

We shall limit our discussion of this complex topic to concepts that are applicable to all the various types of CVD. Usually, at least for other than low pressure CVD, there are two series processes, diffusion and interface reaction, that can limit the rates of growth (interface migration) in CVD. Diffusion is in the vapor phase. Since there are gaseous reactants and products the diffusion may be controlled by the slower of the reactants or the slower of the products. Usually, diffusion is through a laminar layer set up by convective flow arrangements in the CVD reactor. Often, at high temperatures the growth rate is limited by diffusion through the gas phase, while at low temperatures the growth rate is limited by some interface reaction. Figure 11.17 illustrates growth rate versus reciprocal temperature for the CVD of silicon. Since diffusion through the gas is only weakly temperature dependent, whereas the interface reaction is strongly temperature dependent, these two stages can be easily delineated.

Very little is known about the specific steps in the interface reaction or even the identity of the limiting step in the reaction, for any CVD process. The silicon deposition has been studied extensively and still this process is not defined in terms of the identity of the limiting step. Another complication is that reactions can occur in the gas phase that yield intermediate precursors. Thus, often the identity of the precursor responsible for the reaction that donates the substrate atom or molecule is not known. A recent survey of the status of ignorance and knowledge of the mechanism of CVD of silicon is given in reference 10. Despite this ignorance, it is still possible to control and deposit acceptable films via CVD.

Among the surface steps that can control the interface reaction are: decomposition of the gaseous precursor yielding the monomer; diffusion of the monomer over the surface; attachment of the monomer to a kink site at the surface; desorption of a product species from the surface. There is no rule that nature must be simple and CVD is certainly one of the less simple processes to define mechanistically.

We shall learn in the next chapter that under diffusion controlled

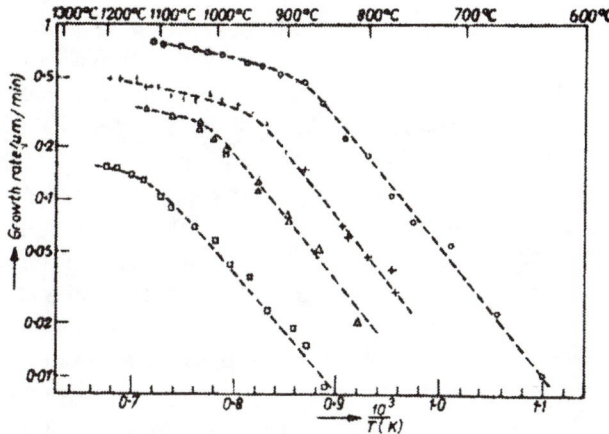

Figure 11.17. Temperature dependence of CVD growth rates of silicon from various source gases. After F.C. Eversteijn, Philips Res. Rept 29, 45(1974).

conditions a planar interface between parent and product phases is morphologically unstable, whereas it is stable under interface controlled conditions. A goal of CVD is to deposit smooth surfaces, which are certainly easier to produce when growth is reaction controlled. Thus, there is interest in defining the transition between these two regimes. This has led to the definition of the CVD number, deduced from equating the rates of growth given by diffusion through the gas laminar layer to that given by the reaction at the surface. This number is

$$CVD = [k_d \delta T_o^2 \ln(T_s/T_b)]/[D_o(T_s - T_b)T_s]$$

where k_d is the mass transfer coefficient at the deposit's surface, cm/s; δ is the boundary layer thickness, cm; $T_{s,b}$ the temperature of the surface and "bulk" gas, i.e. the temperatures at the laminar layer interfaces with the deposit and bulk gas stream, respectively; D_o the diffusion coefficient at the temperature T_o.

7. OSTWALD RIPENING.

Decomposition of a supersaturated solution via the nucleation and growth of precipitates of the product phase yields, at the point where the average concentration in the parent solution is close to equilibrium, a distribution of precipitates that vary in size. We have not treated the dependence during diffusion limited growth of the radius of the precipitate, if it is a sphere, on time. This problem and that of the growth when the diffusion fields of neighboring particles overlap is of interest, but these problems are problems in the solution of the diffusion equations for various boundary conditions, which we will not consider here. In this section, we will treat the interface energy driven growth of large particles at the expense of small particles—the problem known as Ostwald ripening.

Ostwald ripening involves both diffusion and interface reaction and we shall give solutions corresponding to control by one or the other process. For diffusion control the driving force for the reaction is given by the difference in chemical potentials corresponding to equilibrium with particles of radius r and r'. Since the chemical potential $\mu = 2\sigma V/r$ and $\mu' = 2\sigma V/r'$, we can express the driving force as $2\sigma V(1/r - 1/r')$. For simplicity, let us assume that the solution is dilute so that the difference in chemical potentials equals $RT(C - C')/C_{eq}$, where C is the concentration of the solute species in equilibrium with the particle of radius r, etc., and C_{eq} is that for an infinite radius (a planar surface).

A solution to the diffusion problem[11,12] involves an analysis of how the initial distribution function of particle radii varies with time, if at all. In the case that it is valid to assume that the distribution function does not depend on time and that transport is via volume diffusion then the solution to the problem yields for the average particle radius$<r>$

$$<r> = [8\sigma V^2 C_{eq} D/(9RT)]^{1/3} t^{1/3}$$

It is possible to illustrate the derivation of this result using an oversimplified model. If the distribution function of particle radii is independent of time then we can focus on any particle and consider its radius as a function of time. Accordingly, consider the flux between two neighboring particles of radii r and r'. Now the driving force for diffusion between these particles is the chemical potential gradient between them for the solute. We have already shown that this gradient will approximately equal $2\sigma V(r'-r)/(rr'L)$, where L is the distance between the particles. Thus, the flux at the surface of the larger particle will be given by $J = D(C_{eq}/RT)[2\sigma V(r' - r)/(rr'L)]$. Roughly $rr' = <r>^2$. Also, the rate of change of the average radius will equal that of the larger particle by the time independence of the particle distribution curve. Thus, we can write

$$(d\langle r\rangle/dt)/V \;=\; J \;=\; D(C_{eq}/RT)[2\sigma V(r'-r)/(rr'L)] \;=$$

$$[2\sigma VD(C_{eq}/RT)(r'-r)/L]/\langle r\rangle^2$$

This equation integrates to give an equation having the same time depend-ence for r as the rigorous one given before, although the coefficient differs.(The difference $(r'-r)$ may be considered to be independent of $\langle r\rangle$ since the distribution function is independent of time. It represents an average driving force, independent of time. Also, the average diffusion distance between particles is independent of time. Hence, $(r'-r)/L$ is some number.)

Experiment is not quite in accord with this result. Nevertheless, it is worthwhile to generalize this result for the case that either transport occurs via other paths or that Ostwald ripening is reaction controlled. The general result is

$$d^n = d^n_{\,o} + aGt$$

where the values of n for the various rate controlling processes are listed below.

n $= 1$ for viscous flow as the rate controlling process
   $= 2$ " interfacial reaction control
   $= 3$ " volume diffusion in all phases
   $= 4$ " interfacial diffusion as the rate controlling process
   $= 5$ " dislocation pipe diffusion control.

The values of a, and G vary with the process; however, for all diffusion controlled processes G is proportional to the product of σ, D and C, as in the above equation.

Ostwald ripening is of technological importance in the development of high temperature resistant superalloys, which depend upon stable par-ticles to provide time independent strength. Thus, it is not a surprise that the interfaces between the particles in nickel based superalloys can be tailored to have negligible interface energy.

This area is still an active one for research because the predictions of the theory are not in precise agreement with experiment. In particular the main disagreement has to do with the form of the particle distribution function after Ostwald ripening.

8. SINTERING.

Sintering is another process that is interface energy driven. It is a process in which the interstices between contacting particles shrink as a consequence of the tendency to decrease the overall interface energy. There are three stages to the sintering process. In the first stage, necks grow between contacting particles, i.e. the area of contact between two contacting particles increases. This process is illustrated in Figure 11.18. These contact areas between crystalline particles are perforce along grain boundaries since adjacent particles are differently oriented. At the end of the first stage, the growth of these necks leads to the formation of spherical pores connected by grain boundaries. The shrinkage of these pores is the process occurring during the second stage of sintering. The grain boundaries, which had been anchored by the pores, now begin to move and their interaction with the remaining few volume percent of pores leading to the removal of the latter comprises the third stage. Incoherent grain boundaries hasten the sintering process because they are sinks for vacancies. Although surfaces and dislocations are vacancy sinks, as well, they are less important, either because they are far removed from the source of vacancies, or they are too few in number, respectively.

There are a variety of paths for matter transport: vapor transport, volume diffusion, grain boundary diffusion and plastic flow. This fact appears to have been appreciated first by Kucynski[13], who provided a quantitative theory of sintering that has stood the test of time in its major aspects. It appears that volume diffusion is the significant path for many experimental observations of the sintering process. There are two pathways for volume diffusion of matter: from surface of curvature $2/r$ to neck surface of curvature $(1/x - 1/r')$ and from grain boundary between the particles to the same neck surface. The latter pathway appears to be the predominant one for $r' \ll x \ll r$. Let us derive a relation for the rate of increase of x. A simplified model is as follows. Vacancies diffuse from just below the neck surface to the grain boundary, where they annihilate. Their virtual chemical potential at the neck surface is $\sigma\Omega(1/r' - 1/x)$, whereas at the grain boundary it equals

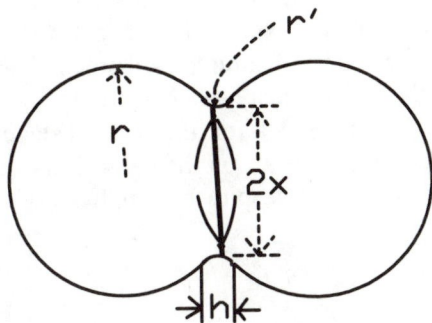

Figure 11.18. Illustrating geometry of neck formed between two spheres in process of sintering.

zero. The corresponding vacancy concentrations are C and C_o, respectively, where $kT(C-C_o)/C_o = \sigma\Omega(1/r' - 1/x)$, $C_o = \exp(-\Delta G_f/kT)$ and ΔG_f is the work to form a vacant site in the lattice by exchange of a subsurface atom with a vacant site on the surface. The diffusion geometry for the vacancies is complicated in that it is radially inwards from the lateral surface of a squat cylinder and axial to one of the two end surfaces. A typical approximation is to assume one dimensional diffusion through a cross-sectional area πx^2 over a diffusion distance L. The value of L used by various investigators varies, but in all cases it is taken to be proportional to x. Suppose $L = Kx$. Then, by equating the flux from one side of the grain boundary to the rate of increase of half the neck volume, we obtain, with the approximation that half the thickness of the neck equals $r'/2$,

$$(1/\Omega)(d/dt)(\pi x^2 r'/2) = (\pi x^2)(D^*\sigma/kTKxr')$$

where $r' << x$, $D^* = D_v C_o$ and D_v is the vacancy diffusion coefficient.

But, by the Pythagorean theorem and $r' << x << r$, $r' = x^2/2r$. Substituting and solving for x as a function of time yields

$$x^5 = (10\sigma\Omega D^* r^2/KkT)t$$

The values of K vary from 1/8 to 1/2 in the literature. (Compare the values in the books of Borg and Dienes, Kingery et al and the paper of Ashby[14].)

It is possible to derive relations between x and t for the other modes of atom transport with the result that the exponent n in the relation $x^n = At$

equals 2, 3, 5 and 7 when the mechanisms are viscous flow, evaporation-condensation, bulk diffusion and surface diffusion, respectively.

Sintering is even more a complicated process than Ostwald ripening, considering the gross changes that occur in the geometry and microstructure. A simple model for the second stage[15] yields a relation in which the pore volume depends on the logarithm of time. This relation seems to be obeyed by the data despite the simplicity of the model on which it is based.

9. DIFFUSION (CHEMICAL) INDUCED GRAIN BOUNDARY MIGRATION (DIGM OR CIGM)).

At temperatures where lattice diffusion is negligible (i.e. diffusion penetration distance is less than 1 Angstrom) it is found that solute diffusion along grain boundaries can induce grain boundary migration.[16-20] As the boundary moves it leaves a higher concentration of the diffusing solute behind it. This phenomenon occurs in both metals and non-metallic materials. The driving force for this process is the lowering in free energy of the system that accompanies the solution of the solute to form a more concentrated solid solution. It is observed at the low temperatures at which lattice diffusion is negligible, because diffusion along the grain boundary can still occur. The solute atoms enter the lattice at the grain boundaries, not by lateral lattice diffusion into the surrounding grains, but as a consequence of trapping at lattice sites as the boundary migrates. Hence, enrichment of the lattice in solute occurs only at the trailing edge of the migrating grain boundary.

A model for the kinetics of boundary migration under DIGM is presented below for the special case of a thin film couple between two pure constituents having the same molar volume V and the free energy composition curve shown in Figure 11.19. As indicated, when the composition in the trailing region behind a boundary is X(atom fraction), then the free energy available to drive the irreversible processes involved is $\Delta g_o / V$, where V is the molar volume and Δg_o is defined in Figure 11.19. Now, Δg_o cannot be the driving force for the boundary migration because part of Δg_o must be dissipated in the process of grain boundary diffusion. We shall consider that Δg_o has only two components: one the driving force for boundary migration, $\Delta g_b / V$, and dissipated irreversibly as the boundary migrates

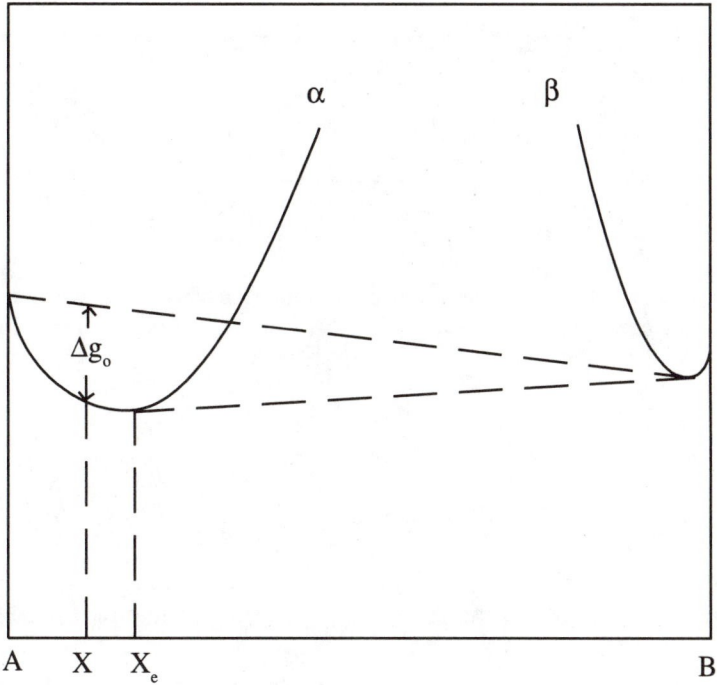

Figure 11.19. Free energy per mole versus mole fraction for two phases

and the other dissipated irreversibly in boundary diffusion.Thus, for the grain boundary illustrated in Figure 11.20 of area w·h and thickness δ and where the partial pressure of B corresponding to the equilibrium concentration of the β phase is maintained on the top face of the thin film

$$[\int_0^h (\Delta g_o /V)dyw]v = [(\Delta g_b /V) \text{ wh}]v + \int_0^h J(y)X(y)w\delta dy$$

where v is the boundary velocity in the x direction. The local flux of B, J(y), along the grain boundary obeys

$$J(y) = LX(y) = L(d\mu_b/dC_b)(-dC_b/dy) = -D_b(dC_b/dy)$$

where D_b is the grain boundary diffusivity for B, μ_b the chemical potential of the solute in the grain boundary and C_b the boundary solute concentration in atoms/volume. Thus,

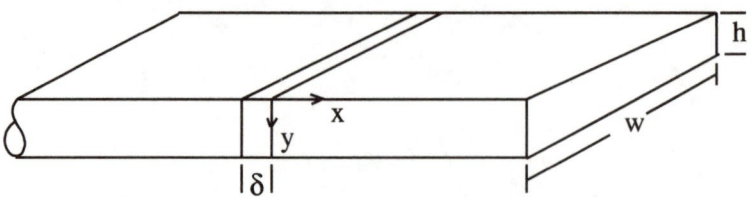

Figure 11.20. Illustrating geometry of thin film with grain boundary of thickness δ lying normal to the boundary. The boundary moves in the x direction with a velocity v and leaves behind it a concentration C=X/V at the position y.

$$\int_0^h J(y)X(y)dy = D_b(d\mu_b/dC_b)\int_0^h (dC_b/dy)^2\, dy.$$

We assume that a steady-state is reached such that the velocity, v, is constant and the boundary composition profile, $C_b(y)$, is independent of time. Hence, by conservation,

$$\delta D_b(d^2C_b/dy^2) = vC$$

where C is the concentration in the lattice adjacent to the boundary at the same y for which the concentration in the boundary is C_b. We also assume that $C_b = sC$, where s is closely given by the equilibrium segregation coefficient, but is not exactly equal to it. Thus, we obtain the second-order differential equation

$$\delta D_b(d^2C_b/dy^2) = vC_b/s.$$

For the boundary conditions that at y = 0, $C_b = sC_e$, where C_e is the concentration in α in equilibrium with β and at y = h, $(dC_b/dy) = 0$, we obtain the solution

$$C_b = (sC_e)\cosh(y'\sqrt{a})/\cosh(\sqrt{a}/2)$$

where y' = (h−y)/2h and a= $(4vh^2)/(s\delta D_b)$. Taking the derivative with respect to y, substituting this derivative in the first equation above along with the relation $(\Delta g_b/V) = v/M$, we obtain

$$v = (<\Delta g_o>/V) / [(1/M) + (C_e^2 h^2 (d\mu_b/dC_b)/3\delta D_b)]$$

where $<\Delta g_o> = \int_o^h (\Delta g_o /V)dy/h$ and which is not less than $\Delta g_{oe}/2$.

  The first term on the right hand side corresponds to reaction control of the interface velocity and the second term to diffusion control of it. The main point to be made is that the driving force for the boundary migration is not Δg_{oe}, but Δg_b. The above analysis, hopefully, will act to clear up a controversy concerning the detailed nature of the driving force. It is not the free energy of mixing, because a major fraction of the latter is used to drive the grain boundary diffusion process. Unfortunately, the above considerations are thermodynamic in origin and do not specify the mechanism of the migration driving force. However, there is no reason to emphasize the lack of understanding of this mechanism, especially when no fuss has been made concerning the mechanism of the driving force for boundary migration in discontinuous precipitation, an analogous process to DIGM.

10. SUMMARY.

  We have considered the motion of a planar interface, either between two terminal phases of a binary system having or not having stable intermediate phases in the phase diagram for the system or between a supersaturated or supercooled parent phase and a product phase. For a multicomponent system this motion involves the coupling between an interface reaction and diffusion. Since these processes are in series, the slower controls. We have obtained relations for the interface velocity and defined the transition between these two controlling regimes. We have also briefly considered the processes of physical vapor deposition, chemical vapor deposition, Ostwald ripening, sintering and DIGM.

REFERENCES.

1. F.C. Larche and J.W. Cahn, Acta Met. 33, 331(1985).
2. G.V. Kidson, J. Nuclear Mater. 3, 21(1961).
3. X.L. Yeh, K. Samwer, and W.L. Johnson, Appl. Phys. Lett.42, 242(1983).

4. R.H. Doremus, RATES OF PHASE TRANSFORMATIONS, Academic Press, NY, 1985, p.116.
5. W.G. Pfann, ZONE MELTING, J. Wiley, NY, 1957;Kruger, NY, 1978.
6. D. Turnbull, J. Physical Chem. 66, 609(1962).
7. M.J. Aziz, J. Appl. Phys.53, 1158(1982); Apply. Phys. Lett.43, 552(1983).
8. B. Caroli, C. Caroli and B. Roulet, Acta Met.34, 1867(1986); J. Cryst. Growth 66 575(1984).
9. J.C. Baker and J.W. Cahn, in SOLIDIFICATION, ASM Seminar, ASM, Metals Park. Ohio(1971).
10. J.M. Jasinski, B.S. Meyerson and B.A. Scott, Ann.Rev. of Phys. Chem. 38, 109(1987).
11. I.M. Lifshitz and V.V. Slyozov, J. Phys. Chem. Sol. 19, 35(1961).
12. C. Wagner, Z. Elektrochem. 65, 581(1961).
13. G. Kuczynski, Trans. A.I.M.E. 185, 169(1949).
14. M. F. Ashby, Acta Met.22, 275(1974).
15. R.L. Coble, J. Appl. Phys. 32, 787(1961).
16. M. Hillert and G.R. Purdy, Acta Met. 26, 333(1978).
17. P.G. Shewmon and G. Meyrick, in DIFFUSION IN SOLIDS, RECENT DEVELOPMENTS, eds. Dayananda and G. Murch, A.I.M.E., Warrendale, Pa., 1985.
18. D.B. Butrymowicz, J.W. Cahn, J.R. Manning and D.E. Newbury, in Advances in Ceramics,6, GRAIN BOUNDARIES AND INTERFACES IN CERAMICS, Am.Ceramics Soc.,1983, p.202.
19. R.W. Baluffi and J.W. Cahn, Acta Met.29, 493(1981).
20. R.S. Hay and B. Evans, Acta Met.35, 2049(1987).

BIBLIOGRAPHY.

Growth of intermediate phases.

V.I. Dybkov, J. Mater. Sc.21, 3078, 3085(1986);22, 4233(1987); J. Phys. Chem. Sol.47, 735(1988).
C. Wagner, Acta Met.7, 99(1969).
S.R. Shatynski, J.P. Hirth, and R.A. Rapp, Acta Met.24, 1071(1975).
D.S.Williams, R.A. Rapp and J.P. Hirth, Met. Trans.12A, 639(1981).
Guan-Xing Li and G.W. Powell, Acta Met. 33, 23(1985).

Vapor deposition.

W.K. Burton, N. Cabrera and F.C. Frank, Phil. Trans(London) A243, 299(1951).

Sintering.

G.C. Kuczynski, MATERIALS SCIENCE RESEARCH 13(Sintering Processes), 1980, Plenum Press, NY.
W.D. Kingery, H.K. Bowen and D.R. Uhlmann,INTRODUCTION TO CERAMICS, 2nd ed., 1976, J. Wiley, NY. This book has a good chapter on sintering processes.

XII-MORPHOLOGICAL INSTABILITY
AND
GROWTH VIA CELLULAR SEGREGATION

INTRODUCTION.

There are phenomena, which are diffusion controlled, but for which the solution to the diffusion problem is indeterminate in the sense that a growth velocity is given as a function of a length parameter of the product phase. In the past, the attempts to solve this problem of indeterminateness have made use of arbitrary criteria that have had no physical basis, as, for example, the criterion of maximum growth velocity, or that of maximum entropy production rate, etc. One of these indeterminate problems concerns the growth of dendrites in solidification. In the early 1960's researchers became aware of the possible role of morphological instability in this mode of growth and the classic paper in this field by Mullins and Sekerka[1] has since inspired a flood of other studies involving morphological instability.

There are two aspects of morphological instability. One involves the morphological instability of a growth form. The other involves the control exerted on the kinetics of the growth process by morphological instability. The first aspect was that treated by Mullins and Sekerka. The second aspect was first treated quantitatively by Langer and Muller-Krumbhaar[2] and they enunciated the significance of "marginal stability" as a principle operating to control the dimension of a growing dendrite during solidification. In their analysis of this problem, they noted the existence of another factor affecting the course of such growth; namely, the effect of fluctuation induced drift of the length parameter, the dendrite tip radius in this case. It is shown in this chapter, that fluctuation induced drift of the length parameter is the factor that provides the basis for explaining the behavior of the length parameter in the various cellular segregation processes: eutectic solidification, eutectoid (pearlitic) growth and discontinuous precipitation.

Thus, in this chapter we will first consider the morphological stability of product interfaces in contact with either supersaturated or supercooled parent phases. Then, we will consider how "marginal stability" exerts its control of the growth morphology in the case of the growth of a dendrite. Finally, we will describe how fluctuation induced drift of the length parameter eliminates the indeterminateness previously existing in the solutions of the diffusion problem for various cellular segregation growth modes.

1. MORPHOLOGICAL STABILITY.

1.1. Physical basis for morphological instability.

Whenever growth of a stable phase occurs by migration of an interface between the stable and metastable phases, the interface is subject to morphological instability when the region adjacent to the interface is either supersaturated or supercooled. This tendency toward morphological instability can be understood qualitatively with the aid of Figure 12.1. This figure shows iso-temperature profiles in a supercooled melt (or iso-concentration profiles in a supersaturated solution) adjacent to a spherical bump on a planar interface.

It is apparent that because of the crowding of the isotherms about the bump that the temperature gradient in the vicinity of the bump is larger in magnitude than it is adjacent to the planar region. Hence, the bump can reject the heat of fusion per unit extension into the melt faster than the planar section can and, consequently, solidify faster than the planar portion of the interface. This qualitative analysis neglects the stabilizing influence of capillarity due to the resistance to increase of the interface area. The influence of capillarity is such that bumps having radii smaller than some critical radius will tend to disappear, whereas bumps larger than this critical radius will grow. Typical products of morphological instabilities are dendrites and whiskers.

These concepts describe the physical nature of the morphological instability problem, at least for the solidification of pure materials into a supercooled liquid. In principle, the same tendency for the morphological instability of a planar interface exists for growth into a supersaturated or

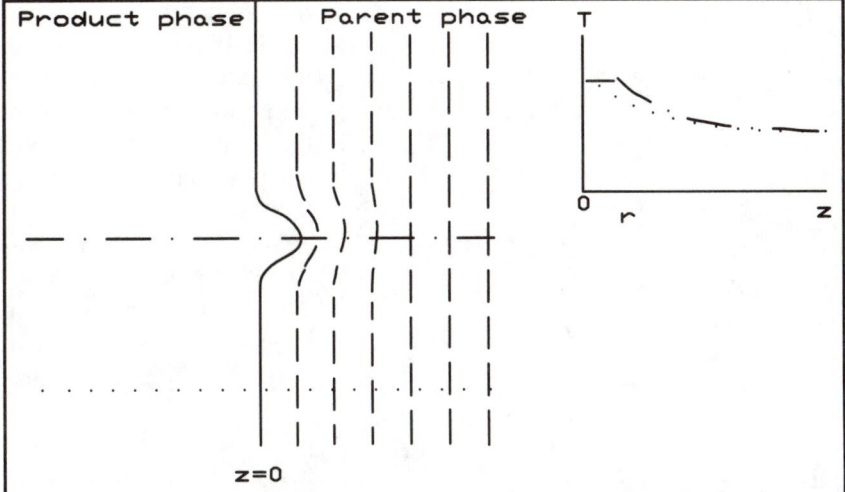

Figure 12.1. Showing the crowding of isotherms (dotted lines) about a bump on a flat interface between an unstable parent phase and a growing product phase. Inset shows plots of temperature versus distance along corresponding lines in main figure.

supercooled phase. Morphological instabilities can arise from other sources as well. Internal stress induced in thin films that are coherent with their substrates can be a source of morphological instability as can surface tension. Also, there are conditions where the reaction front may undergo an oscillation superimposed on its steady state propagation and thereby induce morphological inhomogeneities. The mathematical treatment of all these problems is complex. We shall therefor forego detailed mathematical analysis of morphological instability and instead confine our discussion to the results of such analyses in the remainder of this chapter.

1.2. Liquid/solid interface (solidification).

Consider a solid-liquid interface which originally was planar, but has been perturbed to have the form of a periodic wave, as illustrated in Figure 12.2. This wave has an amplitude δ and a wavenumber $2\pi/\lambda$, where λ is the wavelength. The interface moves with a constant velocity v in the z direction, where z=0 at the interface. We assume a binary system in which m is the slope of the liquidus and k is the distribution coefficient. Also, the concentration gradient ahead of the interface is G_C, and the temperature gradients in the liquid and solid are G_L and G_S. If $(1/\delta)(d\delta/dt)>0$, then the inter-

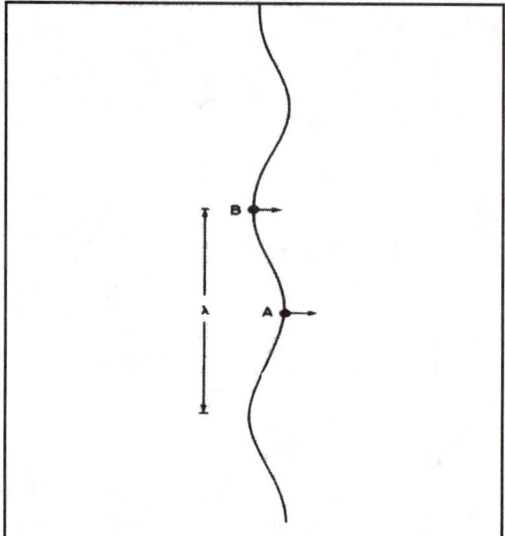

Figure 12.2. Illustrating a perturbed interface with a wave length λ. A and B are the highest and lowest points on the interface, respectively.

face is morphologically unstable. If, on the other hand, this ratio is negative then the periodic perturbation in the planar interface will decay to restore the planar interface.

This problem was solved originally by Mullins and Sekerka[1].The solution is obtained by setting down the partial differential equations that govern the transport of heat in the liquid and the solid and the transport of solute in the liquid. It is assumed that no diffusion occurs in the solid phase. Also, conservation of matter and heat balance at the interface are obeyed. (There is of course a divergence of heat at the interface due to the heat of fusion.) Further, the effect of capillarity on the temperatures and concentrations at the interface is taken into account. The far field boundary conditions are assumed equal to those for a planar interface. The local interface velocities calculated from heat flow relations must equal those evaluated from diffusion considerations. At the same time they also must equal $v + (d\delta/dt)\sin\omega X$, where v is the velocity of the planar interface. The partial differential equations are then solved subject to the boundary conditions and the other conditions. In Appendix 1 we repeat the classical perturbation analysis of Mullins and Sekerka[1]. However, we present here the more general results for the stability criterion due to Trivedi and Kurz[3]. Their result for the stability criterion is

$$-\Gamma\omega^2 - \overline{K}_L G_L \left[\frac{\omega_L - (V/a_L)}{\overline{K}_S\omega_S + \overline{K}_L\omega_L} \right] \qquad - \overline{K}_S G_S \left[\frac{\omega_S + (V/a_S)}{\overline{K}_S\omega_S + \overline{K}_S\omega_S} \right]$$

$$+ mG_C \left[\frac{\omega_C - (V/D)}{\omega_C - (V/D)(1-k)} \right] \begin{array}{l} > 0 \text{ (Morphological Instability)} \\ < 0 \text{ (Morphological Stability)} \end{array}$$

where $\Gamma = \sigma/\Delta S$, σ is the solid-liquid interface energy, ΔS is the entropy of fusion per unit volume, $\omega = 2\pi/\lambda$, where λ is the wave length of the sinusoidal perturbation in shape of the planar interface. Also,

and

$$\overline{K}_L = K_L/(K_S + K_L)$$

$$\overline{K}_S = K_S/(K_S + K_L)$$

where $K_{S,L}$ are the thermal conductivities in liquid(L) and solid(S). Further, $a_{L,S}$ are the thermal diffusivities, V is the velocity of the interface and D is the solute diffusion coefficient in the liquid.

Let us consider this stability criterion for various conditions. We obtain the stability conditions of Mullins and Sekerka by setting $\omega_C = \omega_S = \omega_L = \omega$ in the above relations. In this approximation, the stability criterion becomes

$$\Gamma\omega^2 > -\overline{K}_L G_L - \overline{K}_S G_S + mG_C \left[\frac{\omega_C - (V/D)}{\omega_C - (V/D)(1-k)} \right]$$

At low velocities, the term in the brackets approaches unity and, if the capillarity term is neglected, is the same as the modified constitutional supercooling criterion.

At high velocities or large undercoolings, where $G_S = 0$ is an acceptable approximation and where we also assume $K_S = K_L$ and $a_S = a_L$, then the criterion for stability becomes

$$V^2 > (mG_C D^2/k - G_L a_L^2)/\Gamma$$

By use of the conservation relations at the interface, i.e.

$$DG_C = C_L(1-k)V \text{ and } K_L G_L = -V \Delta H$$

and $a_L = K_L C_p$, and by substitution we obtain

$$V > mC_S(1-k)D/(k^2\Gamma) + a_L \Delta H/(C_p\Gamma)$$

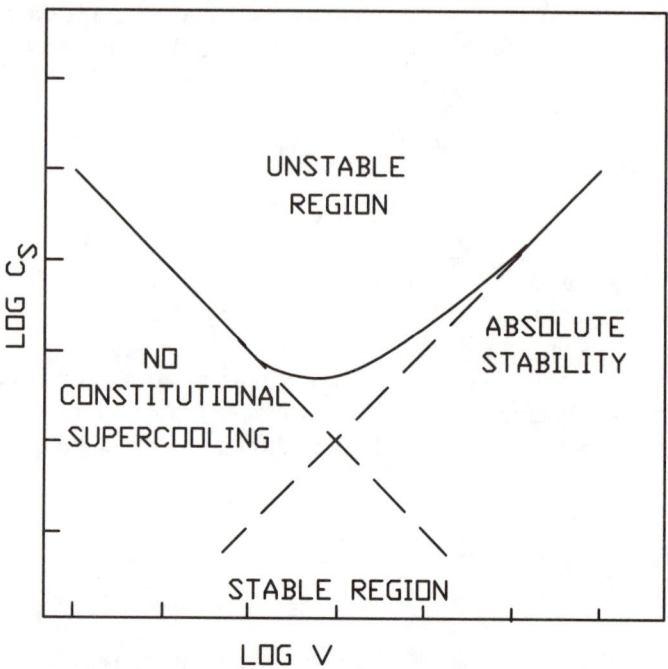

Figure 12.3. Interface stability as revealed in a plot of log solute concentration as a function of log interface velocity.

for the criterion for stability at high interface velocities. The first term on the right hand side corresponds to the result of the Mullins and Sekerka analysis. The second term is a correction due to the Trivedi and Kurz analysis. For the Mullins and Sekerka conditions

$$V < (\Gamma\omega^2 + \bar{K}_L G_L + \bar{K}_s G_s)\, D/\, [mC_s(1 - k)/k]$$

for the criterion for stability at low velocities. In the Trivedi and Kurz analysis the assumptions $\omega_c = \omega_s = \omega_L = \omega$ are not made.

A plot of log solute concentration versus interface velocity is shown in Figure 12.3. As shown, the stable regions at low velocity and high velocity are separated by an unstable region. The morphological stability at low velocity is due to the lack of undercooling, whereas that at high velocity is due to the capillary length $(\Gamma/[mC_s(1-k)/k])$ exceeding the sum of the solute diffusion length (D/V) and the thermal diffusion length (a_L/V).

This result is applicable in a general sense to other processes. Whenever, the sum of the solute and thermal diffusion lengths exceeds the capillary length we may expect a tendency towards morphological instability at sufficient undercooling or supersaturation.

1.3. Vapor/solid interface.

1.3.1. Chemical vapor deposition

The critical condition separating reaction controlled growth from diffusion controlled growth in chemical vapor deposition was investigated in Chaper X. The result was a number called the CVD number. We noted there that the inspiration for the derivation of such a number was the knowledge that morphological instability is possible only in the diffusion controlled regime. A stability analysis for chemical vapor deposition has been carried out by Brekel[4] that is analogous to that for solidification carried out by Mullins and Sekerka. The main difference is that in CVD diffusion and heat flow occur through a laminar layer of constant thickness, h_g, that is large compared to the amplitude of the sinusoidal perturbation in shape of the interface. The temperature and composition at the outer boundary of the laminar layer are maintained constant. Thus, the solutions to the diffusion equations differ to some extent, but not importantly. In the notation used in the previous section, the result for the instability criterion under conditions where the wave length of the perturbation, λ, satisfies $\lambda << h_g$ and $\lambda << d$, where d is the thickness of the substrate, becomes

$$(1/\delta)(d\delta/dt) = V\omega \left[\frac{2G_C - 2C_e^\infty \Gamma \omega^2 + z(G_g + G')}{2(1 + \omega h_g/Nu)G_C + z(G - G' - G'')} \right] > 0$$

where

$$z = C_e^\infty(\Delta H/(RT_o^2)) + (C_o - C_e^\infty)(\Delta E/RT_o^2)$$

$$G = 2K_g G_g/K; \quad G' = 2K_s G_s/K; \quad G'' = 2a\sigma T_o^4/K$$

$$Nu = k_D h_g/D: \quad K = K_s + K_g + K_r; \quad K_r = 4aT_o^3/\omega$$

Also, ΔH is the heterogeneous reaction molar enthalpy ($\Delta H < 0$ for an exothermal reaction) and ΔE is the activation energy for the heterogeneous decomposition reaction

$$AB_g = A_s + B_g$$

Further, C_e^∞ is the equilibrium concentration of AB at a planar interface at temperature T_o, C_o is the actual concentration at the planar interface at temperature T_o, K_i is the thermal conductivity in the i phase, G_c is the concentration gradient, G_g the thermal gradient in the gas phase, G_s is the thermal gradient in the solid phase at the unperturbed interface, a is the emission coefficient, $\sigma = 1.35 \ 10^{-12}$ cal/K^4scm^2, the Stefan-Boltzmann radiation constant, Γ is the capillarity constant($=2\sigma\Omega/kT_o$, where σ is the interface energy and Ω is the atomic volume of the solid, k Boltzmann's constant), ω is the wave number of the sinusoidal morphological perturbation on the planar interface ($\omega = 2\pi/\lambda$, where λ is the wave length of this perturbation), k_D is the mass transfer coefficient or rate constant of the heterogeneous decomposition reaction as given by

$$k_D = k_o \exp(-\Delta E/RT)$$

and D is the diffusivity in the gas phase of the limiting reactant species and, finally, V is the vapor/solid interface velocity.

For infinite Nusselt number the above result for the instability criterion is very similar to that obtained by Mullins and Sekerka and the one we have derived in the previous section. In particular, the parameters $1/z$ and C_e/z play the same roles as the parameters m and T_M play in the solidification instability analysis.

In the expression for $(1/\delta)(d\delta/dt)$ the denominator is positive (if it were negative it would result in the prediction that the capillarity term is not stabilizing, which is contrary to physical sense); in the numerator the term with G_C is positive since the concentration gradient is positive and favors instability; the term containing Γ is negative and represents the stabilizing effect of capillarity; and the term proportional to $z(G + G')$ can be either positive or negative. When the substrate is hot compared to the gas phase (the term $(G + G')$ is negative) and the protrusions are slightly cooler than the valleys of the sinusoidal perturbed surface. In this case and when z is positive then the thermal gradient term is stabilizing against propagation of the morphological perturbation. For diffusion controlled growth (Nu > 1 and $C_o = C_e$) a positive value of z implies a positive value of ΔH (i.e. an endothermic reaction for the decomposition, $AB_g \rightarrow A_s + B_g$). This result can also be understood using the relation for the equilibrium concentration at a curved interface given by

$$C_e = C_e^\infty [1 - (\Delta H/RT_o^2)(T - T_o)].$$

Hence, the equilibrium concentration is slightly lower in the valley as compared to a protrusion and consequently the supersaturation is higher in the valley than above the protrusion and the latter will grow at a slower rate than the valley, i.e. a stabilizing condition. A change in sign of ΔH will reverse this situation because the relative supersaturation also reverses. With a change in sign of the thermal gradients, the valleys become cooler than the protrusions and now endothermal reactions ($\Delta H > 0$) will destabilize the morphology while exothermal ($\Delta H < 0$) reactions stabilize the morphology.

In the event the deposition process is reaction controlled then, because $\Delta E > 0$, the sign of z is also positive for $\Delta H > 0$. For $\Delta H < 0$, the sign of z depends upon the supersaturation, the relative values of C_o and C_e. Hence, no general conclusions on stability can be drawn in this case.

To summarize, morphological instability during chemical vapor deposition is encouraged by a high supersaturation leading to a large positive value of G_c. It is also encouraged when the substrate is hot relative to the gas by an exothermal decomposition reaction, or when the substrate is cold relative to the gas by an endothermal decomposition reaction.

Empirically, morphological instability is quite common in chemical vapor deposition leading to the growth of whiskers. The onset of whisker growth, however, may depend upon other special factors, such as the existence of a special property at the tip of the whisker not existing elsewheres on the surface of the film. For example, in the vapor-liquid-solid mode of whisker formation in CVD, the tip of the whisker has a lower melting point than elsewheres resulting in the formation of a liquid drop on the tip and thus yielding a higher accommodation coefficient at the tip and its more rapid growth, i.e. supersaturation is not the driving force for morphological instability in this case.

1.3.2. Physical Vapor Deposition

There are at least two sources of morphological instability in physical vapor deposition. Shadowing of lower asperities by higher protrusions tends to yield a higher deposition rate on the highest protrusions and their growth at the expense of smaller ones. The finite size of the atoms results in a higher projected area at the surface through the centers of the atoms over protrusions than over valleys for the same area on the surface.[5] Figure 12.4a illustrates this effect, which is sufficient to result after

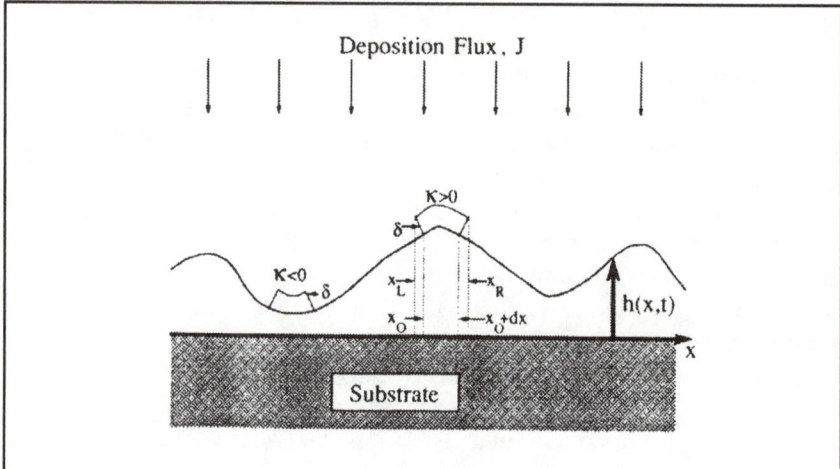

Figure 12.4a. Illustrating the difference in projected length due to surface curvature. K is the surface curvature and δ is the atomic radius. After D.J. Srolovitz, A. Mazor and B.G. Bukliet, J. Vac.Sci.Tech. $\underline{A6}$, 2371(1988).

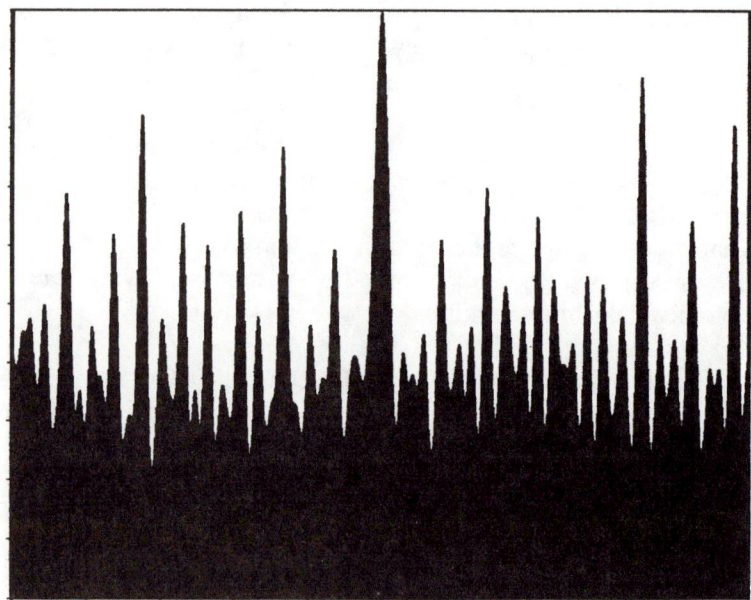

Figure 12.4b. Illustrating the calculated profile developed by deposition in the absence of surface diffusion. After Karunasiri, Bruinsma and Rudnick, Phys.Rev.Lett. $\underline{62}$, 788(1989).

sufficient deposition in a non-planar interface having protrusions, such as those shown in Figure 12.4b. Srolovitz et al[5] and Karunasiri et al[6] have applied perturbation theory to study the morphological stability in physical vapor deposition. In this case, diffusion is a stabilizing influence, tending to smooth out the asperities provoked by the two effects described above. Srolovitz et al predict a transition between Zones I and II and the growth of columnar grains in the latter zone.Karunasiri et al also find that it is possible to grow a flat surface up to certain critical film thickness. The protrusions illustrated in Figure 12.4b are a consequence of an instability with respect to surface curvature in both models. This instability is revealed using a linear stability analysis as per the dendrite stability analysis in the latter model.

Srolovitz[7] has recently discovered that internal stress in a film, as can be induced during epitaxial deposition onto a substrate that has a slightly different lattice parameter from that of the film, is also a source of morphological instability leading to the development of a wavelike surface profile. This instability can be understood physically as follows. For sufficiently small wave length the material above the troughs will be stress-free although the surface energy will be increased by the formation of additional surface area. Hence, the film becomes unstable with respect to such surface morphology perturbation when the wavelengths of the perturbations exceed a critical wavelength that scales inversely as the internal stress. One possible result of such an instability is a distribution of unstressed isolated islands. It is probable that this source of instability operates in most of the other growth processes, as well as in vapor deposition, and it may well be that the onset of Stranski-Krastanov growth in which islands form on a monocrystalline layer is due to this phenomenon.

1.4. Solid/solid interface.

Mullins and Sekerka[8] treated the morphological stability of a growing sphere in a supersaturated field, describing the infinitesimal perturbation in the spherical shape in terms of an expansion in spherical harmonics. The result for the absolute instability criterion is

$$(1/\delta)(d\delta/dt) = \left\{ \frac{D(n-1)}{(C_\beta - C_\alpha(r))r} \right\} \left[\frac{C_o - C_\alpha(r)}{r} - \frac{C_o \sigma v_\beta (n+1)(n+2)}{kTr^2} \right]$$

where

$$C_\alpha(r) = C_\alpha\{1 + [(1 - C_\alpha)/(C_\beta - C_\alpha)][(2\sigma v_\beta)/(kTr\zeta)]$$

and

$$\zeta = 1 + d\ln f_\alpha/d\ln C_\alpha$$

where f_α is the activity coefficient and C_α is the concentration of solute in alpha in equilibrium with a planar precipitate of beta. Also, r is the radius of the spherical precipitate, D is the diffusivity of the solute in alpha, C_o is the far field concentration in alpha, σ is the interfacial energy, v_β is the molar volume of the beta phase, k is Boltzmann's constant and T the absolute temperature.

Thus, since instability will occur when $(1/\delta)(d\delta/dt) > 0$, it is required that $n \geq 2$ and $r > [\sigma v_\beta/(kT(C_o - C_\alpha(r)))][C_o(n+1)(n+2)]$ for the absolute instability. However, a shape change will not occur unless $(1/\delta)(d\delta/dt) > (1/r)(dr/dt)$. The latter inequality requires that $n \geq 3$. If we define

$$r^* = [C_\alpha/(C_o - C_\alpha)][2\sigma v_\beta/kT]$$

and for n=3 we obtain that the growing spherical precipitate will be morphologically unstable when $r > 21r^*$. In this problem, the supersaturation is the driving force for instability and the capillarity provides the stabilizing force.

It is known that the growth of solid phases from supersaturated solids does not often lead to the development of the morphological instability that exists. One reason for the absence of morphological unstable growth forms in this situation is that the growth may be controlled by an interface reaction rather than by diffusion. Another possible explanation is that rapid lateral diffusion of the solute along the interface between parent and product phases decreases the differences in growth velocity between bump and hollow in a planar interface. Still another possible explanation of the lack of morphological instability in solid state growth is that anisotropy of the interface energy, which is more prevalent between solid phases, increases the contribution of the capillarity term to the stabilization of shape perturbations. Finally, still another possible reason may be related to the dependence of the elastic energy induced by the solid state phase transformation on the shape of the product phase.

Summarizing, morphological instability in the shape of an interface between a growing phase and its environment can exist for a variety of reasons. If the environment consists of a supersaturated solution or a supercooled solution then such instability occurs whenever, the diffusion length or the thermal diffusion length exceeds the capillarity length. Also, with a chemical reaction at the interface, as may occur in CVD the effect of the latter on the instability depends upon the sign of the heat of the reaction. In PVD and at low temperature surface curvature drives morphological instability and surface diffusion acts to prevent it. Finally, internal stress can drive morphological instability in thin films. This is not a complete catalogue of all the sources of morphological instability as a complete reading of the applicable literature will demonstrate.

2. MARGINAL STABILITY AND GROWTH.

2.1. Dendritic growth.

It is well known that the solution of the diffusion equation (or the thermal diffusion equation) governing the growth of a stable phase at the expense of an unstable phase in diffusion controlled growth yields an expression for the velocity of the interface as a function of a length parameter of the product. For example, the solution for the diffusion controlled growth of a paraboloid shaped precipitate from a supersaturated field of solute, first given by Ivantsov[9] is an equation for the Peclet number, $p=Vr/2D$, where V is the velocity of the paraboloid tip, r is the tip radius and D the diffusivity in the parent phase. This solution is

$$(C_o - C_\alpha)/(C_\beta - C_\alpha) = p\exp(p)E(p)$$

where E is the integral exponential function $\int_p^\infty (\exp(-u)/u)du$. Thus, p is defined but V and r are not defined, although their product is defined.

In the past, several additional criteria have been suggested to provide a constraint on the value of either V or r. For example, it has been suggested that growth proceeds so as to maximize the growth velocity, to minimize the rate of entropy production, to maximize the rate of decrease of the free energy, etc.. None of these criteria have been given a fundamental justification. Recently, another criterion has been suggested which does stem from fundamental physical underpinnings. In particular, the criterion of marginal stability has been suggested to govern the growth of a dendrite during solidification. Langer and Muller-Krumbhar[2] first suggested that dendrites grow at a critical tip radius because radii smaller than this critical radius tend to increase due to the development of side-branches that interact with the dendrite tip and radii larger than this critical value develop instabilities. Their analysis yielded an equation for this critical tip radius given by

$$r^* = [2a_L d_o/(VK^*)]^{1/2}$$

where $d_o = \sigma C_p / S_f^2$ and the constant K^* was evaluated to have the value $0.025 \pm 0.007 (= 1/(2\pi)^2)$ from a numerical analysis. Huang and Glicksman[10] from an approximate analytical treatment of the marginal stability problem in dendrite growth obtained $K^* = 0.0192$ for a dendrite tip growing in the <100> direction. Kurz proposed another approximation which takes the tip radius to be equal to the shortest wavelength perturbation which would cause the dendrite tip to undergo morphological instability. The latter approximation yields $r^* = 2\pi[D\sigma/(S_f Vk T_o)]^{1/2}$, where

$$T_o = T_L(C_o) - T_S(C_o).$$

The full lines in Figure 12.5 represent solutions to the transport equations for the growth of a dendrite tip. As mentioned above any solution to the transport equations will yield an indeterminate result in that only the product of the growth velocity and the tip radius is determined, but neither the growth velocity nor the tip radius is separately determined. The Ivantsov assumptions have already been mentioned. The other solutions include capillarity and molecular attachment kinetics as additional considerations in the transport equations. The limit to growth at small tip radii is due to capillarity. The open circle correspond to the maximum velocity growth criterion. The closed circle corresponds to the

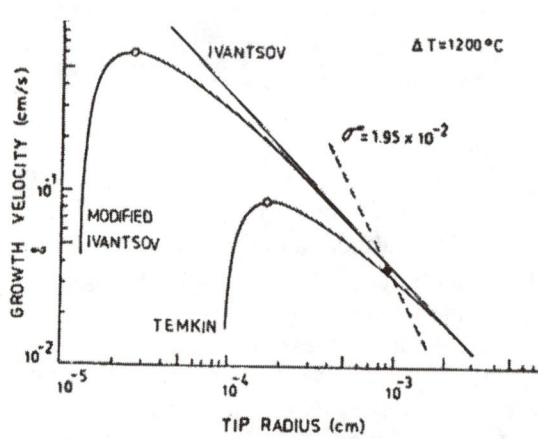

Figure 12.5. Comparison between theoretical predictions and experiment. The solid curves represent solutions to the diffusion equations. The dashed line represents the marginal stability criterion with $K^* = 0.0192$. The experimental point is the solid point at the intersection of the full and dashed lines. After Huang and Glicksman[9].

experimental value of tip radius and dendrite velocity for succinonitrile. The dotted line corresponds to the Langer-Muller-Krumbhar equation for the marginal stability condition with $K^* = 0.0192$. It is apparent that the latter together with the solutions to the transport equations yields agreement with experiment and that the maximum velocity criterion does not. This result has been substantiated with other systems.

Thus, there is reason to believe that marginal stability is a growth criterion that is consistent with both experiment and theory. It is worth repeating that there are two concepts involved in marginal stability. The first concept is that some physical basis exists for a drift in magnitude of the length parameter involved in the Peclet number. The second concept is that there is a critical value of this length parameter at which morphological instabilities develop. We shall see in the next section that these twin pillars of marginal stability may also explain some cellular growth phenomena.

3. GROWTH VIA CELLULAR SEGREGATION.

We include among the types of growth processes that we call cellular: eutectic solidification, eutectoid or pearlitic growth, and discontinuous precipitation. These are cellular growth processes in the sense that the product consists of lamellae, with the plane of the lamellae lying perpendicular to the reaction front. They involve segregation in that atoms must be transported to and from adjacent cells (lamellae) in the reaction front between parent and lamellar product phases in order for the latter to grow.

Reasonable arguments have been presented to conclude that eutectic solidification proceeds under marginal stability conditions.[11-13] However, several problems exist with respect to the generalization of these arguments to the other cellular processes. First among these is the knowledge that eutectoid growth and discontinuous precipitation do not obey the same relations as eutectic solidification experimentally. In particular, it is known that eutectic solidification occurs at front velocities and interlamellar spacings corresponding to those predicted by the maximum velocity criterion, as well as by the criterion for marginal stability, whereas neither eutectoid growth nor discontinuous precipitation obeys the maximum velocity criterion predictions. One possible reason for this difference in behavior is that the path for transport of atoms in these processes differ. In eutectic solidification, this path is a zone in the parent liquid phase adjacent to the lamellar product/parent liquid interface. In discontinuous precipitation it is the interface itself between the parent and lamellar product phases, whereas in eutectoid growth it may be either path.

We present in the following sections a new view of the process that leads to the observed interlamellar spacing, common to all the cellular segregation processes. This new view consists of a generalization of a model for the drift in the interlamellar spacing in eutectic solidification proposed by Langer.[12] First, we consider the case of eutectic solidification and Langer's analysis for the unidirectional drift in interlamellar spacing and the instability that sets in at a critical interlamellar spacing.[12] Then, we extend Langer's drift criterion to the cases of eutectoid growth and discontinuous precipitation. For eutectoid growth we make use of another critical inter-lamellar spacing at which an instability appears, that has come to be known as the "upper catastrophic limit", to suggest that marginal stability is also a criterion for eutectoid growth. We do not provide an analytic basis for the upper catastrophic limit, although we do suggest a physical reason for it. In the case of discontinuous precipitation we suggest that although there exists a unidirectional drift in the interlamellar spacing, the rate of change of the latter becomes very slow above some value of the spacing--sufficiently slow that this spacing remains unchanged during the experimental growth period.

3.1. Eutectic solidification.

First, we present the solutions to the transport equations due to Jackson and Hunt[11] for the case of eutectic solidification under the influence of a temperature gradient that is moving with a constant velocity.

In the above class of eutectic solidification the transport equations have a solution of the form

$$\Delta T(\lambda) = AV\lambda + B/\lambda$$

where

$$A = mP(1 + \zeta)^2 C_o/(\zeta D)$$

$$P = \sum (n\pi)^{-3} \sin^2(n\pi S_\alpha/(S_\alpha + S_\beta))$$

$$B = 2(1 + \zeta)[a_\alpha^L/m_\alpha + a_\beta^L/(\zeta m_\beta)]$$

$$a_\alpha^L = (T_E/L)_\alpha \sigma_\alpha^L \sin\theta_\alpha^L$$

$$a_\beta^L = (T_E/L)_\beta \sigma_\beta^L \sin\theta_\beta^L$$

$$1/m = 1/m_\alpha + 1/m_\beta$$

and V is the reaction front velocity, $\zeta = S_\beta/S_\alpha$ the ratio of the lamellar plate thicknesses in phases β and α, m_i is the slope of the liquidus in equilibrium with the i phase, C_o is the concentration difference at the eutectic temperature, T_E, between the concentrations of the solid phases in equilibrium with the liquid phase at the eutectic composition, D is the diffusivity in the liquid phase, L is the latent heat for the eutectic reaction: liquid-> $\alpha + \beta$, σ is the interfacial energy of the interface between the phases denoted by the subscript and superscript, θ_β^L is the angle between the tangent to the β/L interface and the reaction front at the α/β/L triple junction., and finally λ $=2(S_\alpha + S_\beta)$, i.e. the S values correspond to half the respective plate thickness.

The above result is for the steady-state solidification achieved by having the reaction front move at the same velocity as a moving temperature gradient G, as occurs in directional solidification. Langer[12] provided an analysis of the marginal stability condition for this mode of solidification based on mainly dimensional considerations, which were then justified by a more detailed linear stability analysis by Datye and Langer[13]. We shall reproduce the main arguments of the Langer analysis because we will use them in a discussion of what controls the interlamellar spacing in eutectoid and discontinuous precipitation processes.

Let the position of the interface be described by the function $\eta(x,t)$, which measures the distance of the local interface at x (as defined in Figure 12.6) from its undeformed position relative to the frame of reference moving with the velocity V. The moving temperature gradient has the value G and thus the local undercooling at the displaced interface at position x will be given by

$$G\eta(x,t) = -\Delta T(\lambda)$$

Langer assumed that $\eta(x,t)$ and the vertical displacement y (as defined in Figure 12.6) are coupled by the condition that each lamella must grow in a direction which is locally perpendicular to the solidification front. Thus,

$$\partial y/\partial t = -V \, \partial\eta/\partial x$$

Also, the local lamellar spacing and y are related thru

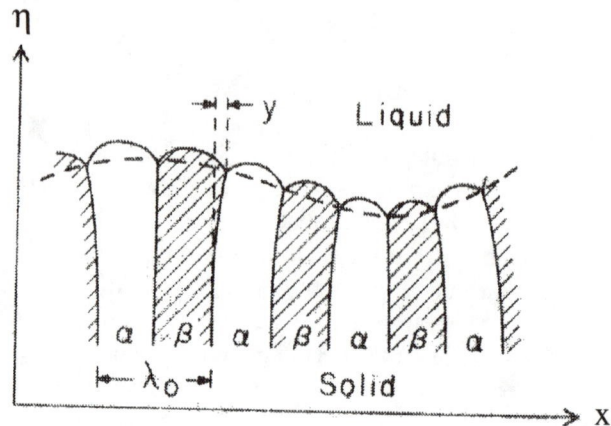

Figure 12.6. Figure defining the coordinates η, y, x at the solidification front. After Langer, Phys. Rev. Lett. **44**, 1023(1980).

$$\lambda(x,t) = \lambda_o(1 + \partial y/\partial x)$$

Taking the appropriate derivatives we obtain

$$\partial(\lambda/\lambda_o)/\partial t = \partial^2 y/\partial t\partial x = -V(\partial^2\eta/\partial x^2) = (V/G)(\partial^2[\Delta T(\lambda)]/\partial x^2)$$

The latter equation can be rewritten into the form

$$\partial\Lambda/\partial\tau = \{\partial/\partial x\} [D(\Lambda)\{\partial\Lambda/\partial x\}]$$

where $\Lambda = \lambda/\lambda_c$ and $D(\Lambda)$ plays the role of a Λ dependent diffusion constant:

$$D(\Lambda) = 1 - (1/\Lambda^2) = (d/d\lambda)\Delta T(\lambda)$$

and λ_c is defined by the relation $(B/AV)^{1/2}$ (i.e. setting the derivative of $\Delta T(\lambda)$ with respect to λ equal to zero and equating the value of λ to λ_c at this condition).

When $\Lambda < 1$, $D(\Lambda) < 0$. Langer noted that when Λ is slightly larger than unity and a fluctuation has caused Λ to drift to slightly subcritical values in some finite region, as illustrated in Figure 12.7, then at the minimum in this curve $\partial\Lambda/\partial x = 0$, $\partial^2\Lambda/\partial x^2 > 0$, and $D < 0$: thus Λ decreases. Also,at $\Lambda(x_1,t) = 1$, it can be shown by differentiating this relation that

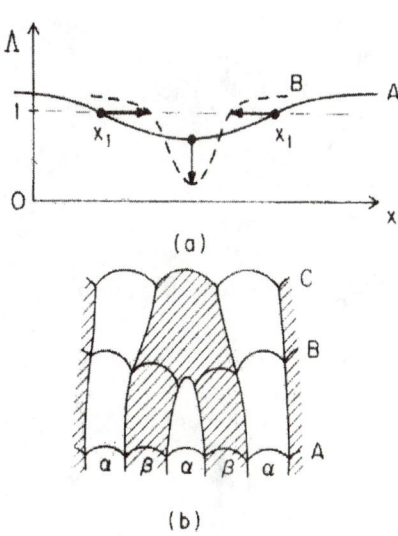

Figure 12.7. After Langer.[12]

$dx_1/dt = -(\partial \Lambda/\partial t)/(\partial \Lambda/\partial x) = -2(\partial \Lambda/\partial x)_{x=x_1}$. Thus, the pair of points labelled x_1 in the figure approach each other. The resulting behavior is indicated by the arrows and dashed curve in Figure 12.7a. All the intensity of an initially diffuse and shallow fluctuation is concentrated at a point. When Λ touches zero, the lamella at that point disappears, and the equation of motion loses its validity. The physical system presumably reverts to a state with fewer lamellae and larger average Λ, as illustrated in Figure 12.7b, from which configuration the entire process must start all over again.

This analysis of the instability in Λ at $\Lambda < 1$ is not sufficient to describe the behavior of the system at $\Lambda > 1$. Langer noted based on an analysis of the drift in Λ due to thermal noise that the direction of the drift in Λ depends upon the sign of $dD/d\Lambda$. In particular, when $dD/d\Lambda > 0$ the drift due to thermal noise is toward decreasing Λ. This result may be understood as follows. A fluctuation that drives Λ downwards persists for a longer time (due to lower D value) than one which goes in the other direction. Thus, the system moves to decreasing values of Λ, where D is smaller. Consequently, the observation that eutectic solidification occurs at the critical value of the interlamellar spacing, where $\Delta T(\lambda)$ is a minimum, may be explained in terms of the drift downwards in Λ, the instability in Λ at $D < 1$ leading to the termination of lamellae and a consequent local increase in Λ to a value of about 2 and the repetition of this cycle. During the downward drift in Λ, because the time spent by the system in an interval $\Lambda + d\Lambda$ increases as Λ decreases, the average value of Λ is much closer to 1 than to 2.

3.2. Extension of Langer's analysis of the drift in interlamellar spacing to eutectoid growth and discontinuous precipitation.

Langer's analysis, which yields a physical basis for the interlamellar spacing chosen by Nature in eutectic solidification, provides the starting point for similar analyses of the factors affecting the operating interlamellar spacing in eutectoid decomposition and discontinuous precipitation. The latter processes involve the transformation at a free boundary of a parent phase to a lamellar product consisting of two phases. The atomic transport required to effectuate the transformation is transverse to the moving free boundary and may occur either ahead of it in the parent phase, in the boundary itself, or behind it in one of the product phases.

We wish to determine how Nature determines the value of λ that operates during eutectoid decomposition and discontinuous precipitation. It seems eminently reasonable that what Langer has found concerning the drift in interlamellar spacing during eutectic solidification may also be applicable to the presently considered processes. Thus, in the following we extend the concepts of Langer to these processes.

Now in both eutectoid decomposition and discontinuous precipitation we may relate the velocity to the free energy available for the boundary migration (i.e. $-\Delta F'$) by

$$V = -M\Delta F'$$

We now make the reasonable assumption that the magnitude of the displacement $\eta(x,t)$ from the planar reaction front moving with the velocity V is proportional to the free energy $-\Delta F'$ (i.e. the larger is $-\Delta F'$ the more energy can be stored in the form of interface energy associated with any curvature produced by the displacement $\eta(x,t)$). Thus,

$$\eta(x,t) = C(-\Delta F'(\lambda))$$

Now, we make use of the remainder of the Langer analysis, in which it is assumed that the growth of each lamella is always perpendicular to the local orientation of the reaction front, i.e.

$$\partial y/\partial t = - V(\lambda)\partial\eta/\partial x$$

$$\lambda = \lambda_o(1 + \partial y/\partial x)$$

where η and y are defined in Figure 12.6..

Hence,

$$\partial(\lambda/\lambda_o)/\partial t=\partial^2 y/\partial t\partial x=-(\partial/\partial x)[V(\lambda)(\partial\eta/\partial x)]=-CM(\partial/\partial x)[\Delta F'(\lambda)(\partial\Delta F'(\lambda)/\partial x)]$$

This relation can be written as

$$\partial(\Lambda)/\partial\tau = (\partial/\partial x)[-\Delta F'(\Lambda)(d\Delta F'(\Lambda)/d\Lambda)(\partial\Lambda/\partial x)]$$

where $\Lambda = \lambda/\lambda_R$ and $\lambda_R = 2\sigma_{\alpha\beta}/\Delta F_o$ while ΔF_o is the free energy difference $(f_\alpha F_\alpha + f_\beta F_\beta) - F_\gamma$ in which the F values are free energies per unit volume at the equilibrium compositions of the product phases and the eutectoid composition of the parent phase and the f values represent the volume fractions of the respective phases.This relation has the form of a conservation relation for Λ, i.e.

$$\partial(\Lambda)/\partial\tau = (\partial/\partial x)[D(\partial\Lambda/\partial x)]$$

where $D = -\Delta F'(\Lambda)(d\Delta F'(\Lambda)/d\Lambda)$ is the Λ dependent diffusion constant. This relation for the diffusion constant is different from that derived by Langer for the case of eutectic solidification. The noise analysis of Langer showed that the drift in Λ depends upon the sign of $dD/d\Lambda$. When this slope is positive the drift is toward decreasing Λ and vice versa. Thus, a determination of the direction of the drift in Λ requires an analysis of the dependence of $\Delta F'$ on Λ.

3.2.1. Dependence of $\Delta F'$ on Λ for eutectoid growth and the drift in Λ.

We shall make use of Cahn's solution to the transport equations for the eutectoid growth problem[14], except that we will not make use of the maximum velocity criterion as Cahn did. His result relating the reaction front velocity V to the interlamellar spacing λ is given by

$$V = -M\Delta F' = -M\{\Delta F_o + C - C[(6/\alpha^{1/2})\tanh(\alpha^{1/2}/4) - (1/2)\text{sech}^2(\alpha^{1/2}/4)) + 2\sigma/\lambda\}$$

where $\alpha = kV\lambda^2/D_b\delta$, $C = (1/2\Omega)(\partial^2 F_M/\partial X_\alpha^2)$, k is the segregation coefficient relating the local concentration in the reaction interface to the local concentration in the product phase at the reaction interface, D_b is the diffusivity along the reaction interface, δ is the thickness of the reaction interface, σ is the specific reaction interface energy and M is the mobility of the reaction front.

Cahn by numerical solution of the above equation showed that α varies from 1 to 2 for reasonable values of ΔF_o,C and M. Thus, the argument of the hyperbolic functions are small enough to justify a linear expansion. We make such an expansion and solve the equation for V, which can then be expressed as a function of $\Lambda = \lambda/\lambda_o$, where $\lambda_o = 2\sigma/\Delta F_o$. The result is

$$V(\Lambda) = - \Delta F_o(1- \Lambda^{-1})/[M + B\Lambda^2\}$$

where $B = kC\lambda_o^2/(32D_b\delta)$. Using $V = - M\Delta F'$ we obtain the desired result

$$\Delta F'(\Lambda) = \Delta F_o(1- \Lambda^{-1})/[1 + B'\Lambda^2]$$

that is applicable to eutectoid growth, where $B' = MB$. Thus, for eutectoid growth the desired expression for $D(\Lambda)$ becomes

$$D(\Lambda) = - \Delta F_o^2(1- \Lambda^{-1})\{B'(3 - 2\Lambda) + \Lambda^{-2}\} / [1 + B'\Lambda^2]^3$$

The dependence of $D(\Lambda)$ on Λ is shown schematically in Figure 12.8 for eutectoid growth and is compared there to the dependence applicable in eutectic solidification. As the figure caption notes the definition of Λ differs between these two phenomena. However, the important point relates to the change in sign of the slope $dD(\Lambda)/d\Lambda$ beyond the maximum in $D(\Lambda)$ for eutectoid growth.

In the case of eutectic solidification, we may recall that the drift in Λ is always towards decreasing Λ, i.e. towards the instability point at $\Lambda = \lambda/\lambda_c = 1$, because the slope $dD(\Lambda)/d\Lambda$ is always positive. In eutectoid growth a similar instability point occurs for $\Lambda \leq \Lambda^*$, where the latter is determined by the relation: $\{B'(3 - 2\Lambda^*) + \Lambda^{*-2}\}= 0$. Also, to the left of the maximum in $D(\Lambda)$ the drift in Λ is toward this instability point. When $B' = 1$, Λ^* is about equal to 53/32 and the maximum in $D(\Lambda)$ occurs at Λ about equal to 2.1.Thus, lamellae that have values of interlamellar spacing corresponding to those to the left of the maximum in $D(\Lambda)$ drift to smaller values of Λ and annihilate according to the mechanism illustrated in Figure 12.7. However, they then yield local interlamellar spacings that correspond to about $2\Lambda^*$, which when $B' = 1$ yields $\Lambda \approx 53/16 = 3.3$. This value of Λ is thus to the right of the maximum in $D(\Lambda)$(i.e. at $\Lambda = 2.1$) and because the slope $dD(\Lambda)/d\Lambda$ is now negative, the drift in Λ is towards increasing values of Λ. Thus, in eutectoid growth and according to the above model the tendency is for the interlamellar spacing to drift towards larger values. The experimental observation is that in eutectoid decomposition the lamellar product grows in the form of colonies in which the interface is convex toward the parent phase and the interlamellar spacing increases in agreement with the expectation we have just developed from the Langer model for drift in Λ.

Why then does the interlamellar spacing not continue to increase indefinitely. It appears that there is a limit to such increase occasioned by the nucleation of new lamellae. This outer limit in λ has been called the upper catastrophic limit by Sundquist.[15] Although there are explanations for the upper catastrophic limit a satisfactory one for the instability at this limit has still not yet been given.We suspect that a detailed model of both the transport and interface reaction that would consider each infinitesimal section of the interface with the general approach of Cahn may provide the desired explanation for the upper catastrophic limit. The instability associated with the latter may be that beyond the upper catastrophic limit it is not possible for the center of the lamella at which solute is discharged to move at the same velocity as the remainder of the reaction front.

If we accept the existence of an upper limit to the interlamellar spacing then the operating interlamellar spacing should be some fraction close to unity of this limit based on the arguments we have given above that

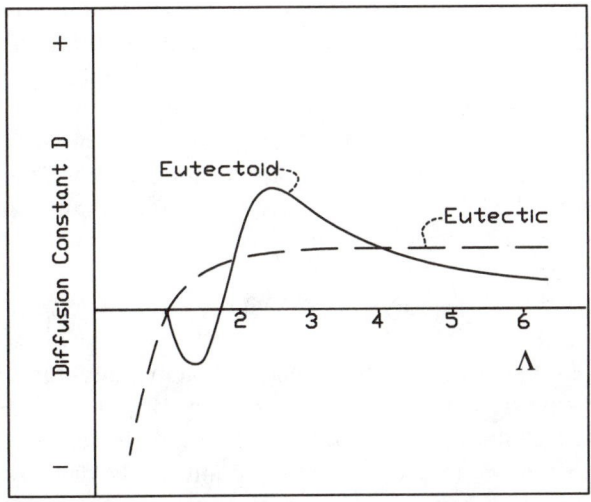

Figure 12.8. Illustrating the difference in behavior of the Λ dependent diffusion constant D between that for a directionally solidified eutectic moving at a fixed velocity and an eutectoid free boundary. The drift in Λ is toward smaller D. Thus, the eutectic always drifts towards smaller Λ, whereas the eutectoid can drift toward larger Λ for values larger than that corresponding to the maximum in D. However, note that the definition of Λ varies for the two cases: $\Lambda = \lambda/\lambda_c$, where λ_c corresponds to the minimum in ΔT in eutectic solidification; $\Lambda = \lambda/\lambda_R$, where λ_R corresponds to $\Delta F(\lambda) = 0$ in eutectoid decomposition.

the lifetime of a given spacing increases as the D value corresponding to that spacing decreases.

3.2.2. Dependence of $\Delta F'$ on Λ for discontinuous precipitation and the drift in Λ.

The description of the factors controlling the interlamellar spacing given in the previous section appears to be acceptable for eutectoid growth. However, the corresponding case of discontinuous precipitation should not obey the same relations if only because the resulting equation for $V(\Lambda)$ cannot be the same. Cahn showed that for the case of discontinuous precipitation the relation for $V(\Lambda)$ becomes

$$V = -M\Delta F' = -M\{\Delta F_o [(3/\alpha^{1/2})\tanh(\alpha^{1/2}/2) - (1/2)\text{sech}^2(\alpha^{1/2}/2)] + 2\sigma/\lambda\}$$

However, now the arguments of the hyperbolic functions are no longer small and expansion of these functions is no longer justifiable. Thus, we must rely on numerical solution of the above relation. Fortunately, Hillert has extended Cahn's model and obtained numerical solutions that yield the $V(\lambda)$ function for various assumed values of the fraction of the drag due to the formation of the α/β interface that is assignable to the α lamellae.[16] These curves are shown in Figure 12.9. They reveal an interesting feature not shown in the corresponding functions applicable to eutectoid growth. Namely, that $V(\lambda)$ approaches a constant value with increasing λ. Since $V(\lambda)$ is proportional to $\Delta F'(\lambda)$ and

$$dD/d\Lambda = -\{\Delta F'(\Lambda)(d^2\Delta F'(\Lambda)/d\Lambda^2) + (d\Delta F'(\Lambda)/d\Lambda)^2\}$$

it is apparent that for those curves that yield a maximum in $V(\lambda)$ in Figure 12.9 and to the right of the inflection point in $V(\lambda)$(i.e., $\Delta F'(\Lambda)$) that $dD/d\Lambda$ is negative and rapidly approaches zero. Thus, any lamellae with interlamellar spacing to the right of the inflection point will be driven to larger interlamellar spacing. However, this rate of change of Λ will rapidly approach zero at some finite value of Λ. Hence, since experimental observations of discontinuous precipitation involve growth periods of finite time, Λ will be limited by the experimental conditions to some maximum value. This process may explain the observed value of Λ found in discontinuous precipitation.

Figure 12.9 shows that for the case where the fraction of the drag due

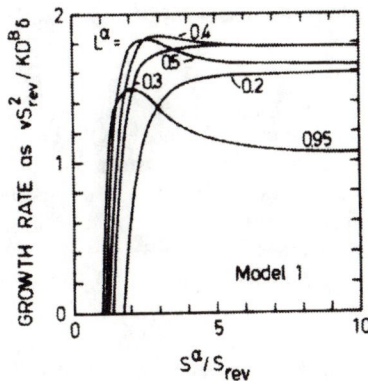

Figure 12.9. After Hillert[16]. The abscissa is equal to our parameter Λ. The ordinate is proportional to the velocity $V(\lambda)$ and to $\Delta F'(\lambda)$ as well. The numbers represent the fraction of the drag due to formation of the lamellar interfaces that is borne by the α lamellae.

to the formation of α/β interface is less than 0.4 then there is no maximum in $V(\lambda)$. The corresponding curves are similar to those applicable to eutectic solidification. However, it appears highly unlikely that this fraction will be less than 0.5.

The considerations governing the interlamellar spacing in eutectoid growth and discontinuous precipitation given in Section 3.2 are new and untested by experiment. Certainly they should be tested and it is hoped that these concepts will inspire experimentalists and theoreticians to renewed attack upon this problem. This author likes the new concepts because they represent a consistent view of the factors that govern pattern formation in moving boundary phenomena. However, despite their esthetic quality they may be inapplicable and only the test of comparison with experiment will provide the necessary data to evaluate their applicability.

Summarizing, we have investigated the phenomenon of morphological instability in a variety of growth environments. The linear stability analysis of Mullins and Sekerka is the starting point for the understanding of most of these phenomena. Also, we have found an approach that, for the first time, provides a consistent and physically reasonable explanation for the value of the length parameter in solutions to the transport relations, in which the growth rate and length parameter are uniquely determined as a product, but not as separate parameters. This approach builds on a concept due to Langer: the drift in the length parameter is driven by fluctuations in a defined direction. The length parameter reached by this drift is unique to the process itself.

.APPENDIX 1- Mullins-Sekerka Analysis of Morphological Instability of a Planar Interface.

Consider a planar interface, moving with a velocity V relative to the bounding phases which are fixed with respect to the laboratory coordinate system, and another coordinate system, to which we will refer certain distances, that moves with the interface. Let the z axis point toward the liquid and the plane $z = 0$ coincide with the interface. Suppose a sinusoidal fluctuation in the shape of the interface described by

$$z = \delta(t)\sin\omega x$$

is introduced. The object of this analysis is to obtain an expression for $d\delta/dt$ to determine whether the perturbation in the shape of the interface grows or decays. This analysis requires the calculation of the local velocity $v(x)$, where x denotes the distance along the interface from some origin, in terms of the values of the local thermal and diffusion gradients. The applicable diffusion equations are

$$\nabla^2 u + (V/D_u)(\partial u/\partial z)$$

where u is a field of either concentration (in the liquid phase) or temperature (in both the liquid and solid phases) and D_u represents the corresponding diffusivity. These equations have to satisfy the far field boundary conditions(the concentration and temperatures that would exist several wave lengths away from the interface for the unperturbed interface.) Also, at the interface local equilibrium is assumed and hence the local temperature at the interface is given by

$$T^* = mc^* + T_M(1 + \Gamma K) = mc^* + T_M(1 + \Gamma\delta\omega^2\sin\omega x)$$

Here K is the average curvature at a point on the interface (positive when concave toward the liquid), $\Gamma = \sigma/L$, where σ is the specific interface energy and L the latent heat of the solvent per unit volume.

The local interface velocity calculated from heat flow considerations must agree with that calculated from diffusion conditions, i.e.

$$v(x) = (1/L)[K_S(\partial T'/\partial z) - K_L(\partial T/\partial z)] = [D/(c^*(k-1))](\partial c/\partial z)^*$$

where the asterisk denotes interface values, k is the distribution coefficient (ratio of equilibrium solute concentration in solid to that in liquid), and the denominator $c^*(k-1)$ gives the difference in concentration between the solid and liquid sides of the interface.

By setting

$$T^* = T_o + a\delta(t)\sin\omega x$$

$$c^* = c_o + b\delta(t)\sin\omega x$$

where the subscript $_o$ denotes values for a flat interface. The solutions that satisfy the diffusion equations and the boundary conditions are then on the liquid side

$$c(x,z) - c_o = (G_c D/V)(1 - \exp[-Vz/D]) + \delta(b - G_c)\sin\omega x \, e^{-\omega^* z}$$

$$T(x,z) - T_o = (G''D''_{th}/V)(1 - \exp[-Vz/D''_{th}]) + \delta(a - G'')\sin\omega x \, e^{-\omega'' z}$$

and on the solid side

$$T'(x,z) - T_o = (G'D'_{th}/V)(1 - \exp[-Vz/D'_{th}]) + \delta(a - G')\sin\omega x \, e^{-\omega' z}$$

where G" and G' are the thermal gradients at the unperturbed flat interface in the liquid and solid, respectively, G_c is the concentration gradient in the liquid at this interface and where

$$\omega^* = (V/2D) + [(V/2D)^2 + \omega^2]^{1/2}$$

$$\omega'' = (V/2D''_{th}) + [(V/2D''_{th})^2 + \omega^2]^{1/2}$$

$$\omega' = (V/2D'_{th}) - [(V/2D'_{th})^2 + \omega^2]^{1/2}$$

The parameters a and b can be obtained by appropriate substitution and are

$$b = \frac{2G_c T_M \Gamma \omega^3 + \omega G_c (H' + H) + G_c [\omega^* - (V/D)](H' - H)}{2\omega m G_c + (H' + H)[\omega^* - (V/D)(1-k)]}$$

$$a = mb - T_M \Gamma \omega^2$$

where $H = (K_L / \overline{K})G$; $H' = (K_S / \overline{K})$ and $\overline{K} = 0.5(K_S + K_L)$.

To determine $d\delta/dt$, we substitute for the temperature gradients in the equation for the local velocity to obtain

$$v(x) = V + (d\delta/dt)\sin\omega x = (\overline{K}/L)(H' + H) + (\overline{K}/L)\omega\{2a - (H' + H)\}\delta\sin\omega x$$

Equating like Fourier components yields

$$V = (\overline{K}/L)(H' + H)$$

$$(d\delta/dt) = (\overline{K}/L)\omega\{2a - (H' + H)\}\delta$$

Substituting for a we then obtain the desired result

$$(1/\delta)(d\delta/dt) = \frac{V\omega\{-2T_M\Gamma\omega^2 [\omega^* - (V/D)(1-k)] - (H' + H)[\omega^* - (V/D)(1-k)] + 2mG_c[\omega^* - (V/D)]\}}{(H' + H)[\omega^* - (V/D)(1-k)] + 2\omega m G_c}$$

REFERENCES.

1. W.W. Mullins and R.F. Sekerka, J. Appl. Phys.35, 444(1964).
2. J.S. Langer and H. Muller-Krumbhaar, Acta Met. 26, 1681;1689;1697(1978).
3. R.Trivedi and W. Kurz, Acta Met. 34, 1663(1986).
4. C.H.J. van den Brekel, Philips J. Res. 33, 20(1978).
5. D.J. Srolovitz, A. Mazor and B.G. Bukiet, J. Vac. Sci. Tech. A6, 2371(1988).
6. R.T.U. Karunasari, R. Bruinsma and J. Rudnick, Phys. Rev. Lett. 62, 788 (1989).
7. D.J. Srolovitz, Acta Met 37, 621(1989).
8. W.W. Mullins and R.F. Sekerka, J.Appl.Phys. 34, 323(1963).
9. G.P. Ivantsov, Dokl.Akad.Nauk SSSR 58, 567(1947).
10.S.C. Huang and M.E.Glicksman, Acta Met. 29,701;717(1981).
11.K.A.Jackson and J.D.Hunt, Trans.Met.Soc. AIME 236, 1129(1966).
12.J.S. Langer, Phys.Rev.Lett. 44, 1023(1980).
13.V. Datye and J.S. Langer, Phys.Rev. B24, 4155(1981).
14.J.W. Cahn, Acta Met. 7, 18(1959).
15.B.E. Sundquist, Acta Met. 16, 1413(1968).
16. M. Hillert, Acta Met. 30, 1689(1982).

BIBLIOGRAPHY.

1. W. Kurz and D.J. Fisher, FUNDAMENTALS OF SOLIDIFICATION, Trans.Tech Publications, Switzerland, 1986.
2. H.Eugene Stanley and N. Ostrowsky, eds., RANDOM FLUCTUATIONS AND PATTERN GROWTH, Proc. NATO Advanced Study Institute, NATO Advanced Science Institute Series E: Applied Sciences 157, Kluwer Academic Publ., Hingham, Mass., 1988.

INDEX

K

Kink at surface step, 295
Kinetics, concepts in, 185
   nucleation, 231
   of interface migration, 247
Kirkendall effect, 203
Kroger-Vink plot, 180
Kronberg-Wilson texture, 256

L

Langer's analysis of
   drift in interlamellar spacing, 326
   marginal stability, 321, 327
Lattice stability energy
   definition, 3
   values(table), 24-6
Liquidus temperature
   average of solidus and, 68
Local equilibrium
   at embryo-host interfaces, 145
   at embryo-host-particle interfaces, 143
   at grain boundary intersections, 124
   at interface in diffusion controlled
      growth, 273
   at liquid/solid interface, 288
Long-range-order, 17, 47

M

Magnetite, 216
Marginal stability, 321, 327
Mass action law, 48, 176
Matano interface, 205
Metastability, 16, 79, 131, 148
Minimum rate of entropy production, 196
Miscibility gap compositions, 42, 44
Monte Carlo method, 51
Morphological instability, 309
   during CVD, 315

during PVD, 317
during solid phase precipitation, 319
in solidification, 311
physical basis for, 310
Mullins-Sekerka, 334

N

Nabarro shape function, 147
Nernst-Einstein relation, 217
Network solid, 62
Non-stoichiometric compounds, 173
Nucleation
   heterogeneous, 238
   homogeneous, 231
   island, 239
   kinetics, 231
   of thin films during deposition, 239
   rate of heterophase, 231
      steady-state, 232
         temperature dependence, 235
   rate of homophase, 240
   versus growth, 237
Nucleus
   free energy of, 141

O

Onsager relations, 194
Order
   long-range, 17
   short-range, 47
Ostwald ripening, 103, 299
Oxides, diffusion in, 215

P

Pairing of vacancies, 164
Partial molar free energy, 43
   definition, 64